Development and Uses of

Biofortified Agricultural Products

Development and Uses of

Biofortified Agricultural Products

Edited by

Gary S. Bañuelos • Zhi-Qing Lin

CRC Press
Taylor & Francis Group
Boca Raton London New York

CRC Press is an imprint of the
Taylor & Francis Group, an informa business

CRC Press
Taylor & Francis Group
6000 Broken Sound Parkway NW, Suite 300
Boca Raton, FL 33487-2742

Printed in the United States of America on acid-free paper
10 9 8 7 6 5 4 3 2 1

International Standard Book Number-13: 978-1-4200-6005-8 (Hardcover)

Library of Congress Cataloging-in-Publication Data

Development and uses of biofortified agricultural products / editors, Gary S. Banuelos, Zhi-Qing Lin. -- 1st ed.
p. cm.
Includes bibliographical references and index.
ISBN 978-1-4200-6005-8 (alk. paper)
1. Crop improvement. 2. Trace elements in plant nutrition. 3. Crops and soils. 4. Trace elements in nutrition. 5. Enriched foods. I. Banuelos, Gary Stephan, 1956- II. Lin, Zhi-Qing. III. Title.

SB106.I47D48 2008
630--dc22 2008019818

Visit the Taylor & Francis Web site at
http://www.taylorandfrancis.com

and the CRC Press Web site at
http://www.crcpress.com

Contents

Foreword

Malnutrition is by far the leading cause of death worldwide. Alarmingly, it results in about 30 million deaths a year as a result of overt nutrient deficiencies as well as chronic diseases (e.g., heart disease, cancer, strokes, and diabetes) associated with inappropriate diets and imbalances in nutrient intake. Incredibly, over three billion people (nearly half the world's population) are afflicted with deficiencies of one or more essential trace elements and vitamins resulting in serious consequences for human health, productivity, happiness, and national development. Until recently, primary interventions to address this global health crisis include dietary micronutrient supplements and food fortification programs. Unfortunately, these types of programs do not reach many of those afflicted (especially resource-poor rural families in developing nations), and in many nations they have not proven to be sustainable. These types of approaches treat the symptoms of malnutrition but do nothing to address the primary causes. The underlying factors responsible for malnutrition are rooted in dysfunctional food systems dependent on agricultural systems that cannot provide all the essential nutrients and health-promoting factors in needed amounts continuously during all seasons. Because agriculture is the primary supplier of all nutrients and health-promoting factors consumed by people, it is logical to suspect that agricultural policies and farming systems are part of the root causes of malnutrition and diet-related diseases. Moreover, the agricultural sector has never had an explicit goal of improving human health, and the nutrition and health communities have never used agriculture as a primary tool to fight malnutrition and diet-related diseases.

Health problems associated with the uneven distribution of micronutrients in the soil and their less-than-efficient uptake by some crop cultivars have become major public health issues in many developing countries. In particular, the urgent needs to improve the bioavailability of essential trace elements (e.g., iron, selenium, zinc, and iodine) and recognize the importance of plant and soil nutrition are conjunctly vitally influential in preserving and improving food quality and, ultimately, human health. To realize these objectives, new strategies, including agronomic and soil management practices, breeding, genetic and molecular insight, and their impact on influencing trace element uptake, accumulation, and bioavailability, have been developed and proposed for producing biofortified agricultural products.

Biofortification (i.e., generating plants that fortify themselves with nutrients and other health promoting factors during their growth) is the first agricultural tool to be embraced by both the agricultural and the nutrition communities to combat malnutrition. Originally, biofortification focused on breeding major staple food crops that would produce edible products enriched in bioavailable amounts of iron, zinc, and provitamin A carotenoids. The concept of biofortification has now expanded to include using soil amendment, fertilizer, and cropping system strategies to improve the nutrient output of farming systems and an ever-widening list of nutritional factors to biofortification in agricultural production. Increasing the output of bioavailable essential trace elements in food crops through adopting agricultural practices that are designed to meet human needs can be accomplished from "off the shelf" technologies, including choice of cropping systems, agronomic practices, and variety selection, as well as through the use of modern genetic engineering technologies. The connection between agriculture and human health must be established if we are to find sustainable solutions to trace-element deficiencies and associated diet-related chronic diseases afflicting the lives and health of massive numbers of people, and the development potential for numerous nations globally. As the world population grows and the demand for food increases, it will become imperative that we perform the research needed to design agricultural systems that not only provide enough food to meet energy needs but also provide healthy foods to prevent nutrient deficiencies and the chronic diseases associated with inappropriate diets and low quality food products. To achieve this goal scientists should have a better understanding of the underlying mechanisms, such as the interactions between agricultural practices, gut microorganisms, and nutrient bioavailability; the processes in the rhizosphere and the availability of essential trace elements to food crops; and effects of soil amendments, organic matter, and farming practices (e.g., cropping systems, tillage, irrigation, etc.) on trace element and vitamin accumulation in edible plant products. The genetic mechanisms that control the accumulation, translocation, and transformation of trace elements in edible plant products need further exploration.

The need to critically analyze and comprehensively synthesize the ever-mounting body of new information on biofortification from a worldwide perspective provided the impetus for the development of this book. It is a compilation of scientific articles provided by the eminent scientists from around the world who are actively engaged in vital aspects of enhancing nutritional quality of food products from the soil up. Some of those international experts have presented at the special symposium entitled "Development and Uses of Biofortified Agricultural Products," as part of the 9th International Conference on the Biogeochemistry of Trace Elements that was held on July 16–17, 2007 in Beijing. This book presents a myriad of research focused on how best to biofortify food crops. The novel research endeavors presented will contribute to the important creation of biofortified agricultural products, and provide insightful opportunities to further our worldwide understanding of nutritionally important trace elements in the context of biochemistry, food chain transfer, and health-related issues. Indeed, an agricultural system that cannot sustain the societies it supports will not be, in itself, sustainable.

Ross M. Welch, Ph.D.
USDA-ARS Plant, Soil and Nutrition Laboratory
Cornell University

Acknowledgments

This book is a product of international cooperation that originated from the inspiration initially provided to me by Prof. Horst Marschner at the University of Hohenheim in Germany. With the support from the USDA-ARS, my efficient technical assistants, and most importantly the nonstop efforts of my coeditor, Dr. Zhi-Qing Lin, we and the participating authors were able to create this book on biofortification. Such endeavors are only possible with financial support, and ours was graciously provided by California State University–Fresno (the Agriculture Research Initiative), Southern Illinois University–Edwardsville (the Graduate School and the College of Arts and Sciences), the Organization for Economic Co-operation and Development Cooperative Research Fellowship Program in Paris, and technical support from Professor Yongguan Zhu of the Chinese Academy of Science. Lastly, both editors are grateful to our families for their continued patience throughout this time-consuming process.

We greatly acknowledge the contribution and effort made by our reviewers: Michele Arienzo, Larry L. Barton, Martin R. Broadley, Rupali Datta, Danika LeDuc, Emmanuel Frossard, Suduan Gao, Yantai Gan, Cynthia Grant, Yibing Ma, Satoshi Mori, Barry Olsen, Thangavel Palaniswamy, Tommy Pflaum, Ruijun Qin, David W. Ramming, Kurt Schluz, and Abderrahmane Tagmount.

The Editors

Gary Bañuelos is a plant/soil scientist at the U.S. Department of Agriculture–Agricultural Research Station Water Management Research Unit, Parlier, California, and adjunct professor at the University of California and California State University–Fresno. Dr. Bañuelos began his work with "green technology" with Professor H. Marschner in Germany, where he initially tested for radioactive vegetables contaminated from the Chernobyl disaster in 1986. Fascinated by the ability of plants to take up contaminates and still survive, Bañuelos focused his research activities on the phytoremediation of soils and waters laden with selenium, boron, and salinity in California. His integrated approach to field phytoremediation involves more than just the selection of crops, but also considers crop rotation with oil crops, irrigation and drainage management, natural chemical transformations within the soil, production of innovative biofortified and biobased products (e.g., biofuel Se-enriched feeds, and organic fertilizer/herbicide), and he strives for creating a sustainable green strategy for managing trace elements in the United States and in other parts of the world. Dr. Bañuelos is the principle author of over 105-refereed technical journals on the topic of phytoremediation. He received his Bachelor of Arts degree in German from California State University, Bachelor of Science in Crop Science, and Master of Science degree in international agriculture from California Poly-Technical University. His diploma in plant biology was earned at the University of Tubingen, Germany, and doctorate in agriculture/plant nutrition at Hohenheim University, Germany, as a National Science Foundation Fellow.

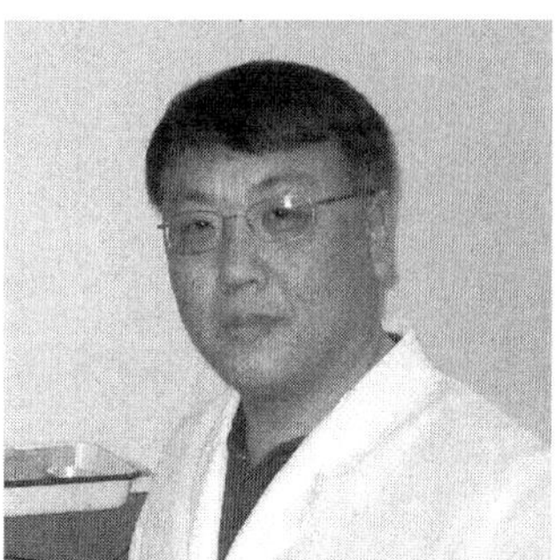

Zhi-Qing Lin is an associate professor in the Department of Biological Sciences and the director of the Environmental Sciences Program at Southern Illinois University at Edwardsville. Dr. Lin started his research on phytotechnology with Prof. Huiyi Huang at the Chinese Academy of Sciences in 1986, and continued with Prof. Norman Terry at the University of California at Berkeley in 1996. His research is related to the biogeochemistry of environmentally important trace elements, with special emphasis on biological accumulation, transformation, and volatilization processes. He is the author or coauthor of over 40 refereed journal and review articles, proceedings, and book chapters. Dr. Lin received his Bachelor of Science degree (1983) in environmental biology from Liaoning University, Master of Science degree (1986, with Professor Huiyi Huang) in Pollution Ecology from the Chinese Academy of Sciences, and Ph.D. degree (1996, with Professor Peter Schuepp) in Environmental Studies from McGill University.

Contributors

Mark G. M. Aarts
Laboratory of Genetics
Wageningen University
Wageningen, The Netherlands
mark.aarts@wur.nl

Majid Afyuni
Isfahan University of Technology (IUT)
Isfahan, Iran

Arden B. Andersen
Crossroads Healing Arts
Goshen, Indiana
drandersen@optihealthguide.com

Gary S. Bañuelos
USDA-ARS Water Management Research Unit
Parlier, California
gary.banuelos@ars.usda.gov

Mônica Batista Benke
Agriculture and Agri-Food Canada
Lethbridge Research Centre
Lethbridge, Alberta, Canada

Stefan Storcksdieck genannt Bonsmann
European Food Information Council
Brussels, Belgium

Gloria S. Cabuslay
International Rice Research Institute
Manila, Philippines

Andrew C. Chang
University of California–Riverside
Riverside, California

Chi Chang
Agriculture and Agri-Food Canada
Lethbridge Research Centre
Lethbridge, Alberta, Canada

Chien-Yi Chen
School of Medical Imaging and Radiological Sciences
Chung Shan Medical University
Taiwan, Republic of China
ccy@csmu.edu.tw

Juan Du
State Key Laboratory of Plant Cell and Chromosome Engineering
Institute of Genetics and Developmental Biology
Chinese Academy of Sciences
Beijing, China

Mingsheng Fan
Key Laboratory of Plant–Soil Interactions, MOE
Key Laboratory of Plant Nutrition, MOA
Department of Plant Nutrition, China Agricultural University
Beijing, China

Emmanuel Frossard
Institute of Plant Sciences
ETH Zurich
Zurich, Switzerland

Xiaopeng Gao
Department of Soil Science
University of Manitoba
Winnipeg, Manitoba
gaox@cc.umanitoba.ca

Yusuf Genc
Molecular Plant Breeding Cooperative Research Centre
Glen Osmond
South Australia, Australia

Robin Graham
School of Agriculture, Food and Wine
University of Adelaide
Glen Osmond
South Australia, Australia

Glenn B. Gregorio
Plant Breeding, Genetics and Biotechnology Division
International Rice Research Institute
Manila, Philippines
g.gregorio@cgiar.org

Xiying Hao
Agriculture and Agri-Food Canada
Lethbridge, Alberta, Canada
haoxy@agr.gc.ca

Ellis Hoffland
Department of Soil Quality
Wageningen University
Wageningen, The Netherlands

Tin Htut
Department of Agricultural Planning
Ministry of Agriculture and Irrigation
Nay Pyi Taw, Myanmar

Guangming Huo
Department of Life Sciences
Nanjing Xiaozhuang College
Nanjing, China

Richard F. Hurrell
Institute of Food Science and Nutrition
Zurich, Switzerland
richard.hurrell@ilw.agrl.ethz.ch

Wen Jiang
China Agricultural University
College of Resources and Environmental Science
Beijing, China
and
Chinese Academy of Agricultural Sciences
Institute of Crop Science
Beijing, China

Amir Khoshgoftarmanesh
Isfahan University of Technology (IUT)
Isfahan, Iran

Pulugurtha Bharadwaja Kirti
Department of Plant Sciences
University of Hyderabad
Hyderabad, India

Thomas W. Kuyper
Department of Soil Quality
Wageningen University
Wageningen, The Netherlands

Jianfen Liang
China Agricultural University
College of Food Science and Nutritional Engineering
Beijing, China

Zhi-Qing Lin
Environmental Sciences Program and Department of Biological Sciences
Southern Illinois University–Edwardsville
Edwardsville, Illinois
zhlin@siue.edu

Hong-Qing Ling
State Key Laboratory of Plant Cell and Chromosome Engineering
Institute of Genetics and Developmental Biology
Chinese Academy of Sciences
Beijing, China
hqling@genetics.ac.cn

Graham Lyons
School of Agriculture, Food and Wine
University of Adelaide
Glen Osmond
South Australia, Australia
graham.lyons@adelaide.edu.au

Antonio Eduardo Furtini Neto
Universidade Federal de Lavras
Lavras, Brazil

Nadella Nirupa
Department of Plant Sciences
University of Hyderabad
Hyderabad, India

Bernd Nowack
Swiss Federal Institute of Material Sciences (Empa)
St. Gallen, Switzerland

Majeti Narasimha Vara Prasad
Department of Plant Sciences
University of Hyderabad
Hyderabad, India
prasad_mnv@yahoo.com

Rainer Schulin
Institute of Terrestrial Ecosystems
ETH Zurich
Zurich, Switzerland
schulin@env.ethz.ch

Francisco Nildo da Silva
Universidade Federal Rural do Semi-Árido
Costa e Silva
Mossoró, Brazil
fnildos@yahoo.com

Maja Slingerland
Crop and Weed Ecology Group
Wageningen University
Wageningen, The Netherlands
maja.slingerland@wur.nl

Tjeerd-Jan Stomph
Crop and Weed Ecology Group
Wageningen University
Wageningen, the Netherlands

Ning Wang
State Key Laboratory of Plant Cell and Chromosome Engineering
Institute of Genetics and Developmental Biology
Chinese Academy of Sciences
Beijing, China

Yutao Wang
Greenhouse and Processing Crops Research Center
Agriculture and Agri-Food Canada
Harrow, Ontario, Canada
and
Institute of Agricultural Resources and Regional Planning
Chinese Academy of Agricultural Sciences
Beijing, China

Jian Wu
Institute of Vegetables and Flowers
Chinese Academy of Agricultural Sciences
Beijing, China

Fusuo Zhang
Key Laboratory of Plant–Soil Interactions, MOE
Key Laboratory of Plant Nutrition, MOA
Department of Plant Nutrition, China Agricultural University
Beijing, China
zhangfs@cau.edu.cn

Liyang Zhang
Department of Life Sciences
Nanjing Xiaozhuang College
Nanjing, China

Tiequan Zhang
Greenhouse and Processing Crops Research Center
Agriculture and Agri-Food Canada
Harrow, Ontario, Canada
zhangt@agr.gc.ca

Chunqin Zou
Key Laboratory of Plant–Soil Interactions, MOE
Key Laboratory of Plant Nutrition, MOA
Department of Plant Nutrition, China Agricultural University
Beijing, China

Yuanmei Zuo
Key Laboratory of Plant–Soil Interactions, MOE
Key Laboratory of Plant Nutrition, MOA
Department of Plant Nutrition
China Agricultural University
Beijing, China

1 The Impact of Trace Elements from Plants on Human Nutrition

A Case for Biofortification

Stefan Storcksdieck genannt Bonsmann and Richard F. Hurrell

CONTENTS

1.1 INTRODUCTION

The trace elements iron (Fe), zinc (Zn), iodine (I), and selenium (Se) are essential micronutrients for various functions in the human body. A diversified diet is the best approach to ensure their adequate supply, yet such a diet may not be available to all population groups at all times. Women of reproductive age and children living in developing countries are especially vulnerable to micronutrient deficiencies because of high demands and poor diet quality (FAO/WHO 2001). In many countries, fortification and supplementation programs have been implemented to combat such deficiencies, for example, wheat flour fortification with Fe in Pakistan, vitamin A supplementation in Niger, use of iodized salt in Morocco, etc. However, these programs may not reach the neediest, the rural poor, as recently reported for Guatemala

(Imhoff-Kunsch et al. 2007). This population group largely relies on subsistence farming, and often cannot afford or lacks access to processed foods. Biofortification is a relatively new approach to improve the nutritional status of populations by enhancing the micronutrient content of their staple plant foods either through conventional breeding or genetic engineering. In addition, agronomic biofortification through fertilization (application to soil, seed, or leaves) helps in increasing plant nutrient content, without changing the plant's genetic makeup.

1.2 IRON (Fe), ZINC (Zn), IODINE (I), AND SELENIUM (Se): HUMAN HEALTH ASPECTS AND FORTIFICATION STRATEGIES

1.2.1 IRON

Over 30% of the world's population is anemic, which is to a large extent due to Fe deficiency (ID) (Stephenson et al. 2000; FAO/WHO 2001). A major cause of ID is low dietary Fe bioavailability, which is typically observed in monotonous plant-based diets consumed by third world populations of low socioeconomic status (Zimmermann et al. 2005; Gibson et al. 2006). Such diets are rich in phytic acid and polyphenols, two well-known inhibitors of Fe absorption (Gillooly et al. 1983; Reddy et al. 1996), but poor in Fe absorption enhancers such as meat (Martinez-Torres and Layrisse 1971) or ascorbic acid (Siegenberg et al. 1991). The most severe form of ID, Fe deficiency anemia (IDA), is common in children and young women in the developing world as well as in developed countries (FAO/WHO 2001). IDA has adverse health effects on pregnancy outcome, infant growth, cognitive performance, immune status, and work capacity (WHO 2002). For pregnant women, anemia contributes to 20% of all maternal deaths. Even mild-to-moderate ID without anemia may lower work capacity and resistance to fatigue (Haas and Brownlie 2001; Brownlie et al. 2004) and impair cognition (Bruner et al. 1996; Stoltzfus et al. 2001).

Bioavailability of nonheme Fe, which constitutes all plant food Fe and over 50% of animal food Fe, usually lies in the range of 2–20% (Monsen and Balintfy 1982). In contrast, heme Fe is absorbed at 15–35%, and its absorption is slightly influenced by other diet constituents (Carpenter and Mahoney 1992). The majority of body Fe, 65–70%, is found in the hemoglobin and myoglobin molecules used for oxygen transport and storage and functions as a cofactor in several enzymes (Narins 1980; Aisen et al. 2001). Storage Fe, which makes up the residual one-third of total body Fe, is sequestered in the proteins ferritin and hemosiderin in the liver, spleen, and bone marrow (Aisen et al. 2001).

Universal Fe fortification is generally recommended for countries in which the risk of ID is high for all population groups other than adult men and postmenopausal women. The latest WHO guidelines on food fortification with micronutrients give recommendations for most vehicles and, more importantly, a way to calculate fortificant levels (Allen et al. 2006). Several efficacy and effectiveness trials have demonstrated the beneficial impact of food fortification on Fe status in infants, children, and women. Foods that have successfully been used for Fe fortification include rice in India (Moretti et al. 2006), fish and soy sauce in Vietnam and China (Fidler et al. 2003; Chen et al. 2005), wheat and maize flour in Venezuela (Layrisse et al. 1996), infant formulae in the United States (Committee on Nutrition 1999), milk in Chile (Olivares et al. 1989), and maize-meal porridge in South Africa (Faber et al. 2005). Absorption of Fe from cereal- and legume-based foods may also be increased by the addition of ascorbic acid and ethylenediaminetetraacetic acid (EDTA), or both, or by the degradation or removal of phytic acid (Hurrell et al. 2004). Phytic acid is highest in whole-grain flours and can be decreased considerably by removing its hull (Reddy et al. 1982). Iron absorption is still low, however, even from porridges made from low-extraction flours, because small amounts of phytate inhibit Fe absorption. Phytic acid has been completely degraded in weaning cereals by adding commercial exogenous phytases (Davidsson et al. 1997), or by activating the native phytases by a combination of soaking, germinating, and fermenting (Marero et al. 1991). Whereas large-scale fortification of centrally processed flour or salt can be an effective means to supply Fe to the urban poor, reaching poor populations in remote rural areas is much more difficult. It is here that biofortification can make its most important contribution to alleviating malnutrition.

1.2.2 Zinc

Although it is difficult to biochemically define mild-to-moderate Zn deficiency in individuals, it appears likely that Zn deficiency is common in many developing countries (Gibson 1994; Sandstead 1999; Shrimpton et al. 2002). Because absorbable forms of Fe and Zn are found in many of the same foods, high rates of Fe deficiency in Southeast Asia and sub-Saharan Africa suggest the probable widespread occurrence of Zn deficiency in the same populations. In a meta-analysis by Brown et al. (2002), Zn supplementation was found to have a positive effect on linear growth and weight gain (but not weight for height) in prepubertal children. The response is especially marked in children who are stunted or underweight, probably because their habitual Zn intakes were inadequate. About one in three preschool children in developing countries have stunted growth (de Onis 2000), and a proportion of this growth failure is likely due to Zn deficiency.

Foods that are good sources of bioavailable Zn, such as animal products, are generally not available to poorer populations in the developing world due to their high cost or limited supply. Whole-grain cereals and legumes have fairly high amounts of Zn, but its absorption is inhibited by the high phytate levels in these foods. Based on national food balance data to estimate the prevalence of Zn deficiency in various regions of the world, it can be predicted that populations in Southeast Asia and sub-Saharan Africa have low daily intakes of Zn and are at risk for Zn deficiency (IZiNCG 2004). The co-occurrence of Fe and Zn deficiencies argues for a combined approach to Fe and Zn biofortification in foods. Although a high Fe:Zn ratio may inhibit Zn absorption when given as a supplement without food, there is no significant interaction between Fe and Zn when given with food (Lonnerdal 2000; IZiNCG 2004).

Zinc is required by more than 100 enzymes in the body; many bodily functions are affected by its deficiency, including physical growth, immune competence, reproductive function, and neural development (Hotz and Brown 2001). Zinc deficiency may impair neurobehavioral function in children (Grantham-McGregor and Ani 2001), and may increase the risk for adverse outcomes of pregnancy (King 2000).

Addition of Zn to infant formulas and Zn fortification of cereal flours have been implemented in several countries, including Mexico, Indonesia, and South Africa. Although a number of trials have been conducted, there is no evidence so far that Zn fortification can effectively improve Zn status in humans (Gibson 2006). In addition, several efficacy trials of Zn in complementary food supplements and multimicronutrient-fortified beverages for schoolchildren have not shown any clear benefits (Abrams et al. 2003; Ash et al. 2003; Gibson 2006). The design of fortification programs or efficacy trials is hampered by the lack of a good Zn status measure and by the limited information on absorption and sensory properties of different fortification compounds.

1.2.3 Iodine

Iodine deficiency—mainly caused by its low dietary supply—continues to be one of the major global public health problems with about 1.5 billion people worldwide at risk of deficiency-related disorders (Lyons et al. 2004). Populations living in areas where the soil has been leached of I by past glaciation or repeated heavy rainfall experience I deficiencies (de Benoist et al. 2004). Crops grown in this type of soil have low I content, and consequently fail to provide adequate amounts of dietary I to humans.

Inadequate iodine intake may lead to goiter, cretinism, mental retardation, hypothyroidism, and several other developmental abnormalities collectively referred to as the I deficiency disorders (IDD) (Zimmermann 2006). Urinary excretion data indicate widespread I deficiency in sub-Saharan Africa, India, and Eastern Europe (WHO 2006), but studies in developed countries such as Germany or Australia have also reported large segments of the population (20–76%) with suboptimal I status (Gunton et al. 1999; McDonnell et al. 2003; Buhling et al. 2003).

Iodine is essential for the synthesis of the thyroid hormones triiodothyronine and thyroxine, which regulate key biochemical reactions such as protein synthesis, energy metabolism, and basal metabolic rate (Institute of Medicine 2001b). Dietary I is usually reduced in the gut prior to near-complete absorption. The thyroid selectively concentrates circulating I for hormone synthesis, the excess I largely being eliminated via urine (Vanderpas 2006). Despite its high bioavailability, I utilization may be impaired by plant-derived substances called *goitrogens* that are mainly found in cassava and *Brassica* vegetables. These compounds block thyroidal I uptake and its incorporation into thyroid hormone precursors (Vanderpas 2006). Similarly, humic substances found in shallow or polluted wells may also interfere with the thyroidal iodination process (Institute of Medicine 2001b).

Iodized salt is the prime example of food fortification with I. Table salt is used across all cultures and often processed centrally, thus providing an ideal vehicle that provides global coverage and is easy to fortify. The WHO recommends iodization of all salt for human and livestock consumption, the so-called universal salt iodization, for the control of IDD (WHO 1999). Iodine may be added to foods as iodide or iodate, usually in the form of potassium salts. For iodized salt, preference should be given to potassium iodate because it is more stable than potassium iodide. Other foods used for fortification with I include bread, water, milk, sugar, and fish sauce.

Milk has become an important source of I in Northern Europe, the United Kingdom and the United States as a result of the use of iodophors by the dairy industry rather than through I fortification. In Finland, animal fodder is enriched with I, leading to increased I content of animal-derived foods (Allen et al. 2006).

Despite the relative success of iodized salt in large parts of the world, areas of severe I deficiency remain, for example, in Xinjiang province of China (Zhu et al. 2003), and strategies other than traditional fortification are needed to combat IDD there.

1.2.4 Selenium

Selenium deficiency is estimated to affect several hundred million individuals (Combs 2001). Poor Se status has been associated with compromised immune functions as well as an increased risk of cancer (e.g., prostate and lung) and cardiovascular disease (Lyons et al. 2004; Rayman 2005). Selenium deficiency, which is endemic in some regions of Northern China, Korea, Tibet, and Siberia, may cause osteoarthropathy (Kashin-Beck disease) in preadolescence or adolescence, or cardiomyopathy (Keshan disease) in children (Institute of Medicine 2001a). The latter seems to require an additional stress such as an infection or chemical exposure (Ge et al. 1983).

Food Se content depends on the Se content of the soil where the animal was raised or the plant was grown (Institute of Medicine 2001a). Low-Se soils are common in New Zealand, Denmark, Eastern Europe, the United Kingdom, central Siberia, and a belt from north-east to south-central China (Lyons et al. 2004). Accordingly, epidemiological data indicate suboptimal Se intake in several European countries such as the Czech Republic, the United Kingdom, and Germany, whereas parts of the United States, Canada, Japan, and Venezuela well exceed the Se intake range associated with optimal plasma glutathione peroxidase function (Rayman 2005). Large areas of Africa and Southeast Asia are also likely to be Se deficient, but this needs to be confirmed by further analytical data.

Selenium at nutritional doses can prevent cancer initiation, and at supranutritional doses may slow down cancer progression (Rayman 2005). The nutritional effect appears to be related to a reduction in oxidative stress mediated by the selenoprotein enzyme glutathione peroxidase, especially when other antioxidants are lacking. At supranutritional doses, anticancer effects of Se may be mediated by methylselenol, which retards cancer development (progression and metastasis) by preventing the formation of new blood capillaries (angiogenesis) needed to feed the tumor (Rayman 2005).

Absorption of Se occurs either as organic selenomethionine/selenocysteine or as inorganic selenite/selenate, and is not regulated by the body. Selenomethionine synthesized by plants is the major dietary source of Se, whereas the inorganic salts are mostly used for food fortification and in supplements (Institute of Medicine 2001a). Selenium generally assumes a biological function in the form of redox-active selenocysteine, which constitutes the active site of enzyme systems such as glutathione peroxidases known to detoxify hydrogen peroxide and (phospho)lipid hydroperoxides, thioredoxine reductase involved in DNA synthesis and ascorbic acid regeneration, or iodothyronine deiodinases needed for thyroid hormone activation (Institute of Medicine 2001a; Lyons et al. 2004).

Few foods have been fortified with Se so far. Sodium selenite ($NaSeO_3$) and selenate ($NaSeO_4$) are generally considered the most suitable fortificants for table salt, infant foods, margarine, and sports drinks. They display a Se bioavailability of ~50% (Institute of Medicine 2001a), and have shown comparable utility in milk-based infant formulas (Van Dael et al. 2001). Selenomethionine and other organic forms of Se are similarly well absorbed but better retained, which may be a cause for concern from a toxicological viewpoint.

Salt fortification with sodium selenite ($NaSeO_3$) (15 mg kg^{-1}) has been introduced successfully in parts of China that had endemic Se deficiency in 1983. As a result, average daily Se intakes increased from 11 to 80 μg with a concomitant decline of ~90% in the prevalence of Keshan disease (Cheng and Qian 1990; Allen et al. 2006). In preterm infants, fortification of infant formulas with sodium selenate ($NaSeO_4$) to three times the basal level improved Se status and plasma glutathione peroxidase after 12 weeks, compared to preterm infants receiving unfortified formula (10 μg Se L^{-1}) (Tyrala et al. 1996).

In general, fortification of foods with Se is a delicate task due to a very narrow safety margin. Daily Se intakes well below 50 μg or above 400 μg may have adverse health consequences such as increased risk of cancer or cardiovascular disease (Varo et al. 1994; Reilly 1998; Institute of Medicine 2001a). Stability of the fortificant is another issue because organic Se is highly volatile and sensitive to heating, and may impair the sensory qualities of food, namely odor and color (Higgs et al. 1972; Reilly 1998; Allen et al. 2006).

1.3 BIOFORTIFICATION WITH Fe, Zn, I, AND Se

Biofortification has been defined as the "development of micronutrient-dense staple crops using the best traditional breeding practices and modern biotechnology" (Nestel et al. 2006). Agronomic biofortification through fertilization provides an additional approach to improve the micronutrient content of plants grown and animal foods raised on soils so treated. Likewise, staple plant seeds and grains may be enriched in micronutrient metals to enhance disease resistance and crop yield on micronutrient-poor soils. Crop yield in general must not be compromised by biofortification in order to ensure acceptance by farmers with any newly bred varieties. Further questions to be addressed concern meaningful increases in micronutrient content, sufficient bioavailability translating into improved nutritional status, consumer acceptance, and cost effectiveness (Nestel et al. 2006).

Folate; vitamins A, C, and E; provitamin A carotenoids; and elements such as Fe, Zn, I, and Se have so far been considered for biofortification. Whereas biofortification of rice with provitamin A ("Golden Rice") has been transferred from the laboratory to field trials already in Asia, the application of biofortification is still a relatively new concept in Africa and largely confined to breeding orange-fleshed sweet potatoes in Southern Africa. The International Crops Research Institute for Semi-Arid Tropics in India has recently identified pearl millet with increased Zn content but applicability of the crop in Africa has not yet been investigated.

1.3.1 Iron

Cultivars of wheat, bean, cassava, maize, rice, and yam show considerable variation in Fe content (Welch et al. 2000; Frossard et al. 2000; Haas et al. 2005; Genc et al. 2005; Nestel et al. 2006), which suggests that selective breeding may be able to increase Fe content of staple foods. A fortification level commonly used in wheat flour of 40 mg Fe kg^{-1} would be useful (Flour Fortification Initiative 2004). However, such a level may be difficult to attain in milled wheat, because most of the Fe is removed during the milling process. This problem was evident when the efficacy of a rice cultivar high in Fe was tested in a 9-month feeding trial in Philippine women consuming either "high-Fe" rice (3.21 mg Fe kg^{-1}) or a local variety (0.57 mg Fe kg^{-1}) (Haas et al. 2005). Possibly because the high-Fe rice added only ~1.5 mg Fe d^{-1} to the diet, there was no clear benefit on Fe status. High concentrations of phytate and polyphenols may account for the low Fe absorption from other cereals and legumes, many of which have high native Fe content (Hurrell 2002). Donangelo et al. (2003) compared Fe bioavailability in humans from two varieties of red bean: an Fe-rich genotype (containing 65% more Fe) and a lower-Fe genotype. Only a small amount of Fe was absorbed from both cultivars, probably due to their high phytate and polyphenol content. Decreasing the content of these strong inhibitors in high-Fe cultivars may be necessary for them to have a positive impact on Fe absorption for human nutrition. Genotypes of maize, barley, and rice have been identified that are low-phytic-acid mutants, with phytic acid phosphorus content reduced by up to two-thirds (Raboy 2001). Although such reductions may improve Fe absorption from diets containing small amounts of meat and ascorbic acid (Tuntawiroon et al. 1990), it may be necessary to theoretically decrease phytic acid by > 90% to usefully increase Fe absorption from the monotonous cereal-based diets available in many developing countries (Hallberg et al. 1989).

Because of these limitations, genetic engineering may prove to be the most effective way to provide a useful amount of absorbable Fe in plant foods (Zimmermann and Hurrell 2002; Lonnerdal 2003). The Fe content of rice can be increased two- to threefold by introducing the ferritin gene from soybean (Goto et al. 1999) or *Phaseolus vulgaris* (Lucca et al. 2001). Iron uptake from soils may be increased by introducing a ferric reductase gene into plant root systems (Samuelsen et al. 1998). To reduce the phytic acid content of rice, Lucca et al. (2001) introduced a phytase from *Aspergillus fumigatus* that was developed to withstand food processing. Although phytase activity increased sevenfold, the phytase proved to be unstable and was destroyed when the rice was cooked. Overall, these studies suggest that Fe content can be increased in staple foods by plant breeding or genetic engineering or both. The challenge now is to show that it can be increased to nutritionally useful levels and that the additional Fe is bioavailable.

1.3.2 Zinc

Zinc fertilization and breeding for Zn accumulation represent the main strategies to increase its content in cereal crops. A screening of different genotypes yielded Zn contents in the range of 7.5–96.7 mg kg^{-1} for rice, 4.5–56.5 mg kg^{-1} for wheat, and 11–95 mg kg^{-1} for maize (Yang et al. 2007), which justifies further efforts to increase

Zn density through breeding. In a recent pilot trial at the University of Adelaide (Lyons et al. 2007), researchers examined the effect of agronomic biofortification with a micronutrient mix (Zn, Se, I, and Cu) on cassava root and leaf concentrations of these micronutrients. Cassava Zn content was largely determined by soil Zn and soil pH, and was shown to increase from 20 to 61 mg kg^{-1} dry matter (DM) in roots and from 30 to 78 mg kg^{-1} DM in leaves (Lyons et al. 2007). Field trials to investigate agronomic biofortification of cassava with Zn, Se, I, and Cu are being conducted in Colombian sites, which vary in soil pH, as well as other soil and climatic characteristics. Included in these feed trials are sites low in available Zn, where a yield response to Zn biofortification could be expected.

Fe, Zn, and phytate concentrations were determined in 76 varieties of sorghum from Benin in order to assess the impact of genetic and environmental effects on the composition of the grains, and to identify farmers' varieties with high potential Fe and Zn bioavailability (Kayode et al. 2006). In contrast to their Fe content, it appeared that none of the varieties provided adequate amounts of Zn to meet the nutritional requirements of sorghum consumers. However, strategies to silence genes controlling phytate production could be applied or phytase could be introduced as described for Fe in the previous section to improve Zn absorption.

One concern regarding agronomic fortification with Zn is that it is only suitable for crops grown on Zn-poor soils. It is probable that the edible parts of plants will increase in Zn content only up to the level which would be expected if they were grown on Zn-adequate soils. These levels, however, are likely to be too low to prevent Zn deficiency in human populations consuming cereal- and legume-based diets. For these populations, it would be necessary to increase Zn levels by genetic engineering to decrease phytate levels, or to introduce phytases, which would degrade phytic acid either during preparation or digestion of food.

1.3.3 Iodine

Biofortification strategies with I have focused on supplementation of irrigation water or soil fertilizer, whereas genotypic variation and accumulation potential seem to be less useful in improving I density in grains (Mackowiak and Grossl 1999; Dai et al. 2004; Genc et al. 2005). Cereals and vegetables represent major sources of dietary I, and they provide 52–70% of total daily intake in China and > 90% in developed countries as exemplified by the United Kingdom. Dai et al. (2004) analyzed I uptake into pot-grown pakchoi, spinach, water spinach, carrots, celery, and onions from soils treated with potassium iodate at 0, 1, or 5 mg I kg^{-1} soil. Among leafy vegetables, spinach exhibited the highest potential to accumulate I in its edible parts (> 10 mg kg^{-1}, fresh weight), whereas carrots, celery, and onions displayed somewhat lower I levels (~1 mg kg^{-1}, fresh weight). However, crop yield was significantly reduced for spinach and pakchoi at the highest I dose. Iodine toxicity and lower biomass yield were also reported in spinach (Zhu et al. 2003) and rice plants (Mackowiak and Grossl 1999) grown in hydroponic culture, with iodate being less detrimental than iodide. Notably, only the highest dose of iodate (100 μM) improved rice seed I content over control plants (Mackowiak and Grossl 1999). Iodide treatment produced comparable figures at 10 μM, but totally inhibited seed growth at 100 μM. Nutritionally relevant I enrichment of edible parts was achieved with spinach, and this seems to be much less likely with rice or wheat (Zhu et al. 2003).

Taken together, these findings indicate the need to carefully consider the form of I and its dosage levels, as well as target plants under field conditions. Nevertheless, I-deficient soils remain a major cause of low I content in plant foods, and it needs to be addressed before plant breeding strategies can become relevant. Iodine is perhaps less of a priority than Fe, Zn, or Se as salt iodization has proven extremely effective in improving the I status of human populations worldwide and is relatively cheap and easy to implement.

1.3.4 Selenium

The Se content of foods usually is a function of soil Se levels, which in turn may be improved by the application of Se-fortified fertilizers. The effectiveness of such an agronomic biofortification is best exemplified by field data from Finland. Low Se content in agricultural soils (0.2–0.3 mg kg^{-1}) led the Finnish government to introduce mandatory fortification of fertilizers with Se in 1984 (Varo et al. 1994). $NaSeO_4$ was added at 16 mg kg^{-1} to fertilizers used for grain production, and within one year these additions raised spring wheat Se content from 20 to over 200 μg kg^{-1} dry matter (DM). Winter grain varieties were less responsive, probably due to immobilization of Se after reduction from selenate to selenite in the soil. Nevertheless, this approach has helped in increasing daily Se intake of the population from 25–60 μg in 1974 to an average of 90 μg in 1992, with a corresponding increase in serum Se levels (Reilly 1998). In fact, Se fertilization was so effective that fortification levels were reduced to 6 mg kg^{-1} in 1991 (Varo et al. 1994) and are now set at 10 mg kg^{-1} (Lyons et al. 2003). The significant decline in rates of cancer and cardiovascular disease observed in Finland since 1985 may be partly attributed to improved Se status (Varo et al. 1994).

In the United Kingdom and other European countries, Se status of the population has dropped continuously upon replacement of Se-rich wheat from the United States and Canada with locally produced wheat of low Se content. Based on the Finnish experience, Se fortification of fertilizers has been recommended to balance the resulting dietary Se deficiency (Rayman 1997; Broadley et al. 2006).

Foliar application of Se solution onto pea plants has been shown to increase seed Se content up to 30-fold over control seeds (Smrkolj et al. 2006), and yielded 743 μg Se kg^{-1} compared to 21 μg Se kg^{-1}. In addition, leaf and stem Se concentrations of plants grown from these enriched pea seeds were 7- and 13-fold higher than their untreated counterparts. However, not all plant foods respond similarly to Se fertilization. For example, Eurola and colleagues (1991) reported low Se concentrations in tomatoes (30 μg kg^{-1} DM), whereas cruciferous vegetables were shown to effectively accumulate Se (600–700 μg kg^{-1} DM). Preliminary studies with Se-accumulating soybeans and *Brassica* vegetables showing a 15-fold variation in Se content also indicate some potential for breeding and genetic engineering strategies to improve plant Se uptake and retention (Combs 2001).

1.4 CONCLUSIONS

Low dietary Fe, Zn, I, and Se concentrations are common in the developing world with populations consuming predominantly plant-based diets. Deficiencies in these micronutrients are common and lead to a range of major health problems affecting mainly women and children. With the possible exception of iodized salt, industrially fortified foods rarely reach the rural poor and, for these populations, biofortification would be a more sustainable strategy to combat micronutrient malnutrition. Concentrations of Fe and Zn can be increased in staple foods by either plant breeding or genetic engineering or both; however, the latter has much more potential to provide the amounts lacking in the diet. Additionally, phytic acid reduction in the edible parts would substantially improve both Fe and Zn absorption. Farmers can grow improved varieties independently in continuous crop cycles through vegetative propagation without loss of the trait. Productivity is even expected to increase because micronutrients are as essential for plant growth and disease resistance as they are for human health and well-being. Plant varieties with sufficient micronutrients have been shown to grow deeper roots in mineral-deficient soils, and are better at tapping subsoil water and minerals, making them more drought tolerant. Zinc levels in food crops grown on Zn-deficient soils can also be usefully increased by agronomic fertilization. Similarly, the use of fertilizers seems to be the most suitable biofortification strategy for combating I and Se deficiencies as I- and Se-poor soils are the major causative factor for their low dietary intakes. The challenge is now to demonstrate that Fe, Zn, I, and Se in biofortified foods can be increased to nutritionally meaningful levels and that the additional nutrients are more bioavailable.

REFERENCES

Abrams, S.A., A. Mushi, D.C. Hilmers et al. 2003. A multinutrient-fortified beverage enhances the nutritional status of children in Botswana. *Journal of Nutrition* 133(6): 1834–1840.

Aisen, P., C. Enns, and M. Wessling-Resnick. 2001. Chemistry and biology of eukaryotic iron metabolism. *International Journal of Biochemistry & Cell Biology* 33(10): 940–959.

Allen, L., B. de Benoist, O. Dary et al. 2006. Guidelines on food fortification with micronutrients. WHO/FAO: Geneva.

Ash, D.M., S.R. Tatala, E.A. Frongillo, Jr. et al. 2003. Randomized efficacy trial of a micronutrient-fortified beverage in primary school children in Tanzania. *American Journal of Clinical Nutrition* 77(4): 891–898.

Broadley, M.R., P.J. White, R.J. Bryson et al. 2006. Biofortification of UK food crops with selenium. *Proceedings of the Nutrition Society* 65(2): 169–181.

Brown, K.H., J.M. Peerson, J. Rivera et al. 2002. Effect of supplemental zinc on the growth and serum zinc concentrations of prepubertal children: A meta-analysis of randomized controlled trials. *American Journal of Clinical Nutrition* 75(6): 1062–1071.

Brownlie IV, T., V. Utermohlen, P.S. Hinton et al. 2004. Tissue iron deficiency without anemia impairs adaptation in endurance capacity after aerobic training in previously untrained women. *American Journal of Clinical Nutrition* 79(3): 437–443.

Bruner, A.B., A. Joffe, A.K. Duggan et al. 1996. Randomised study of cognitive effects of iron supplementation in non-anaemic iron-deficient adolescent girls. *Lancet* 348(9033): 992–996.

Buhling, K.J., J. Schaff, H. Bertram et al. 2003. Supply of iodine during pregnancy—An inventory in Berlin, Germany. *Zeitschrift fuer Geburtshilfe und Neonatologie* 207(1): 12–16.

Carpenter, C.E. and A.W. Mahoney. 1992. Contributions of heme and nonheme iron to human nutrition. *Critical Reviews in Food Science and Nutrition* 31(4): 333–367.

Chen, J., X. Zhao, X. Zhang et al. 2005. Studies on the effectiveness of NaFeEDTA-fortified soy sauce in controlling iron deficiency: A population-based intervention trial. *Food and Nutrition Bulletin* 26(2): 177–189.

Cheng, Y.Y. and P.C. Qian. 1990. The effect of selenium-fortified table salt in the prevention of Keshan disease on a population of 1.05 million. *Biomedical and Environmental Sciences* 3(4): 422–428.

Combs, G.F., Jr. 2001. Selenium in global food systems. *British Journal of Nutrition* 85(5): 517–547.

Committee on Nutrition. 1999. Iron Fortification of Infant Formulas. *Pediatrics* 104(1): 119–123.

Dai, J.L., Y.G. Zhu, M. Zhang et al. 2004. Selecting iodine-enriched vegetables and the residual effect of iodate application to soil. *Biological Trace Element Research* 101(3): 265–276.

Davidsson, L., P. Galan, F. Cherouvrier et al. 1997. Bioavailability in infants of iron from infant cereals: Effect of dephytinization. *American Journal of Clinical Nutrition* 65(4): 916–920.

de Benoist, B., M. Andersson, I. Egli et al. 2004. Iodine status worldwide—WHO Global Database on Iodine Deficiency. Geneva: WHO.

de Onis, M. 2000. Measuring nutritional status in relation to mortality. *Bulletin of the World Health Organization* 78(10): 1271–1274.

Donangelo, C.M., L.R. Woodhouse, S.M. King et al. 2003. Iron and zinc absorption from two bean (*Phaseolus vulgaris* L.) genotypes in young women. *Journal of Agricultural and Food Chemistry* 51(17): 5137–5143.

Eurola, M.H., P.I. Ekholm, M.E. Ylinen et al. 1991. Selenium in Finnish foods after beginning the use of selenate-supplemented fertilisers. *Journal of the Science of Food and Agriculture* 56(1): 57–70.

Faber, M., J.D. Kvalsvig, C.J. Lombard et al. 2005. Effect of a fortified maize-meal porridge on anemia, micronutrient status, and motor development of infants. *American Journal of Clinical Nutrition* 82(5): 1032–1039.

FAO/WHO. 2001. *Human Vitamin and Mineral Requirements*. FAO: Rome. 195–221.

Fidler, M.C., L. Davidsson, T. Walczyk et al. 2003. Iron absorption from fish sauce and soy sauce fortified with sodium iron EDTA. *American Journal of Clinical Nutrition* 78(2): 274–278.

Flour Fortification Initiative. 2004. Wheat Flour Fortification: Current Knowledge and Practical Applications, In *Summary Report of an International Technical Workshop*. Cuernavaca, Mexico.

Frossard, E., M. Bucher, F. Machler et al. 2000. Potential for increasing the content and bioavailability of Fe, Zn and Ca in plants for human nutrition. *Journal of the Science of Food and Agriculture* 80(7): 861–879.

Ge, K., A. Xue, J. Bai et al. 1983. Keshan disease-an endemic cardiomyopathy in China. *Virchows Archiv. A, Pathological Anatomy and Histopathology* 401(1): 1–15.

Genc, Y., J.M. Humphries, G.H. Lyons et al. 2005. Exploiting genotypic variation in plant nutrient accumulation to alleviate micronutrient deficiency in populations. *Journal of Trace Elements in Medicine and Biology* 18(4): 319–324.

Gibson, R.S. 1994. Content and bioavailability of trace elements in vegetarian diets. *American Journal of Clinical Nutrition* 59(5 Suppl): 1223S–1232S.

Gibson, R.S. 2006. Zinc: The missing link in combating micronutrient malnutrition in developing countries. *Proceedings of the Nutrition Society* 65(1): 51–60.

Gibson, R.S., L. Perlas, and C. Hotz. 2006. Improving the bioavailability of nutrients in plant foods at the household level. *Proceedings of the Nutrition Society* 65(2): 160–168.

Gillooly, M., T.H. Bothwell, J.D. Torrance et al. 1983. The effects of organic acids, phytates and polyphenols on the absorption of iron from vegetables. *British Journal of Nutrition* 49(3): 331–342.

Goto, F., T. Yoshihara, N. Shigemoto et al. 1999. Iron fortification of rice seed by the soybean ferritin gene. *Nature Biotechnology* 17(3): 282–286.

Grantham-McGregor, S. and C. Ani. 2001. A review of studies on the effect of iron deficiency on cognitive development in children. *Journal of Nutrition* 131(2S-2): 649S–666S and discussion 666S–668S.

Gunton, J.E., G. Hams, M. Fiegert et al. 1999. Iodine deficiency in ambulatory participants at a Sydney teaching hospital: is Australia truly iodine replete? *Medical Journal of Australia* 171(9): 467–470.

Haas, J.D. and T. Brownlie, IV. 2001. Iron deficiency and reduced work capacity: A critical review of the research to determine a causal relationship. *Journal of Nutrition* 131(2): 676S–690S.

Haas, J.D., J.L. Beard, L.E. Murray-Kolb et al. 2005. Iron-biofortified rice improves the iron stores of nonanemic Filipino women. *Journal of Nutrition* 135(12): 2823–2830.

Hallberg, L., M. Brune, and L. Rossander. 1989. Iron absorption in man: ascorbic acid and dose-dependent inhibition by phytate. *American Journal of Clinical Nutrition* 49(1): 140–144.

Higgs, D.J., V.C. Morris, and O.A. Levander. 1972. Effect of cooking on selenium content of foods. *Journal of Agricultural and Food Chemistry* 20(3): 678–680.

Hotz, C. and K.H. Brown. 2001. Identifying populations at risk of zinc deficiency: The use of supplementation trials. *Nutrition Reviews* 59(3 Pt 1): 80–84.

Hurrell, R.F. 1997. Preventing iron deficiency through food fortification. *Nutrition Reviews* 55(6): 210–222.

Hurrell, R.F. 2002. Fortification: Overcoming technical and practical barriers. *Journal of Nutrition* 132(4): 806S–812S.

Hurrell, R.F., S. Lynch, T. Bothwell et al. 2004. Enhancing the absorption of fortification iron. A SUSTAIN Task Force report. *International Journal for Vitamin and Nutrition Research* 74(6): 387–401.

Imhoff-Kunsch, B., R. Flores, O. Dary et al. 2007. Wheat flour fortification is unlikely to benefit the neediest in Guatemala. *Journal of Nutrition* 137(4): 1017–1022.

Institute of Medicine. 2001a. *Dietary Reference Intakes for Vitamin C, Vitamin E, Selenium, and Carotenoids. Dietary Reference Intakes,* IOM. Washington, D.C.: National Academy Press.

Institute of Medicine. 2001b. *Dietary Reference Intakes for Vitamin A, Vitamin K, Arsenic, Boron, Chromium, Copper, Iodine, Iron, Manganese, Molybdenum, Nickel, Silicon, Vanadium, and Zinc. Dietary Reference Intakes*, IOM. Washington, D.C.: National Academy Press.

IZiNCG. 2004. Assessment of the risk of zinc deficiency in populations and options for its control. *Food and Nutrition Bulletin* 25(1, Suppl. 2): S91–S204.

Kayode, A.P., A.R. Linnemann, J.D. Hounhouigan et al. 2006. Genetic and environmental impact on iron, zinc, and phytate in food sorghum grown in Benin. *Journal of Agricultural and Food Chemistry* 54(1): 256–262.

King, J.C. 2000. Determinants of maternal zinc status during pregnancy. *American Journal of Clinical Nutrition* 71(5S Suppl): 1334S–1343S.

Layrisse, M., J.F. Chaves, Mendez-Castellano et al. 1996. Early response to the effect of iron fortification in the Venezuelan population. *American Journal of Clinical Nutrition* 64(6): 903–907.

Lonnerdal, B. 2000. Dietary factors influencing zinc absorption. *Journal of Nutrition* 130(5S Suppl): 1378S–1383S.

Lonnerdal, B. 2003. Genetically modified plants for improved trace element nutrition. *Journal of Nutrition* 133(5): 1490S–1493S.

Lucca, P., R. Hurrell, and I. Potrykus. 2001. Approaches to improving the bioavailability and level of iron in rice seeds. *Journal of the Science of Food and Agriculture* 81(9): 828–834.

Lyons, G., J. Stangoulis, and R. Graham. 2003. High-selenium wheat: Biofortification for better health *Nutrition Research Reviews* 16(1): 45–60.

Lyons, G.H., J.C.R. Stangoulis, and R.D. Graham. 2004. Exploiting micronutrient interaction to optimize biofortification programs: The case for inclusion of selenium and iodine in the HarvestPlus program. *Nutrition Reviews* 62(6): 247–252.

Lyons, G.H., H. Ceballos, Y. Genc et al. 2007. Agronomic biofortification of cassava with zinc and other micronutrients to improve human health. http://www.fertilizer.org/ifa/publicat/PDF/2007_zinccrops2007_lyons.pdf. (accessed May 05, 2007).

Mackowiak, C.L. and P.R. Grossl. 1999. Iodate and iodide effects on iodine uptake and partitioning in rice (*Oryza sativa L.*) grown in solution culture. *Plant and Soil* 212(2): 135–143.

Marero, L.M., E.M. Payumo, A.R. Aguinaldo et al. 1991. Antinutritional factors in weaning foods prepared from germinated cereals and legumes. *Food Science and Technology-Lebensmittel-Wissenschaft and Technologie* 24(2): 177–181.

Martinez-Torres, C. and M. Layrisse. 1971. Iron absorption from veal muscle. *American Journal of Clinical Nutrition* 24(5): 531–540.

McDonnell, C.M., M. Harris, and M.R. Zacharin. 2003. Iodine deficiency and goitre in schoolchildren in Melbourne, 2001. *Medical Journal of Australia* 178(4): 159–162.

Monsen, E.R. and J.L. Balintfy. 1982. Calculating dietary iron bioavailability: Refinement and computerization. *Journal of the American Dietetic Association* 80: 307–311.

Moretti, D., M.B. Zimmermann, S. Muthayya et al. 2006. Extruded rice fortified with micronized ground ferric pyrophosphate reduces iron deficiency in Indian schoolchildren: A double-blind randomized controlled trial. *American Journal of Clinical Nutrition* 84(4): 822–829.

Narins, D. 1980. Absorption of Nonheme Iron. In *Biochemistry of Nonheme Iron*, A. Bezkorovainy (Ed.), New York: Plenum Press.

Nestel, P., H.E. Bouis, J.V. Meenakshi et al. 2006. Biofortification of staple food crops. *Journal of Nutrition* 136(4): 1064–1067.

Olivares, M., T. Walter, E. Hertrampf et al. 1989. Prevention of iron deficiency by milk fortification. The Chilean experience. *Acta Paediatrica Scandinavica. Supplement* 361: 109–113.

Raboy, V. 2001. Seeds for a better future: "Low phytate" grains help to overcome malnutrition and reduce pollution. *Trends in Plant Science* 6(10): 458–462.

Rayman, M.P. 1997. Dietary selenium: Time to act. *British Medical Journal* 314(7078): 387–388.

Rayman, M.P. 2005. Selenium in cancer prevention: A review of the evidence and mechanism of action. *Proceedings of the Nutrition Society* 64(4): 527–542.

Reddy, M.B., R.F. Hurrell, M.A. Juillerat et al. 1996. The influence of different protein sources on phytate inhibition of nonheme-iron absorption in humans. *American Journal of Clinical Nutrition* 63(2): 203–207.

Reddy, N.R., S.K. Sathe, and D.K. Salunkhe. 1982. Phytate in legumes and cereals. *Advances in Food Research* 28: 1–92.

Reilly, C. 1998. Selenium: A new entrant into the functional food arena. *Trends in Food Science and Technology* 9(3): 114–118.

Samuelsen, A.I., R.C. Martin, D.W. Mok et al. 1998. Expression of the yeast FRE genes in transgenic tobacco. *Plant Physiology* 118(1): 51–58.

Sandstead, H.H. 1999. Improving study design. *American Journal of Clinical Nutrition* 70(1): 110.

Shrimpton, R., R. Gross, I. Darnton-Hill et al. 2002. Effect of supplemental zinc on the growth and serum zinc concentrations of prepubertal children: A meta-analysis of randomized controlled trials. *American Journal of Clinical Nutrition* 75(6): 347–349.

Siegenberg, D., R. D. Baynes, T. H. Bothwell et al. 1991. Ascorbic acid prevents the dose-dependent inhibitory effects of polyphenols and phytates on nonheme-iron absorption. *American Journal of Clinical Nutrition* 53(2): 537–541.

Smrkolj, P., M. Germ, I. Kreft et al. 2006. Respiratory potential and Se compounds in pea (*Pisum sativum* L.) plants grown from Se-enriched seeds. *Journal of Experimental Botany* 57(14): 3595–3600.

Stephenson, L.S., M.C. Latham, and E.A. Ottesen. 2000. Global malnutrition. *Parasitology* 121 Suppl: S5–S22.

Stoltzfus, R.J., J.D. Kvalsvig, H.M. Chwaya et al. 2001. Effects of iron supplementation and anthelmintic treatment on motor and language development of preschool children in Zanzibar: Double blind, placebo controlled study. *British Medical Journal* 323(7326): 1389–1393.

Tuntawiroon, M., N. Sritongkul, L. Rossander-Hulten et al. 1990. Rice and iron absorption in man. *European Journal of Clinical Nutrition* 44(7): 489–497.

Tyrala, E.E., M.W. Borschel, and J.R. Jacobs. 1996. Selenate fortification of infant formulas improves the selenium status of preterm infants. *American Journal of Clinical Nutrition* 64(6): 860–865.

Van Dael, P., L. Davidsson, R. Munoz-Box et al. 2001. Selenium absorption and retention from a selenite- or selenate-fortified milk-based formula in men measured by a stable-isotope technique. *British Journal of Nutrition* 85(2): 157–163.

Vanderpas, J. 2006. Nutritional epidemiology and thyroid hormone metabolism. *Annual Review of Nutrition* 26(1): 293–322.

Varo, P., G. Alfthan, J.K. Huttunen et al. 1994. Nationwide Selenium Supplementation – Effects on Diet, Blood and Tissue Levels, and Health. In *Selenium in Biology and Human Health,* R.F. Burk (Ed.), New York: Springer Verlag. 197–218.

Welch, R.M., W.A. House, S. Beebe et al. 2000. Genetic selection for enhanced bioavailable levels of iron in bean (*Phaseolus vulgaris* L.) seeds. *Journal of Agricultural and Food Chemistry* 48(8): 3576–3580.

WHO. 1999. *Progress Towards the Elimination of Iodine Deficiency Disorders (IDD).* Geneva: WHO.

WHO. 2002. *The World Health Report 2002—Reducing Risks, Promoting Healthy Life.* Geneva: WHO.

WHO. 2006. World Health Organization Global Database on Iodine Deficiency.

Yang, X.E., W.R. Chen, and Y. Feng. 2007. Improving human micronutrient nutrition through biofortification in the soil-plant system: China as a case study. *Environmental Geochemistry and Health* 29(5): 413–428.

Zhu, Y.G., Y.Z. Huang, Y. Hu et al. 2003. Iodine uptake by spinach *(Spinacia oleracea L.)* plants grown in solution culture: Effects of iodine species and solution concentrations. *Environment International* 29(1): 33–37.

Zimmermann, M.B. and R.F. Hurrell. 2002. Improving iron, zinc and vitamin A nutrition through plant biotechnology. *Current Opinion in Biotechnology* 13(2): 142–145.

Zimmermann, M.B., N. Chaouki, and R.F. Hurrell. 2005. Iron deficiency due to consumption of a habitual diet low in bioavailable iron: A longitudinal cohort study in Moroccan children. *American Journal of Clinical Nutrition* 81(1): 115–121.
Zimmermann, M.B. 2006. The influence of iron status on iodine utilization and thyroid function. *Annual Review of Nutrition* 26(1): 367–389.

2 Biofortification in the Food Chain, and Use of Selenium and Phyto-Compounds in Risk Reduction and Control of Prostate Cancer

Graham Lyons, Yusuf Genc, and Robin Graham

CONTENTS

2.1 INTRODUCTION

Malnutrition is easily the most important cause of global mortality, with over 50% of all deaths annually due to diet-related diseases. Dietary deficiencies of the micronutrients iron (Fe), zinc (Zn), selenium (Se), and iodine (I) as well as various vitamins are widespread globally, affecting well over half of the world's population, and often occur concurrently (WHO 2003). Selenium, Fe, Zn, and vitamins A, B, and C have immunomodulating functions, and thus influence the susceptibility of a host to infectious diseases, and their courses and outcomes (Bhaskaram 2002; Failla 2003). The global crisis in malnutrition is due to dysfunctional food systems that fail to deliver sufficient essential nutrients. A food systems paradigm encompasses an agriculture that aims not only at productivity and sustainability, but also at adequate nutrition (Graham et al. 2001). Biofortification of staple crops with micronutrients by breeding for higher density of micronutrients in edible parts or by fertilization is an important component of a food systems strategy to address widespread dietary deficiency in human populations. Breeding (*genetic* biofortification) may be more suitable to enhance provitamin A carotenoids and Fe, whereas a fertilizer strategy (*agronomic* biofortification) may be more effective for Se, I, and Zn.

In the first part of this chapter, we investigate the genotypic variation in wheat (*Triticum aestivum* L.) and other cereals for grain Se density, and present two case studies of agronomic biofortification: Se-biofortified wheat and cassava biofortified with Zn, Se, and I. Current food systems research being conducted under the *HarvestPlus* Biofortification Challenge Program is described in the second section. Finally, evidence is advanced for the role of Se and several phytochemicals in both risk reduction and control of prostate cancer, a major cause of death in men in developed countries. Incidence and mortality rates of prostate cancer are likely to increase with the adoption of more *western* diets by the burgeoning middle-class in many developing countries.

2.2 CASE STUDY 1: BIOFORTIFICATION OF WHEAT WITH SELENIUM

2.2.1 Selenium: Its Importance and Uneven Distribution

Selenium is profoundly important for human health, being involved (mainly through the activity of selenoenzymes) in antioxidant, antiinflammatory, anticancer, antiheart disease, antiviral, and antiaging activities, as well as in fertility, thyroid, brain, and immune functions. Its importance is exemplified by its status as the only micronutrient to be specified in the human genome (as selenocysteine, the 21st amino acid) (Rayman 2002). It is difficult to quantify the number of Se-deficient people in the world (i.e., those who consume insufficient Se daily to support maximal expression of the key selenoenzymes), but it is likely to be in the range of 500–1000 million (Combs, 2001).

Selenium in a food system depends mainly on the levels of available Se in soils used for agriculture. It is ubiquitous in soils but very unevenly distributed, hence the variable Se status observed in different populations and their subgroups. Selenium availability in soils depends upon soil pH, redox potential, calcium carbonate level, cation exchange capacity (Bañuelos and Schrale 1989; Broadley et al. 2006), and organic carbon, iron, and aluminum levels (Ylaranta 1983). Dry climate, low organic matter, high temperature, high soil pH, and aeration are likely to increase the selenate–selenite ratio in the soil and hence the availability of Se to plants, as selenate is less strongly adsorbed to soil minerals (Combs 2001).

Low-Se areas include much of Eastern Europe, sub-Saharan Africa, a belt through the middle of China, and New Zealand. Higher-Se countries include Canada, Japan, Venezuela, and the United States (Reilly 1996). Selenium deficiency in Australia usually occurs on low-pH soils in higher rainfall areas, including southwest Western Australia, South Australia's Mount Lofty Ranges and Kangaroo Island, and the southern coast of Queensland (Reilly 1996; Judson and Reuter 1999). The Se status of the Australian population is estimated to be above the hypothetical global mean, with most individuals likely to ingest sufficient Se to maximize the activity of the important antioxidant selenoenzyme, glutathione peroxidase, in blood (Figure 2.1). This value of around 100 μg Se L^{-1} may be lower in Western Australia and Tasmania. There is also some evidence that Se in Australian food systems may have declined in recent decades (Figure 2.2; Lyons et al. 2003 and 2004a).

Wheat is one of the most important global dietary sources of Se. The sensitivity of a population's Se status to changes in its source of wheat is illustrated by the United Kingdom, where blood Se levels have declined by around 50% since changing from high-Se Canadian- and U.S.-grown wheat to low-Se European wheat in the mid-1980s (Rayman 2000). In Australia, it is estimated that nearly half of the Se intake of most adults is provided by wheat (Barrett et al. 1989).

2.2.2 Genetic Biofortification of Wheat with Selenium: Is It Feasible?

A strategy of breeding a staple crop such as wheat with the ability to fortify itself with micronutrients offers a sustainable, cost-effective alternative to other strategies such as individual supplementation and fertilization. Genetically fortified crops would be more likely to reach those most in need and have the added advantage of requiring no change in current consumer behavior to be effective (Graham et al. 2001). Sufficient genotypic variation in the trait under selection is necessary for conventional breeding to be successful. For example, a twofold range in micronutrient concentration in genotypes grown under identical conditions and with similar yield would be considered plenty. Breeding criteria for micronutrient-enriched staple

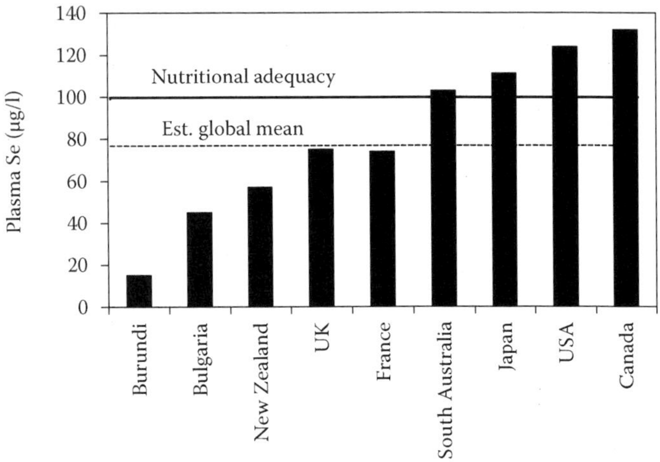

FIGURE 2.1 Comparative plasma Se concentrations from around the world. (Data from Marinov et al. 1998. *Akush Ginekologia (Sophia)* 37: 15–16; Robinson et al. 1997. *British Journal of Nutrition* 77: 551–563; Bates et al. 2002. *European Journal of Clinical Nutrition* 56(9): 873–881; Terrier et al. 1995. *Biological Trace Element Research* 47: 279–285; Lyons et al. 2004a. *Medical Journal of Australia* 180(8): 383–386; Imai et al. 1990. *Nutrition Research* 10: 1205–1214; Burke et al. 1992. *Photodermatology, Photoimmunology and Photomedicine* 9: 1–6; Nomura et al. 2000. *Cancer Prevention Biomarkers* 9: 883–887; Benemariya et al. 1993. *Science of the Total Environment.* 136: 49–76.)

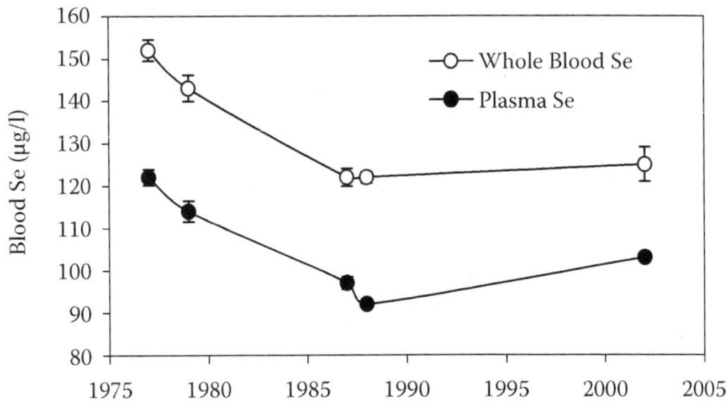

FIGURE 2.2 Trend in whole blood and plasma selenium concentrations in South Australians, 1977–2002. Sample sizes: 1977: n = 117; 1979: n = 30; 1987: n = 199; 1988: n = 200; 2002: n = 288; total n = 834. (From Lyons et al. 2004a. *Medical Journal of Australia* 180(8): 383–386. With permission.)

crops include (1) maintenance of crop productivity (i.e., no yield penalty), (2) stability of micronutrient enrichment traits across various edaphic and climatic zones, (3) demonstration of significant effects of enriched micronutrients on human health, (4) demonstration of bioavailability of enriched micronutrients for human nutrition, and (5) consumer acceptance (Welch and Graham 2004). Transgenic breeding strategies also have an important role to play in micronutrient biofortification, as shown by the development of *Golden Rice* (Potrykus 2003). Current plant biotechnology projects are investigating improved protein and mineral nutrients in cassava and rice (Sautter et al. 2006).

2.2.2.1 Wheat-Selenium Concentration Can Be Highly Variable

To determine whether sufficient genotypic variation in grain Se density in wheat exists to permit a breeding approach to increase the Se concentration, surveys, and field trials were conducted by the University of Adelaide and Centro Internacional de Mejoramiento de Maiz y Trigo (International Maize and Wheat Improvement Center, CIMMYT) in Australia and Mexico, respectively. Diverse genotypes, including ancestral and wild relatives of wheat, landraces, modern commercial bread and durum wheats, doubled haploid populations, synthetic wheats, triticale, rye, and barley were analyzed for Se. Grain Se concentrations varied 144-fold, from 5 $\mu g\ kg^{-1}$ (grown on high-Fe lateritic soil on Kangaroo Island, South Australia) to 720 $\mu g\ kg^{-1}$ (grown on soil with high pH, low Fe, low S, and low organic matter at Minnipa, Eyre Peninsula, and South Australia). The grand mean concentration for Se was 155 ± 15 $\mu g\ kg^{-1}$, and the median was 100 $\mu g\ kg^{-1}$. Selenium in most wheat grown in South Australia from 2000 to 2002 was in the range 70–280 $\mu g\ kg^{-1}$ (Lyons et al. 2005a). This study provisionally places South Australian wheat higher than wheat from New Zealand, the United Kingdom, most of Europe, and China, but lower than Canada and the United States for grain Se concentration.

2.2.2.2 Genotypic Variation of Selenium Density in Modern Wheat Cultivars Is Difficult to Detect

The genotype × environment studies showed that most of the variation in grain Se concentration was due to variations in available soil Se. Total soil Se was found to be an unsuitable indicator of availability. Although it is commonly stated in the literature that a total soil Se concentration of up to 600 $\mu g\ kg^{-1}$ is considered to be deficient (Gupta and Gupta 2000), this value may need to be revised, as our study found at least a 100-fold variation in Se concentration in wheat grain, for example, 5 $\mu g\ kg^{-1}$ at Kangaroo Island and 720 $\mu g\ kg^{-1}$ at Minnipa, in soils testing < 200 $\mu g\ kg^{-1}$ of total Se. This observation suggests a wide variation in available Se concentration in these two soil types. The Minnipa site grew the highest-Se wheat in the survey and could not be considered Se deficient. South Australian soils are renowned for their microspatial variability in available micronutrient concentration, and this study shows this to be particularly so for Se (Lyons et al. 2005a). Spatial variation in available soil Se, even within an individual site, was found to be large, making detection of genotypic differences in Se uptake efficiency difficult. For example, a sixfold variation in grain Se concentration from 110 to 690 $\mu g\ kg^{-1}$ was found at Bordertown between samples from four replicate plots of a single wheat cultivar grown together in one field. In trials conducted in Mexico, the soil Se variation, although not as large as in South Australia, was still substantial, as indicated by a 27-fold variation in grain Se concentration, from 9 to 244 μg kg^{-1} (Lyons et al. 2005a).

No significant genotypic variation in grain Se density among modern commercial bread or durum wheat (*T. turgidum* L. var. durum), triticale (*X. Triticosecale* Wittmack) or barley (*Hordeum vulgare* L.) varieties was detected in this study. However, diploid wheat (*Aegilops tauschii* L.) and rye (*Secale cereale* L.) were significantly higher in grain Se concentration than other small-grain cereals in separate field trials, and rye was significantly higher in foliar Se content than two wheat landraces in a short-term hydroponic trial (Lyons et al. 2005a). Although genotypic differences may exist in modern wheat varieties, they are likely to be insignificant in comparison with background soil variation, at least in South Australia and Mexico. Field sites that are spatially very uniform in available soil Se and pot trials with varying Se concentrations (including low concentrations of available Se) would be needed to allow comparison of grain Se concentration and content in order to accurately assess genotypic variation. Moreover, enhanced uptake efficiency for selenate (the most soluble and mobile form of Se) may be of limited value in wheat grown on soils of very low available Se, where most Se is present as selenite, selenide, and elemental Se (Cary and Allaway 1969).

2.2.3 Agronomic Biofortification of Selenium in Wheat

2.2.3.1 Selenate Effective in Increasing Grain Selenium Concentration

Agronomic biofortification of Se in wheat was investigated in Australia as an alternative to breeding to increase grain Se concentration. The study included trials in the field, glasshouse, and growth chamber. Selenium was applied to the soil in aqueous solution to ensure accurate dose delivery. It was found that Se applied as sodium selenate to the soil at seeding was more effective than postanthesis foliar application, even on soils of variable pH (6.6–8.6 pH H_2O), Fe (7–66 mg kg^{-1} DTPA Fe), S (2.4–4.4 mg kg^{-1}), and organic carbon (0.8–1.3%) content. Soil application of selenate (SeO_4) at rates of 4 and 120 g Se ha^{-1}, increased grain Se concentration progressively from 0.062 to 8.33 mg kg^{-1}. This was a 133-fold increase at Charlick, the site with the lower baseline Se level, lower pH and higher Fe, S, and carbon, whereas the foliar application of selenate at the highest application rate increased grain Se concentration from 0.062 to 1.24 mg kg^{-1}, a 20-fold increase, at the same Charlick site (see Figure 2.3; Lyons et al. 2004b).

Earlier studies in Europe and North America likewise demonstrate the effectiveness of agronomic biofortification using sodium selenate, and these have been reviewed thoroughly by Broadley et al. (2006). In a higher rainfall year, it is likely that the foliar Se would have been more effective than observed in this trial, as early senescence in the form of extensive leaf yellowing was evident as early as anthesis. The addition of selenate to N–P–K fertilizers for use on crops and pastures has been practiced in Finland since 1984, and has been an effective method to increase the entire population's Se status (Aro et al. 1995; Hartikainen 2005). Although it does not appear that Se is an essential micronutrient for higher plants, benefits of Se addition have been

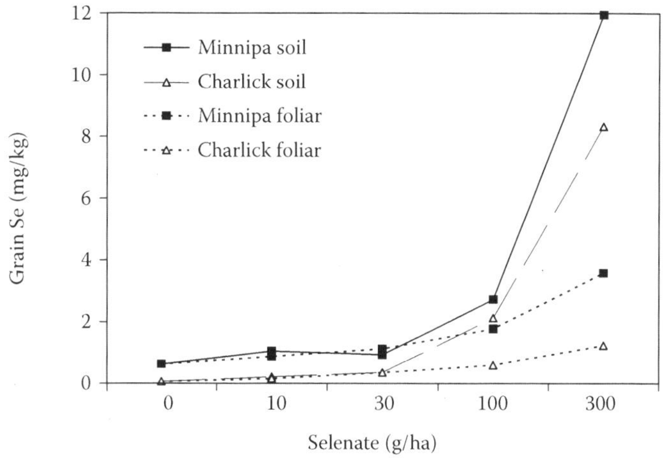

FIGURE 2.3 Grain selenium concentration of wheat grown at two sites in South Australia in 2002. Selenium was applied as sodium selenate either to the soil at seeding or as a foliar spray after flowering. Means are presented from four replications. The two sites have different soil characteristics, including pH: Minnipa 8.6; Charlick 6.6. (From Lyons et al. 2005b. *Journal of Trace Elements in Medicine and Biology* 19: 75–82. With permission.)

observed, including reduction in oxidative stress, delayed senescence, increased accumulation of starch and sugars, and enhanced growth (Hartikainen 2005).

Sulfur was found to inhibit Se uptake. A relatively low application of 30 kg S ha^{-1} at seeding was associated with a 16% reduction in grain Se concentration in a field trial (Lyons et al. 2004b), which supports the findings of other studies. These include the United Kingdom study by Adams et al. (2002), who found a negative correlation between grain Se and S, and between grain Se and the soil S application rate. The effect is due to the competitive inhibition by S of Se uptake via the main S transporter (Läuchli 1993). Gypsum (calcium sulphate, which is applied at rates of up to 10 t ha^{-1} to treat sodic soils) and high-S fertilizers, such as single superphosphate, ammonium sulphate, and potassium sulphate, are likely to reduce Se concentration in crops. Many of Australia's low-Se soils occur in coastal areas, where S deposition from rainfall can be as high as 20 kg ha^{-1} yr^{-1} (Blair et al. 1997), which may reduce Se uptake by crops grown in these areas. Researchers in the United Kingdom point out that the decline of Se in soils, crops, and humans may also be due to reduced use of phosphate rock and single superphosphate, in which Se can occur at up to 55 mg kg^{-1} (Broadley et al. 2006).

Selenium phytotoxicity studies conducted in the field, glasshouse, and laboratory showed that wheat is tolerant of relatively high levels of available Se in soil. A critical tissue concentration (in whole tops harvested at 30 days) for Se toxicity of around 325 mg kg^{-1} was found in a trial using wheat grown in UCMix (a type of growth medium developed by the University of California) (Barker et al. 1998) containing 2.6 mg kg^{-1} of soil Se as selenate for 30 days. This trial suggested that toxicity would not occur at selenate application rates between 10 and 200 g ha^{-1} Se for the biofortification for wheat (Lyons et al. 2005c).

2.2.3.2 Sprouting Biofortification

Field biofortification with Se is quite a wasteful process, with at best 20% recovery in grain when applied to wheat in the field. Selenium not taken up by crops soon after application is usually unavailable later in the same season or in subsequent seasons due to its adsorption onto soil particles. This observation supports the need to continue screening crop varieties for germplasm, which is efficient at Se uptake and accumulation, even though, as discussed earlier, environmental variation in available Se tends to outweigh genotypic variation.

It can be argued that Se is a relatively scarce resource that should be used efficiently, especially as demand increases (Haug et al. 2008). Furthermore, Se is not easy to recycle. Its use as a minor component in many products makes it difficult (or impossible) to recycle efficiently, and only around an estimated 15% of refined Se comes from secondary sources (Reilly 1996; STDA 2006). However, the impending scarcity argument can also be applied to certain other micronutrients, notably Zn. Current global Se reserves are estimated at around 170,000 t, whereas current total usage is around 500 t per year (USGS 2005). For example, if all of the 1.8 million hectares of wheat in the United Kingdom were fertilized with Se at 10 g ha^{-1}, this rate would equate to just 18 t Se per year (USGS 2005).

Sprouting biofortification is an alternative to Se application in the field; it allowed 100% Se recovery when sodium selenite was used for sprouting rye. After 5 days, the biomass (seed residue, shoots, and roots) was dried, powdered, and found to contain 60 mg Se kg^{-1}. The total Se was speciated and found to be entirely composed of organic Se forms, mainly selenomethionine. Thus, during sprouting there was complete biotransformation of inorganic Se (as selenite) into organic forms. This Se-enriched powder (4 mg Se kg^{-1}) was then added to normal rye flour to make rye bread, with no detrimental effects/flavors in the bread (Bryszewska et al., 2005).

Gibson et al. (2006) have shown that barley can be biofortified with Se to produce beer with increased levels of organic Se. In an early trial conducted by this group, field biofortification resulted in beer containing 30–100 μg Se L^{-1}. Malting of barley involves sprouting; hence, the same Se biofortification method (as described earlier) can be used to produce high-Se beer. Poschl and Seitz (2004) hypothesized that beer with enhanced levels of Se and/or Zn would be less damaging to the liver, for example, by reducing the formation of or binding the carcinogenic alcohol metabolite acetaldehyde. Further trials, in which biofortified barley is being used to make beer are being conducted by a brewer in South Korea, a low Se country.

2.2.3.3 Benefits of Agronomically Biofortified Selenium-Wheat

Selenium is an ideal micronutrient for agronomic biofortification. Selenate is readily taken up by plants growing on most soils of pH 5.5–9.0. It is transported easily throughout the plant, it accumulates in edible parts, and it is converted to organic forms, mainly selenomethionine, which are suitable for humans. Selenium is relatively evenly distributed throughout cereal grain, and thus can be abundant in milled products such as white flour and polished rice. Moreover, Se does not bind appreciably to phytate or polyphenols, and it is usually highly bioavailable (Lyons et al. 2003).

Findings from a clinical trial conducted in Adelaide in 2005 suggested that biofortified Se would be more bioavailable than exogenous selenomethionine when subjected to strong heating during processing (Lyons, unpublished). A follow-up speciation study using isotope-dilution HPLC coupled with ICP mass spectrometry showed that selenomethionine in biofortified wheat is resistant to heat-induced oxidation, whereas selenomethionine added to wheat postharvest is not (Kirby et al. 2006).

In human health terms, Se is best known for its antioxidant and anticancer-related activities, and evidence for these effects is strong. Most anticancer studies have used Se in the selenite form because it is easily obtainable and showed significant bioactivity against a range of cancers from the earliest trials. However, Finley and Davis (2001) examined the inhibitory effect of different Se forms on colon cancer carcinogenesis using a rat model. They found naturally high-Se wheat to be the most effective Se form against colon cancer precursors, resulting in a 48% reduction in aberrant crypts, compared with 0, 8, and 36% for selenite, selenomethionine, and Se-biofortified broccoli, respectively.

2.2.4 Commercialization of Selenium-Biofortified Wheat

The research projects discussed in the following text demonstrate that agronomic biofortification of cereals with Se is effective, inexpensive, and provides desirable bioavailable forms of Se. The *HarvestPlus* biofortification group at the University of Adelaide has collaborated with a South Australian milling/food company, Laucke Flour Mills, to develop and market Se-biofortified wheat products in Australia and the United Kingdom. The products were launched in 2005 and include an expanded-grain wafer (which contains 3 mg Se kg^{-1}), wholemeal flour, a bread "pre-mix" for home baking, and *BioMax* bread made by Country Life Bakery in Melbourne (the latter three products contain around 1.2 mg Se kg^{-1}). People who would be most likely to benefit from these products include male smokers, people over 65 years, and those at increased risk of various cancers (notably prostate cancer) and RNA viral diseases (Luty-Frackiewicz et al. 2002; Akbaraly et al. 2005). Novel wheat (or other cereal) products that contain enhanced levels of organic Se due to agronomic biofortification could be considered as *functional foods*, which are likely to experience strong consumer demand as evidence builds for selenium's crucial role in human health. To encourage this type of production, growers who produce high-Se wheat could receive substantial premiums. For example, it is reported that growers in South Dakota have been paid up to $1000 per tonne by European buyers for high-Se durum wheat (R.M. Welch, personal communication).

2.3 CASE STUDY 2: AGRONOMIC BIOFORTIFICATION OF CASSAVA WITH ZINC, SELENIUM, AND IODINE

2.3.1 Introduction

Cassava (*Manihot esculenta* Crantz) is an important staple crop, especially for resource-poor populations in sub-Saharan Africa. However, it is usually not regarded as a particularly nutritious crop because the storage roots contain only around 1.5% protein (Nassar and Sousa 2007) and are generally low in micronutrients. The leaves, on the other hand, are highly nutritious, containing 21–39% protein (El-Sharkawy 2004) and high levels of carotenoids and other beneficial phytocompounds. Moreover, surveys conducted by our University of Adelaide group in Africa and the Solomon Islands suggest that cassava would be a more efficient accumulator of Zn in its leaves than are sweet potato, maize, sorghum, and banana. The leaves are commonly eaten by people in the Democratic Republic of the Congo and other parts of Africa. Many soil types on which cassava is grown are deficient in Zn, Se, and I, and these deficiencies are also prevalent, often concurrently, in the human populations in these areas (Vanderpas et al. 1990; Oldfield 1999).

Recent genotype–environment interaction studies show that variation in Zn concentration in cassava roots is due mostly to soil available Zn and soil pH levels (CIAT 2006). Although genotypic variation for Zn has been reported (Chavez et al. 2005), results were not conclusive. There may have been some narrow genotypic variation, but environmental interaction was large. Therefore, breeding for higher Zn in cassava may not be feasible, although it has been shown that Zn fertilization can be highly effective in overcoming Zn deficiency, to which cassava is susceptible (Asher et al. 1980).

As for Zn, agronomic biofortification of cassava with Se and I may be preferable to breeding for higher concentrations if, as for wheat, the genotypic variation of Se and I density in edible parts is low (Lyons et al. 2005a). If cassava could be efficiently and inexpensively biofortified agronomically, it could become a valuable source of dietary Zn, Se, and I, in addition to its important role as a provider of dietary energy. Overall, little research has been conducted in this area. It is not yet known whether Zn, Se, and I applied to soil can elevate the concentrations of these micronutrients sufficiently in cassava roots and leaves to improve significantly the Zn, Se, and I status of consumers, without causing toxicity to the plants. The following pilot trial was conducted in 2005 to try to answer this question.

2.3.2 Growth Chamber Pilot Trial

A pilot agronomic biofortification trial with cassava was conducted in a growth chamber at the University of Adelaide. The chamber conditions were set as follows: 12-h photoperiod; day temperature 25–28°C; night temperature 12–18°C; light intensity 545 μmol m^{-2} s^{-1}; and mean relative humidity 60%. Stems of a high-yielding, subtropical, low-cyanogen cultivar, *M Aus 10*, bred at the University of Queensland, were planted in UCMix growth medium that is based on sand and "coco peat," and contains all essential nutrients, with pH (H_2O) of 6.0 (Barker et al. 1998). After 6 weeks, the plants (n = 40) were transplanted to large plastic pots (51 cm diameter, 70 kg total weight, including 12% water). Six months after planting, six treatment plants were biofortified by application of Zn, Se, copper (Cu), and I to the growth medium (as zinc sulphate, sodium selenate, copper sulphate, and potassium iodate, respectively) at concentrations shown in Table 2.1. Copper was included in case the applied Zn resulted in Cu deficiency.

Storage roots and youngest mature leaves were harvested 9 months after planting and were analyzed by ICP-optical emission spectrometry (ICP-OES) for Zn, Cu, and other minerals and ICP-mass spectrometry (ICP-MS) for Se and I. There were four untreated control replications and six treatment replications. Data analyses were performed on treatment means by t-test (significant at $P < 0.05$).

There was no difference in yield between controls and biofortified plants. However, agronomic biofortification resulted in large increases in concentration of Zn, Se, and I in both storage roots and leaves, along with increased Cu in roots (Table 2.1). No toxicity symptoms were observed. The most significant effect was a measured 82-fold increase in Se concentration in leaves. The 19-fold increase in root I concentration was unexpected, as I is considered to be transported almost exclusively in the xylem and is thus easily accumulated in leaves but not in grain or tubers (Mackowiak and Grossl 1999). The mean Se concentration in roots and leaves of 5 and 25 mg kg^{-1}, respectively, and the I level of 192 mg kg^{-1} in leaves would be considered too high for human consumption. However, these concentrations would be diluted as the plants continued to grow, especially considering that roots are normally consumed after 11 to 16 months' growth.

To ensure lower concentrations in leaves, lower doses of Se and I were used in field trials, described subsequently. Transfer and accumulation of both volatile Se and I from biofortified plants to control plants was observed. No effect of biofortification by Zn, Se, and I application was evident for other mineral nutrients, including Fe, manganese (Mn), boron (B), nickel (Ni), calcium (Ca), magnesium (Mg), sodium (Na), potassium (K), phosphorus (P), and S in the roots and leaves. In a small concurrent trial, soil application of micronutrients was more effective in increasing the concentrations of Zn, Se, and I in both roots and leaves than foliar application.

TABLE 2.1

Concentrations of Agronomically Biofortified Micronutrients in Storage Roots and Leaves of Cassava (mg kg^{-1} DM: Mean [SD])

		Roots		Leaves		
Micronutrient	**Concentration[a]**	**Control**	**Treated**	**Control**	**Treated**	**Significance**
Zinc	10	20 (4)	61 (23)	30 (3)	78 (14)	$P < 0.05$ (both[b])
Selenium	2	0.2 (0.1)	5.0 (0.4)	0.31 (0.2)	25 (7)	$P < 0.001$ (both)
Copper	2	0.9 (0.1)	2.3 (0.9)	1.2 (0.4)	1.5 (0.2)	$P < 0.05$ (roots)
Iodine	4	0.035	0.65	10 (3)	192 (89)	$P < 0.01$ (both)

[a] Concentration in UC mix growth medium (mg kg^{-1}).

[b] Both roots and leaves.

2.3.3 FIELD TRIALS

The successful agronomic biofortification of cassava with micronutrients in the growth chamber trial led to field trials whereby Centro Internacional de Agricultura Tropical (CIAT) investigated agronomic biofortification of cassava with Zn, Se, I, and Fe at several sites in Columbia, with varied soil pH, and other soil and climatic characteristics. Included were sites low in available Zn, at which a yield response to Zn biofortification could be expected. Further benefits of this multiple-micronutrient approach could include exploitation of synergies, for example, Se/I and Se/Zn interactions (Lyons et al. 2004c). We are waiting for results

2.4 A FOOD SYSTEMS STRATEGY TO OVERCOME MICRONUTRIENT MALNUTRITION

2.4.1 *HarvestPlus* Proof-of-Concept Trials

Addressing the deficiencies of a food system at the village level in a subsistence economy by biofortification has not been studied systematically. Can biofortification of staple crops provide sufficient increases in concentrations of bioavailable micronutrients in edible parts to overcome micronutrient deficiencies in human populations? We hypothesize that by correcting the nutrient imbalances/deficiencies in plants, micronutrient malnutrition people who consume these crops will be largely alleviated. Certainly, in the case of Se (applied with N-P-K fertilizer in Finland since 1984) and I (applied to irrigation water in far western China), the agronomic biofortification strategy has already been successful (Cao et al. 1994; Hartikainen 2005). Graham et al. (2007) make the important point that low-input agriculture is not practical in today's world because the human population is well in excess of what can be sustained without modern inorganic fertilizers. Hence, even subsistence farmers need some fertilizers to compete effectively. Deficiencies of vitamins and certain other nutrients need to be addressed by dietary diversification, and/or by changes in food processing and preparation in these proof-of-concept studies.

Food system studies, which involve improved crop varieties (e.g., orange-fleshed sweet potato and quality protein maize), increased crop diversity, specific fertilizers, and intercropping (e.g., with Faba beans and chickpeas on high-pH soils), are planned for regions deficient in Se, I, and Zn in China and Africa. In China, trials have commenced on the loess plateau of Shaanxi province, where the osteoarthropathy, Kashin-Beck disease, is prevalent. This is a debilitating arthritic condition that affects mainly the knee and finger joints. Etiological factors include Se and I deficiencies and, possibly, fungal contamination of food crops and the presence of fulvic acids in drinking water (Sudre and Mathieu 2001). In Africa, the Great Lakes region in the vicinity of Lakes Victoria and Tanganyika where HIV/AIDS is prevalent is targeted for this program.

2.4.1.1 Shaanxi Province, China

The loess soil types, which cover large areas of central China, can contain relatively high concentrations of macro- and micronutrients, but micronutrients are generally of low availability due largely to high soil pH of 8.2–9.4 (Table 2.2). The village of Mazhuang, which is located around 100 km north of the Northwest Agriculture and Forestry University, Yangling was identified as a suitable place to conduct field and food system trials. The village has around 250 inhabitants, and subsistence agriculture predominates, with apples as a cash crop; other similar villages exist nearby that could act as controls for an intervention study. The area has the highest prevalence of Kashin-Beck disease in China.

It is possible that biofortification of food crops with Se, along with increased consumption of leafy vegetables (which would increase I intake), could reduce the incidence of Kashin-Beck disease. The I levels in leaves in this region are around 650–700 μg kg^{-1}, a substantial level but very little I reaches grain or potato tubers. Table 2.3 shows actual and target levels of Zn, Se, and I in grain collected from Mazhuang village.

Ironically, at Mazhuang (with its calcareous soils), some children show signs of Ca deficiency. The villagers' maize-dominated diet can be improved by an increase in the consumption of beans (which grow well there) and tomatoes, as well as the leafy vegetables noted earlier. Available Zn and B are also relatively low, and hence fertilization with these minerals (along with N-P-K in certain trial plots) may increase crop yields. Interestingly, deficiencies of Fe and Zn in apples in this area have been alleviated by banging metal nails into the tree trunks.

The Mazhuang project will commence with agronomic trials conducted in the glasshouse and field to evaluate different methods of agronomic biofortification in maize, wheat, soybean, potato, cabbage, and *Phaseolus* bean. A range of different legumes will also be tested, initially in a glasshouse, for their ability to acidify the rhizosphere and hence release nutrients and micronutrients for utilization by intercropped/rotated food crops.

The next stage would involve agronomic biofortification of food crops of the village, using soil-applied Zn, Se, and I, added with the standard inorganic P and K fertilizer commonly used in the area. It is important that this intervention involves as little change/disruption to the farming practices/village life as possible. The aim is to achieve an efficient, sustainable, nutritious food system for the region. A similar village, with crops grown on the same soil type but not biofortified, will serve as a control. The nutrition and health of the villagers who consume the biofortified crops will be compared to that of the control villagers who consume their usual, nonbiofortified crops. It is necessary to determine whether the biofortified micronutrients are provided in sufficient quantities and are bioavailable enough to make a significant improvement in the nutrition, and hence the health of the people. The control village is necessary to observe possible seasonal changes in nutritional status/health that would have a confounding effect on results from the treatment village.

2.4.1.2 The Great Lakes, East Africa

A large proportion of the population in East Africa suffers from micronutrient malnutrition. It is estimated that Fe deficiency anemia affects 70% of children, whereas over 50% of children are likely to be Zn and/or vitamin A deficient, and around 30% of the total population may be I and/or Se deficient. An extensive Se-deficient belt

TABLE 2.2
Comparison of Total Nutrient and Micronutrient Levels in Topsoils from Yongshou (Central China) and Minnipa (Eyre Peninsula, South Australia)

	Mineral Nutrient/Micronutrient Concentration (mg kg^{-1})														
Site	Fe	Mn	B[a]	Cu	Co	Ni	Zn	Ca	Mg	Na	K	P	S	Pb	Se
Yongshou	26000	540	< 2	22	12	31	62	90000	9600	400	7600	530	200	21	0.69
Minnipa	8500	69	7	3.6	2.9	5.5	13	21000	2600	75	2400	8	110	< 10	0.08

[a] Actual B levels will be higher than this, as some B is lost due to acid digestion.

TABLE 2.3

Concentration of Selected Micronutrients (Zn, Fe, Se, and I) in Grain of Maize, Wheat, and Soybean and in Potato Tubers from Mazhuang, Yongshou, Shaanxi Province, China, Collected in 2006

	Micronutrient			
	Zn	Fe	Se	I
	mg kg^{-1}		μg kg^{-1}	
Maize	20 (40)[a]	23	11 (400)	< 10 (20)[b]
Wheat	17 (40)	62	14 (400)	< 10 (20)
Soybean	53 (75)	56	13 (400)	20 (50)
Potato	20 (35)	22	16 (400)	< 10 (50)

[a] Means of 2–3 samples. (In brackets: target concentration for agronomic biofortification: not applicable to Fe).

[b] Iodine target level for the leaves of these crops is 1 mg kg^{-1}.

extends from the central Democratic Republic of the Congo to Rwanda. Concurrent deficiencies of I, Zn, Fe, and other mineral micronutrients, along with vitamins A and B12, are common in subsistence populations within this area (WHO 2003; Graham et al. 2007). Such areas having multiple deficiencies in soils, crops, and people, would be suitable for a trial to prove the concept of a food systems approach, similar to the one described earlier in China. The Lake Kivu area, which is elevated, densely populated, and a target zone for intervention under the sub-Saharan Africa Challenge Program, is of particular interest. Important crops here include bananas, sweet potatoes, potatoes, cassava, sorghum, maize, *Phaseolus* beans, peas, and coco yams (Graham et al. 2007). Our basic premise is that if we are able to balance mineral nutrients in crops, we will go a long way toward balancing the nutrients in humans. Initial plant tissue surveys on sorghum, maize, beans, potato, sweet potato, cassava, and bananas have shown the most common deficiencies to be B, Ca, K, Mn (except at low pH sites), P, Se, and Zn (although low Zn levels are species-specific). Cassava leaves are consistently higher in Zn concentration than the leaves of other food crops tested, especially banana. For example, at one site near Lake Kivu, the Zn concentration in cassava leaves was 100 mg kg^{-1}, whereas banana leaves contained 10 mg kg^{-1}.

Various fertilizer treatments will be applied in a series of field trials, which will be conducted on soil types identified as being typical of the target villages, and will include the major food crops of the region. In addition, soil acidity may need to be controlled to improve the availability of nutrients/micronutrients to the plants. Given a normal season, it would be expected that the treatment of N-P-K fertilizer plus micronutrients (e.g., Zn, B, Se, and I) would produce higher yield and greater density of most of these micronutrients in edible parts of the crops. Thus, the treatment could be seen as a combined fertilizer/agronomic biofortification approach. Additionally, as with the China project, improved crop varieties that have higher levels of other nutrients deficient in the population (e.g., orange-fleshed sweet potato for beta-carotene [pro-vitamin A] and beans for Fe) could be trialed alongside locally adapted varieties.

These projects have the potential to provide an example of a sustainable food systems strategy that could be extended to many parts of the world. There is increasing interest in providing fertilizers to African smallholder farmers, and this has been achieved realistically by the Farm Input Promotions Africa Ltd (FIPS-Africa) organization run by Dr. Paul Seward in Nairobi, Kenya. FIPS sells and provides application instructions for a mixed macro-/micronutrient fertilizer produced by Athi River Mining Co., Kenya, in amounts as small as 200 g per packet (a few grains per plant) so as to be affordable. This fertilizer, together with other low-cost inputs, has reportedly increased maize yields in some areas up to fivefold (Seward 2006). The Rockefeller Foundation, one of the FIPS funders, is planning to replicate the FIPS model in other African countries, with an emphasis on an extensive retail network at small town/village level, and with retailers/distributors trained to provide correct advice on fertilizer use.

2.5 NUTRITIONAL RISK REDUCTION AND CONTROL OF PROSTATE CANCER

2.5.1 Introduction

In this section, the role of Se and various phyto-compounds against prostate cancer is discussed to illustrate the importance of dietary components in the reduction of risk and in the control of an important chronic disease. Prostate cancer is the second most common cause of male cancer-related deaths in the United States (Lowe and Frazee 2006) and the United Kingdom (Jian et al. 2007). Due to the effect of different levels of screening on incidence rates of prostate cancer, mortality rates provide a more reliable guide to the risk of advanced prostate cancer in different populations. Mortality rates are high in the Caribbean, central and southern Africa, northern and western Europe, Australia, New Zealand, and North and South America and low in China, India, Southeast Asia, and Japan (Parkin et al. 2005). Black men appear to be at higher risk of prostate cancer (Parkin et al. 2005), and its incidence is increasing in developing countries as people shift from traditional diets to more calorie-rich "western" diets (Jian et al. 2007).

2.5.2 Chemoprevention: Potential for Risk Reduction and Early-Stage Treatment

The higher risk of prostate cancer in black men notwithstanding, it has been estimated that around 80% of all chronic diseases are linked to lifestyle, mainly diet, and cannot be explained by genetic differences (Wong et al. 2005). This observation is well illustrated by cancer studies involving migrants: migration exposes individuals to the same cancer risk as other inhabitants of the adopted country, so that usually within a generation the migrants exhibit similar incidence and mortality rates as others in the adopted country (Locatelli et al. 2004). There is an urgent need to link more closely agriculture and food processing with human health in order to reduce the global burden of diet-related diseases (Graham et al. 2007).

The concept of *chemoprevention* can be defined as a strategy that aims to reduce the incidence and burden of cancer by the development of agents to prevent, reverse, or delay the carcinogenic process (Sporn and Suh 2002). Prostate cancer's high rate of occurrence and long lead time make it an ideal disease for pharmacologic or nutritional chemoprevention (Lowe and Frazee 2006). Given prostate cancer's usually late age of onset, any delay in carcinogenesis will lead to a reduction in its incidence (Thorpe et al. 2007). The long and complex process of carcinogenesis presents numerous opportunities for clinical interventions both in prevention and treatment of premalignant neoplasms—in the case of prostate cancer, *prostate intraepithelial neoplasia* (PIN) (Gupta 2007).

Naturally occurring agents that may decrease the risk of prostate cancer include Se, vitamin E, n-3 fatty acids, lycopene, epigallocatechin-3-gallate (EGCG, from green tea), curcumin (from turmeric), and punicalagin and ellagic acid (from pomegranate) (Whanger 2004; Chan et al. 2005; Bemis et al. 2006; Jiminez del Rio et al. 2006; Narayanan 2006; Thompson 2007). The preclinical data for these agents indicate strong efficacy against prostate cancer cells in vitro, with molecular targets ranging from cell-cycle regulation to induction of apoptosis (programmed cell death), and animal models support the notion that these agents could play a chemopreventive role in humans (Gupta 2007; Von Low et al. 2007).

Growing evidence supports the concept of chemopreventive agent combinations against prostate cancer (Neill and Fleshner 2006; Gupta 2007), which target multiple molecular sites and provide either additive or synergistic effects, such as those demonstrated for green tea and lycopene (Jian et al. 2007), Se and vitamin E (Pak et al. 2002; Peters et al. 2007a), and tomato and broccoli (Canene-Adams 2007). It appears that synergies also exist between antioxidants/micronutrients/phyto-compounds and both chemotherapeutic drugs and radiotherapy for cancer. Recent studies show that Se and certain phyto-compounds can both increase the effectiveness of chemotherapeutic drugs against various cancers and reduce their toxic side effects (Bjorkhem-Bergman et al. 2002; Chisholm et al. 2004; Falsaperla et al. 2005; Theobald 2006; Simone et al. 2007). It is unrealistic to expect a "silver bullet" cure for a condition as multicausal, complex, and heterogeneous as prostate cancer. The anticancer actions of these nutritional components involve diverse mechanisms which, when combined, act on the development of prostate cancer from initiation to progression.

2.5.3 Effectiveness of Selenium as a Chemopreventive Agent Compared to Other Naturally Occurring Compounds

Next, we examine in more detail the antiprostate cancer effects of four of the most promising naturally occurring chemopreventive agents/foods: Se, lycopene/tomato, epigallocatechin-3-gallate/green tea, and punicalagin and ellagic acid/pomegranate. Nearly 50 years of evidence from research ranging from in vitro cell studies to randomized clinical trials supports selenium's anticancer effects, which appear to be strongest against prostate cancer. Reviews of Se's anticancer effects include those of El-Bayoumy (2001) and Whanger (2004). Mechanisms of Se's anticancer action include antioxidant effects (which include stabilizing DNA), neutralization of carcinogens (e.g., cadmium), cell-cycle arrest, immune enhancement, apoptosis, antiangiogenesis, and inhibition of tyrosine kinase (Whanger 2004). For example, polythiols (sulfur-containing proteins) can react with fibrinogen (a blood protein) to form an insoluble polymer around cancer cells that masks tumor antigens and allows cancer cells to escape immune recognition and destruction by natural killer cells. Sodium selenite can oxidize cell membrane thiols, thus preventing the formation of this coat. In addition, Se appears to have a particular role against cancer cells that have become resistant to drugs: selenite was found to be even more effective against doxorubicin-resistant lung cancer cells than against nonresistant lung cancer cells, and there was no effect on normal lung cells (Bjorkhem-Bergman et al. 2002).

Epidemiological and clinical studies that have provided evidence for selenium's efficacy against prostate cancer in recent years include the following: Inuit (Eskimo) men have among the highest Se and n-3 fatty acid intake globally and the lowest incidence and prevalence of prostate cancer, including the microfocal (early-stage) form (Dewailly et al. 2003). The Nutritional Prevention of Cancer Trial, United States, followed approximately 1300 people, mostly men, for 6 years. The treatment group received 200 μg Se per day as high-Se yeast. A protective effect (52% reduction in prostate cancer incidence) was found in men with a baseline plasma Se concentration less than 110 $\mu g\ L^{-1}$ (Duffield-Lillico et al. 2003). In the Baltimore Longitudinal Study on Ageing, low baseline Se status was associated with a fourfold increase in risk of advanced prostate cancer (Brooks et al. 2001). In a case-control study of men with a defective allele for the important mitochondrial antioxidative enzyme, Mn superoxide dismutase, those in the lowest quartile for combined intake of Se, vitamin E, and lycopene (the red carotenoid in tomato) had a tenfold higher risk of advanced prostate cancer than those in the top quartile of intake (Li et al. 2005). As noted earlier, several studies have described the synergistic effect of Se and vitamin E against prostate cancer, and this is currently being tested in the Selenium and Vitamin E Cancer Prevention Trial (SELECT) of over 35,000 men that is being conducted in the United States, Canada, and Puerto Rico, and is expected to report findings in 2013 (National Cancer Institute 2005).

Epidemiological studies have suggested that a diet high in tomatoes or tomato products is protective against prostate cancer (Giovannucci 2002; Campbell et al. 2004), but some studies have shown no effect (Peters et al. 2007b). Tomatoes are the main dietary source of lycopene (Campbell et al. 2004). Laboratory studies provide support for the effectiveness of lycopene against prostate cancer (Barber et al. 2006; Kanagaraj et al. 2007). Animal trials show varied results, including effectiveness for tomato extracts in combination with broccoli, but no significant effect for pure lycopene (Canene-Adams et al. 2007), and reduction in PC-346C human prostate cancer cell volume in nude mice due to a combined lycopene and vitamin E treatment, but not from lycopene alone (Limpens et al. 2006).

Lycopene is present in human prostate tissue in substantial concentrations (Khachik et al. 2002), which supports the plausibility of a direct effect on the prostate, and the combined evidence from human intervention trials for lycopene and/or tomatoes/tomato products against prostate cancer appears (unusual for a natural chemopreventive agent) to outweigh that from animal studies. In a pilot phase 11 clinical study, natural lycopene reduced prostate-specific antigen (PSA) velocity over one year (Barber et al. 2006), and a daily serving of tomatoes or tomato products, but not pure lycopene, increased the resistance of mononuclear leukocytes to DNA strand breaks induced by reactive oxygen species (considered to be an early biomarker of increased cancer risk) (Ellinger et al. 2006). Men newly diagnosed with prostate cancer who consumed a high lycopene tomato extract for just 3 weeks before radical prostatectomy had lower PSA, less PIN tissue, and smaller tumors than controls (Kucuk et al. 2002). A trial of 40 men with high-grade PIN found a 10% prostate cancer incidence (and a mean 42% reduction in PSA level) in those who consumed 8 mg lycopene daily for a year, compared to a 30% incidence (and a 23% increase in PSA) in the control group (Mohanty et al. 2005). Another trial with tomato sauce supplementation demonstrated a significant uptake of lycopene into prostate tissue, reduced DNA damage in both leukocyte and prostate tissue, and reduced PSA (Bowen et al. 2002). However, not all clinical trials have demonstrated positive tomato/lycopene effects (Jatoi et al. 2007).

Mechanisms of lycopene's antiprostate-cancer effects include protection of plasma lipoproteins, lymphocyte DNA, and serum proteins against oxidative damage; reduction of PSA; upregulation of connexion expression (Basu and Imrhan 2007); induction of apoptosis (Bowen et al. 2002; Kucuk et al. 2002); reduction in IGF-1 receptor expression; and an increase in insulin growth factor-binding protein-3 (IGFBP-3) (Kanagaraj et al. 2007).

Several clinical trials have shown that products such as tomato paste may be more effective than pure lycopene against prostate cancer (Ellinger et al. 2006), which agrees with findings from animal trials, and suggests that other tomato components may act to enhance the anticancer effects of lycopene or have direct effects themselves. Synergistic effects of lycopene with other nutrients in reducing biomarkers of oxidative stress and carcinogenesis have also been demonstrated (Li et al. 2005; Basu and Imrhan 2007). There is a need for larger, well-designed randomized clinical trials, ideally including a combined treatment (e.g., tomato paste with sodium selenite and natural vitamin E appears particularly promising) to compare with tomato paste and pure lycopene and placebo, to determine whether tomato products and/or lycopene are really effective against prostate cancer.

Catechins are a class of flavonoid that is particularly concentrated in green tea (*Camellia sinensis*); (-)epigallocatechin-3-gallate (EGCG) is the most abundant and is considered to be the component most responsible for the health benefits of green tea. The effects of EGCG on prostate cancer and breast cancer have been thoroughly reviewed by Stuart et al. (2006). Mechanisms of EGCG's antiprostate-cancer effects include decrease in IGF-1, increase in IGFBP-3, cell-cycle arrest, induction of apoptosis, and inhibition of various processes required for angiogenesis and metastasis, including vascular endothelial growth factor (VEGF) (Stuart et al. 2006).

Several clinical trials have been conducted to evaluate the effectiveness of green tea extracts against prostate cancer. A strong positive effect was evident in a chemoprevention trial with 60 men with high-grade PIN, who took three 200 mg capsules of green tea extract daily for 1 year. At the end of the trial, only one tumor was diagnosed in the treatment group (3% incidence) compared to nine tumors in the placebo group (30% incidence) (Bettuzzi et al. 2006). These researchers have recently developed a gene signature profile that distinguishes benign prostate disease from that which will develop into prostate cancer, and if so, whether it is likely to respond to treatment with green tea catechins (Bettuzzi et al. 2007). Green tea extracts were not successful in trials that used men with advanced prostate cancer (Jatoi et al. 2003; Choan et al. 2005). Some of the participants in the study by Jatoi et al. (2003) experienced side effects, which Stuart et al. (2006) attributed to caffeine, and they recommend that future trials should use decaffeinated green tea or pure EGCG. However, it has been found that decaffeination can reduce EGCG concentration by as much as 90% (Henning et al. 2006). Thus, either pure EGCG or green tea at a lower dose than the 6 g daily used in the study by Jatoi et al. (2003) may be more suitable, and ideally using a study group at high risk of prostate cancer (e.g., with high-grade PIN), rather than men with advanced disease.

Finally, potential synergies with other chemopreventive agents are worth exploring. For example, trials could include a treatment using green tea catechins and pomegranate polyphenols (Adhami and Mukhtar 2006), or Se-biofortified green tea (Xu et al. 2007). Pomegranate (*Punica granatum*) has strong antioxidant (Gil et al. 2000) and antiinflammatory activity (Afaq et al. 2005). Its anticancer effects have only recently been studied but are particularly promising. Research to date has been reviewed by Adhami and Mukhtar (2006) and Lansky and Newman (2007). These authors report laboratory and animal studies that have found strong effects of various extracts of pomegranate fruit against prostate cancer, including cell-cycle arrest, promotion of apoptosis, and inhibition of metastasis and angiogenesis.

Pomegranate juice, peel, and oil, along with ellagic acid, which is considered to be one of the major biologically-active components, have all shown anticancer effects, and fermented juice polyphenols look especially promising (Adhami and Mukhtar 2006; Lansky and Newman 2007). Indeed, Lansky and Newman (2007) state, "the phytochemistry and pharmacological actions of all *Punica granatum* components suggest a wide range of clinical applications for the treatment and prevention of cancer …." Other researchers place more emphasis on the role of the ellagitannin, punicalagin, which is concentrated in the rind (Jiminez del Rio et al. 2006; Sestili et al. 2007). Its antioxidant action is expressed through its scavenging reactions and also its metal chelation ability (Kulkarni et al. 2007). A bioavailability study found that the bioefficacy of pomegranate ellagitannins/polyphenols should be attributed to colonic microflora metabolites rather than to the polyphenols themselves (Cerda et al. 2004).

Only one clinical trial of pomegranate and cancer has been reported to date. This was a phase II, two-stage intervention trial (not placebo-controlled) for men ($n = 48$) with rising PSA after surgery or radiotherapy. The men consumed 8 oz of pomegranate juice daily until disease progression. The mean PSA doubling time significantly increased with treatment from a mean of 15 months at baseline to 54 months posttreatment. In vitro assays comparing the effect of baseline and posttreatment serum on the growth of prostate cancer cells showed a 12% decrease in cell proliferation, a 17% increase in apoptosis, and a 23% increase in serum nitric oxide (Pantuck et al. 2006).

2.6 CONCLUSIONS

2.6.1 What Next? The Need for More Well-Designed Clinical Trials

It is clear that the natural agents discussed previously, along with numerous others, have potential as chemopreventives for prostate and other cancers. Before they can be recommended as useful chemopreventive strategies for patients or individuals at high risk of prostate cancer, their activity must be confirmed in rigorously designed randomized controlled clinical trials. Some of these should include combinations of the most promising dietary agents and ideally involve participants either with susceptible genotypes or with premalignant neoplasms, such as high-grade PIN. Indeed, numerous men would fit both categories. This would provide a more targeted and efficient approach than using long-term trials involving large numbers of participants, many of whom may not be at high risk of the disease.

A good example of this targeted, "holistic" type of randomized trial using a high-risk population is that reported by Ornish et al. (2005). This was an RCT consisting of 93 men with clinically diagnosed prostate cancer (Gleason score < 7; PSA 4–10) who had elected not to undergo any conventional treatment. The treatment group made comprehensive lifestyle changes (including a vegetarian diet and dietary supplements of Se, vitamin E, vitamin C, fish oil, and soy protein; 30 min exercise daily; and stress management techniques, including yoga and meditation for 60 min daily), whereas the control group underwent "usual care." PSA decreased 4% in the treatment group and increased 6% in the control group, and LDL-cholesterol decreased 23% in the treatment group, with no decrease in the controls. Perhaps most interesting of all the findings was from an in vitro study: the growth of LNCaP prostate cancer cells was inhibited almost eight times more by serum from the experimental group than by serum from the controls (70% vs. 9%, $P < 0.001$).

Important aspects of experimental design include the form and dose rate of the potential chemopreventive. In the case of green tea, tomato, and pomegranate, there is evidence that extracts of the whole product/food (and in the case of tomato, involvement of heat processing) may be more effective as chemopreventives than the best-known pure active components, namely, EGCG, lycopene, and ellagic acid, respectively. It could be argued, for example, that taking accumulated evidence into account, the SELECT trial would be more likely to provide positive findings if treatments of 200 μg Se as sodium selenite per day and 150 IU of d-alpha tocopherol succinate were used rather than 200 μg Se as l-selenomethionine and 400 IU of dl-alpha tocopherol acetate (Gupta 2007).

ACKNOWLEDGMENTS

The authors gratefully acknowledge funding from *HarvestPlus*, Washington, DC (G. Lyons and R. Graham) and the Molecular Plant Breeding Co-operative Research Centre, Australia (Y Genc). *HarvestPlus* (www.HarvestPlus.org) is a global alliance of agriculture and nutrition research institutions working to increase the micronutrient density of staple food crops through biofortification and the views expressed do not necessarily reflect those of *HarvestPlus*. Analytical support from Waite

Analytical Services and the University of Adelaide, South Australia, is gratefully acknowledged, along with editorial assistance from Dr. Martin Broadley, University of Nottingham.

REFERENCES

Adams, M. L., E. Lombi, F.-J. Zhao, and S. P. McGrath. 2002. Evidence of low selenium concentrations in UK bread-making wheat grain. *Journal of the Science of Food and Agriculture* 82: 1160–1165.

Adhami, V. M. and H. Mukhtar. 2006. Polyphenols from green tea and pomegranate for prevention of prostate cancer. *Free Radical Research* 40(10): 1095–1104.

Afaq, F., M. Saleem, C. G. Krueger, J. D. Reed, and H. Mukhtar. 2005. Anthocyanin- and hydrolysable tannin-rich pomegranate fruit extract modulates MAPK and NF-kappaB pathways and inhibits skin tumorigenesis in *CD-1* mice. *International Journal of Cancer* 113: 423–433.

Akbaraly, N. J., J. Arnaud, I. Hininger-Favier, V. Gourlet, A. M. Roussel, and C. Berr. 2005. Selenium and mortality in the elderly: Results from the EVA study. *Clinical Chemistry* 51(11): 2117–2123.

Aro, A., G. Alfthan, and P. Varo. 1995. Effects of supplementation of fertilizers on human selenium status in Finland. *Analyst* 120: 841–843.

Asher, C. J., D. G. Edwards, and R. H. Howeler. 1980. *Nutritional Disorders of Cassava*. St. Lucia, Queensland: Department of Agriculture, University of Queensland.

Bañuelos, G. and G. Schrale. 1989. Plants that remove selenium from soils. *California Agriculture* May–June 19–20.

Barber, N. J., X. Zhang, and G. Zhu et al. 2006. Lycopene inhibits DNA synthesis in primary prostate epithelial cells in vitro, and its administration is associated with a reduced prostate-specific antigen velocity in a phase-11 clinical study. *Prostate Cancer Prostatic Disease* 9(4): 407–413.

Barker, S. J., B. Stummer, L. Gao, I. Dsipain, P. J. O'Connor, and S. E. Smith. 1998. A mutant in *Lycopersicon esculentum* Mill. with highly reduced VA mycorrhizal colonisation: Isolation and preliminary characterization. *Plant Journal* 15: 791–797.

Barrett, J., C. Patterson, C. Reilly, and U. Tinggi. 1989. Selenium in the diet of children with phenylketonuria. In *Nutrient Availability: Chemical and Biological Aspects,* D.A.T. Southgate, I.T. Johnson, and G.R. Fenwick, (Eds.), London: Royal Society of Chemistry. 281–283.

Basu, A. and V. Imrhan. 2007. Tomatoes versus lycopene in oxidative stress and carcinogenesis: Conclusions from clinical trials. *European Journal of Clinical Nutrition* 61(3): 295–303.

Bates, C. J., C. W. Thane, A. Prentice, H. T. Delves, and J. Gregory. 2002. Selenium status and associated factors in a British National Diet and Nutrition Survey: Young people aged 4–18 y. *European Journal of Clinical Nutrition* 56(9): 873–881.

Bemis, D. L., A. E. Katz, and R. Buttyan. 2006. Clinical trials of natural products as chemopreventive agents for prostate cancer. *Expert Opinion in Investigative Drugs* 15(10): 1191–1200.

Benemariya, H, H. Robberecht, and H. Deelstra. 1993. Daily dietary intake of copper, zinc, and selenium by different population groups in Burundi, Africa. *Science of the Total Environment* 136: 49–76.

Bettuzzi, S., M. Brausi, F. Rizzi, G. Catagnetti, G. Peracchia, and A. Corti. 2006. Chemoprevention of human prostate cancer by oral administration of green tea catechins in volunteers with high-grade prostate intraepithelial neoplasia: A preclinical report from a one-year proof-of-principle study. *Cancer Research* 66: 1234–1240.

Bettuzzi, S., F. Rizzi, and L. Belloni. 2007. Clinical relevance of the inhibitory effect of green tea catechins (GtCs) on prostate cancer progression in combination with molecular profiling of catechin-resistant tumors: An integrated view. *Polish Journal of Veterinary Science* 10 (1): 57–60.

Bhaskaram, P. 2002. Micronutrient malnutrition, infection, and immunity: An overview. *Nutrition Reviews* 60: S40–S45.

Bjorkhem-Bergman, L., K. Jonsson, L. C. Eriksson et al. 2002. Drug-resistant human lung cancer cells are more sensitive to selenium cytotoxicity: Effects on thioredoxin reductase and glutathione reductase. *Biochemical Pharmacology* 63(10): 1875–1884.

Blair, G. J., G. C. Anderson, M. Crestani, and D. C. Lewis. 1997. The sulfur status of soils and the S accessions in rainfall at trial sites in the National RPR Project. *Australian Journal of Experimental Agriculture* 37: 995–1001.

Bowen, P., L. Chen, and M. Stacewicz-Sapuntzakis et al. 2002. Tomato sauce supplementation and prostate cancer: Lycopene accumulation and modulation of biomarkers of carcinogenesis. *Experimental Biology and Medicine* 227 (10): 886–893.

Broadley, M. R., P. J. White, and R. J. Bryson et al. 2006. Biofortification of UK food crops with selenium. *Proceedings of the Nutrition Society* 65: 169–181.

Brooks, J. D., E. J. Matter, and D. W. Chan et al. 2001. Plasma selenium level before diagnosis and the risk of prostate cancer development. *Journal of Urology* 166(6): 2034–2038.

Bryszewska, M. A., W. Ambroziak, A. Diowksz, M. J. Baxter, N. J. Langford, and D. J. Lewis. 2005. Changes in the chemical form of selenium observed during the manufacture of a selenium-enriched sourdough bread for use in a human nutrition study. *Food Additives and Contaminants* 22(2): 135–140.

Burke, K. E., R. G. Bedford, G. F. Combs, I. French, and D. R. Skeffington. 1992. The effect of topical L-selenomethionine on minimal erythema dose of ultraviolet irradiation in humans. *Photodermatology, Photoimmunology and Photomedicine* 9: 1–6.

Campbell, J. K., K. Canene-Adams, B. L. Lindshield, T. W. M. Boileau, S. K. Clinton, and J. W. Erdman Jr. 2004. Tomato phytochemicals and prostate cancer risk. *Journal of Nutrition* 134: 3486S–3492S.

Canene-Adams, K., B. L. Lindshield, S. Wang, E. H. Jeffrey, S. K. Clinton, and J. W. Erdman Jr. 2007. Combinations of tomato and broccoli enhance antitumor activity in Dunning R3327-H prostate adenocarcinomas. *Cancer Research* 67 (2): 836–843.

Cao, X. Y., X. M. Jiang, and A. Kareen et al. 1994. Iodination of irrigation water as a method of supplying iodine to a severely iodine-deficient population in Xinjiang, China. *Lancet* 344: 107–110.

Cary, E.E. and W.H. Allaway. 1969. The stability of different forms of selenium applied to low-selenium soils. *Soil Science Society of America Proceedings* 33: 571.

Cerda, B., J. C. Espin, S. Parra, P. Martinez, and F. A. Tomas-Barberan. 2004. The potent in vitro antioxidant ellagitannins from pomegranate juice are metabolised into bioavailable but poor antioxidant hydroxy-6H-dibenzopyran-6-one derivatives by the colonic microflora of healthy humans. *European Journal of Nutrition* 43(4): 205–220.

Chan, J. M., P. H. Gann, and E. L. Giovannucci. 2005. Role of diet in prostate cancer development and progression. *Journal of Clinical Oncology* 23(32): 8152–8160.

Chavez, A. L., T. Sanchez, and G. Jaramillo et al. 2005. Variation of quality traits in cassava roots evaluated in landraces and improved clones. *Euphytica* 143: 125–133.

Chisholm, K., B. J. Bray, and R. J. Rosengren. 2004. Tamoxifen and epigallocatechin gallate are synergistically cytotoxic to MDA-MB-231 human breast cancer cells. *Anticancer Drugs* 15(9): 889–897.

Choan, E., R. Segal, and D. Jonker et al. 2005. A prospective trial of green tea for hormone refractory prostate cancer: an evaluation of the complementary/alternative therapy approach. *Urologic Oncology* 23: 108–113.

CIAT (International Centre for Tropical Agriculture). 2006. Improved cassava for the developing world. In *Annual Report Project IP3.* Cali, Colombia: CIAT.

Combs, G. F. 2001. Selenium in global food systems. *British Journal of Nutrition* 85: 517–547.

Dewailly, E., G. Mulvad, and H. Sloth Pedersen et al. 2003. Inuit are protected against prostate cancer. *Cancer, Epidemiology, Biomarkers, and Prevention* 12(9): 926–927.

Duffield-Lillico, A. J., B. L. Dalkin, and M. E. Reid et al. 2003. Selenium supplementation, baseline plasma selenium status and incidence of prostate cancer: An analysis of the complete treatment period of the Nutritional Prevention of Cancer Trial. *British Journal of Urology International* 91(7): 608–612.

El-Bayoumy, K. 2001. The protective role of selenium on genetic damage and on cancer. *Mutation Research* 475: 123–139.

Ellinger, S., J. Ellinger, and P. Stehle. 2006. Tomatoes, tomato products and lycopene in the prevention and treatment of prostate cancer: Do we have the evidence from intervention studies? *Current Opinion in Clinical Nutrition and Metabolic Care* 9(6): 722–727.

El-Sharkawy, M. A. 2004. Cassava biology and physiology. *Plant Molecular Biology* 56: 481–501.

Failla, M. L. 2003. Trace elements and host defense: Recent advances and continuing challenges. *Journal of Nutrition* 133: 1448S–1451S.

Falsaperla, M., G. Morgia, A. Tartarone, R. Ardito, and G. Romano. 2005. Support ellagic acid therapy in patients with hormone refractory prostate cancer (HRPC) on standard chemotherapy using vinorelbine and estramustine phosphate. *European Urology* 47(4): 449–454.

Finley, J. W. and C. D. Davis. 2001. Selenium from high-selenium broccoli is utilized differently than selenite, selenate, and selenomethionine, but is more effective in inhibiting colon carcinogenesis. *Biofactors* 14: 191–196.

Gibson, C., Y. H. Park, K. H., and Myoung, M. K. et al. 2006. The biofortification of barley with selenium. In *Proceedings of the Institute of Brewing and Distilling (Asia Pacific Section)* Convention, Hobart Tasmania, 19–24 March 2006.

Gil, M. I., F. A. Tomas-Barberan, B. Hess-Pierce, D. M. Holcroft, and A. A. Kader. 2000. Antioxidant activity of pomegranate juice and its relationship with phenolic composition and processing. *Journal of Agricultural and Food Chemistry* 10: 4581–4589.

Giovannucci, E. 2002. A prospective study of tomato products, lycopene, and prostate cancer risk. *Journal of the National Cancer Institute* 94(5): 391–398.

Graham, R. D., R. M. Welch, and H. E. Bouis. 2001. Addressing micronutrient malnutrition through enhancing the nutritional quality of staple foods: principles, perspectives, and knowledge gaps. *Advances in Agron*omy 70: 77–142.

Graham, R. D., D. A. Saunders, and I. Ortiz-Monasterio et al. 2007. Sustainable, nutritious food systems. *Advances in Agronomy* 92: 1–74.

Gupta, S. 2007. Prostate cancer chemoprevention: Current status and future prospects. *Toxicology and Applied Pharmacology* 224(3): 369–376.

Gupta, U. C. and S. C. Gupta. 2000. Selenium in soils and crops, its deficiencies in livestock and humans: Implications for management. *Communications in Soil Science and Plant Analysis* 31(11–14): 1791–1807.

Hartikainen, H. 2005. Biogeochemistry of selenium and its impact on food chain quality and human health. *Journal of Trace Elements in Medicine and Biology* 18(4): 309–318.

Haug, A., R. D. Graham, O. A. Christophersen, and G. H. Lyons. 2008. How to use efficiently the world's scarce selenium resources to increase the selenium concentration in food. *Microbial Ecology in Health and Disease* 19: 209–228.

Henning, S. M., W. Aronson, and Y. Niu et al. 2006. Tea polyphenols and theaflavins are present in prostate tissue of humans and mice after green and black tea consumption. *Journal of Nutrition* 136(7): 1839–1843.

Imai, H., T. Suzuki, H. Kashiwazaki, T. Takamoto, T. Izumi, and K. Moji. 1990. Dietary habit and selenium concentrations in erythrocyte and serum in a group of middle-aged and elderly Japanese. *Nutrition Research* 10: 1205–1214.

Jatoi, A., N. Ellison, and P. A. Burch et al. 2003. A Phase II clinical trial of green tea in the treatment of patients with androgen independent metastatic prostate carcinoma. *Cancer* 97: 1442–1446.

Jatoi, A., P. Burch, and D. Hillman et al. 2007. A tomato-based, lycopene-containing intervention for androgen-independent prostate cancer: Results of a Phase II study from the North Central Cancer Treatment Group. *Urology* 69 (2): 289–294.

Jian, L., A. H. Lee, and C. W. Binns. 2007. Tea and lycopene protect against prostate cancer. *Asia Pacific Journal of Clinical Nutrition* 16(Suppl 1): 453–457.

Jiminez del Rio, M., A. Ramazanov, S. Sikorski, Z. Ramazanov, and I. Chkhikvishvili. 2006. A new method of standardization of health-promoting pomegranate fruit (*Punica granatum*) extract. *Georgian Medical News* 140: 70–77.

Judson, G. J. and D. J. Reuter. 1999. Selenium. In *Soil Analysis: an Interpretation Manual*, K. I. Peverill, L. A. Sparrow, and D. J. Reuter, (Eds.), Melbourne, Australia: CSIRO Publishing. 325–329.

Kanagaraj, P., M. R. Vijayababu, B. Ravisankar, J. Anbalagan, M. M. Aruldhas, and J. Arunakaran. 2007. Effect of lycopene on insulin-like growth factor-1, IGF binding protein-3 and IGF type-1 receptor in prostate cancer cells. *Journal of Cancer Research and Clinical Oncology* 133(6): 351–359.

Khachik, F., L. Carvalho, P. S. Bernstein, G. J. Muir, D. Y. Zhao, and N. B. Katz. 2002. Chemistry, distribution, and metabolism of tomato carotenoids and their impact on human health. *Experimental Biology and Medicine* 227: 845–851.

Kirby, J., G. Lyons, M. Karkkainen, and M. McLaughlin. 2006. HPLC-(ID)-ICPMS for the determination of selenium species in biofortified grains and biscuits. In *INTERACT 2006. Royal Australian Chemical Institute & Australasian Society for Ecotoxicology*, Perth, Western Australia, 24–28 September, 2006.

Kucuk, O., F. H. Sarkar, and Z. Djuric et al. 2002. Effects of lycopene supplementation in patients with localized prostate cancer. *Experimental Biology and Medicine (Maywood)* 227(10): 881–885.

Kulkarni, A. P., H. S. Mahal, S. Kapoor, and S. M. Aradhya. 2007. In vitro studies on the binding, antioxidant, and cytotoxic actions of punicalagin. *Journal of Agricultural and Food Chemistry* 55(4): 1491–1500.

Lansky, E. P. and R. A. Newman. 2007. *Punica granatum* (pomegranate) and its potential for prevention and treatment of inflammation and cancer. *Journal of Ethnopharmacology* 109(2): 177–206.

Läuchli, A. 1993. Selenium in plants: Uptake, functions, and environmental toxicity. *Botanica Acta*. 106: 455–468.

Li, H., P. W. Kantoff, and E. Giovannucci et al. 2005. Manganese superoxide dismutase polymorphism, prediagnostic antioxidant status, and risk of clinical significant prostate cancer. *Cancer Research* 65(6): 2498–2504.

Limpens, J., F. H. Schroder, and C. M. de Ridder et al. 2006. Combined lycopene and vitamin E treatment suppresses the growth of PC-346C human prostate cancer cells in nude mice. *Journal of Nutrition* 136(5): 1287–1293.

Locatelli, I., P. Lichtenstein, and A. I. Yashin. 2004. The heritability of breast cancer: A Bayesian correlated frailty model applied to Swedish twins data. *Twin Research* 7(2): 182–191.

Lowe, J. F. and L. A. Frazee. 2006. Update on prostate cancer chemoprevention. *Pharmacotherapy* 26(3): 353–359.

Luty-Frackiewicz, A., Z. Jethon, and L. Januszewska. 2002. Effect of smoking and alcohol consumption on the serum selenium level of Lower Silesian population. *Science of the Total Environ*ment 285(1–3): 89–95.

Lyons, G. H., J. C. R. Stangoulis, and R. D. Graham. 2003. High-selenium wheat: Biofortification for better health. *Nutrition Research Re*views 16: 45–60.

Lyons, G. H., G. J. Judson, J. C. R. Stangoulis, L. T. Palmer, J. A. Jones, and R. D. Graham. 2004a. Trends in selenium status of South Australians. *Medical Journal of Australia* 180(8): 383–386.

Lyons G. H., J. Lewis, M. F. Lorimer et al. 2004b. High-selenium wheat: Agronomic biofortification strategies to improve human nutrition. *Food, Agriculture, and Environment* 2(1): 171–178.

Lyons, G. H., J. C. R. Stangoulis, and R. D. Graham. 2004c. Exploiting micronutrient interaction to optimize biofortification programs: the case for inclusion of selenium and iodine in the *HarvestPlus* program. *Nutrition Reviews* 62(6): 247–252.

Lyons G. H., I. Ortiz-Monasterio, J. Stangoulis, and R. Graham. 2005a. Selenium concentration in wheat grain: Is there sufficient genotypic variation to use in breeding? *Plant and Soil* 269: 369–380.

Lyons, G. H., G. Judson, I. Ortiz-Monasterio, Y. Genc, J. C. R. Stangoulis, and R. D. Graham. 2005b. Selenium in Australia: Selenium status and biofortification of wheat for better health. *Journal of Trace Elements in Medicine and Biology* 19: 75–82.

Lyons, G. H., J. C. R. Stangoulis, and R. D. Graham. 2005c. Tolerance of wheat (*Triticum aestivum* L.) to high soil and solution selenium levels. *Plant and Soil* 270: 179–188.

Mackowiak, C. L. and P. R. Grossl. 1999. Iodate and iodide effects on iodine uptake and partitioning in rice (*Oryza sativa* L.) grown in solution culture. *Plant and Soil* 212: 135–143.

Marinov, B., K. Tsachev, and V. Koleva. 1998. The concentration of selenium in the maternal serum in cases of missed abortion. *Akush Ginekologia (Sophia)* 37: 15–16.

Mohanty, N. K., S. Saxena, U. P. Singh, N. K. Goyal, and R. P. Arora. 2005. Lycopene as a chemopreventive agent in the treatment in the treatment of high-grade prostate intraepithelial neoplasia. *Urologic Oncology* 23(6): 383–385.

Narayanan, B. A. 2006. Chemopreventive agents alter global gene expression pattern: Predicting their mode of action and targets. *Current Cancer Drug Targets* 6(8): 711–727.

Nassar, N. M. and Sousa, M. V. 2007. Amino acid profile in cassava and its interspecific hybrid. *Genetics and Molecular Research* 6(2): 292–297.

National Cancer Institute. 2005. Selenium and Vitamin E Cancer Prevention Trial (SELECT): Questions and Answers. NCI, US National Institutes of Health. www.cancer.gov/cancertopics/factsheet/Prevention/SELECT (accessed August 23, 2007).

Neill, M. G. and N. E. Fleshner. 2006. An update on chemoprevention strategies in prostate cancer for 2006. Current Opinion in Urology 16(3): 132–137.

Nomura, A. M., J. Lee, G. H. Stemmermann, and G. F. Combs, 2000. Serum selenium and subsequent risk of prostate cancer. *Cancer Prevention Biomarkers* 9: 883–887.

Oldfield, J. E. 1999. *Selenium World Atlas*. Grimbergen, Belgium: Selenium Tellurium Development Association.

Ornish, D., G. Weidner, and W. R. Fair et al. 2005. Intensive lifestyle changes may affect the progression of prostate cancer. *Journal of Urology* 174: 1065–1069.

Pak, R. W., V. J. Lanteri, J. R. Scheuch, and I. S. Sawczuk. 2002. Review of vitamin E and selenium in the prevention of prostate cancer: Implications of the selenium and vitamin E chemoprevention trial. *Integrated Cancer Therapies* 1(4): 338–344.

Pantuck, A. J., J. T. Leppert, and N. Zomorodian et al. 2006. Phase II study of pomegranate juice for men with rising prostate-specific antigen following surgery or radiation for prostate cancer. *Clinical Cancer Research* 12(13): 4018–4026.

Parkin, D. M., F. Bray, J. Ferlay, and P. Pisani. 2005. Global cancer statistics, 2002. *CA: A Cancer Journal for Clinicians* 55: 74–108.

Peters, U., C. B. Foster, N. Chatterjee et al. 2007a. Serum selenium and risk of prostate cancer—A nested case-control study. *American Journal of Clinical Nutrition* 85(1): 209–217.

Peters, U., M. F Leitzmann, and N. Chatterjee et al. 2007b. Serum lycopene, other carotenoids, and prostate cancer risk: A nested case-control study in the prostate, lung, colorectal, and ovarian cancer screening trial. *Cancer Epidemiology Biomarkers Prevention* 16(5): 962–968.

Poschl, G. and H. K. Seitz. 2004. Alcohol and cancer. *Alcohol and Alcoholism* 39(3): 155–165.

Potrykus, I. 2003. Nutritionally enhanced rice to combat malnutrition disorders of the poor. *Nutrition Reviews* 61(Suppl. 1): S101–S104.

Rayman, M. P. 2000. The importance of selenium to human health. *Lancet* 356: 233–241.

Rayman, M. P. 2002. The argument for increasing selenium intake. *Proceedings of the Nutrition Society* 61: 203–215.

Reilly, C. 1996. *Selenium in Food and Health*. Blackie Academic and Professional: London.

Robinson, M. F., C. D. Thomson, P. Jenkins, G. Luzhen, and P. D. Whanger. 1997. Long-term supplementation with selenate and selenomethionine: urinary excretion by New Zealand women. *British Journal of Nutrition* 77: 551–563.

Sautter, C., S. Poletti, P. Zhang, and W. Gruissem. 2006. Biofortification of essential nutritional compounds and trace elements in rice and cassava. *Proceedings of the Nutrition Society* 65(2): 153–159.

Sestili, P., C. Martinelli, and D. Ricci et al. 2007. Cytoprotective effect of preparations from various parts of *Punica granatum* L. fruits in oxidatively injured mammalian cells in comparison with their antioxidant capacity in cell-free systems. *Pharmacological Research* 56(1): 18–26.

Seward, P. 2006. Small farmers in Kenya increase yields up to five times with the "Maxi Maize Production" planting string. In *Fertilizers Agriculture*, October, 2006, 4. Paris, France: International Fertilizer Industry Association.

Simone, C. B., N. L. Simone, V. Simone, and C. B. Simone. 2007. Antioxidants and other nutrients do not interfere with chemotherapy or radiation therapy and can increase kill and increase survival, Part 2. *Alternative Therapies in Health and Medicine* 13(2): 40–47.

Sporn, M. B. and N. Suh. 2002. Chemoprevention: An essential approach to controlling cancer. *National Review of Cancer* 2: 24–31.

STDA, Selenium Tellurium Development Association. 2006. http://www.stda.net/se-te.htm. (accessed August 14, 2007).

Stuart, E. C., M. J. Scandlyn, and R. J. Rosengren. 2006. Role of epigallocatechin gallate (EGCG) in the treatment of breast and prostate cancer. *Life Sciences* 79: 2329–2336.

Sudre, P. and F. Mathieu. 2001. Kashin-Beck disease: from etiology to prevention or from prevention to etiology? *International Orthopaedics* 25(3): 175–179.

Terrier, F., H. Hida, A. Favier, and P. Marconnet. 1995. Muscle GSH-Px activity after prolonged exercise, training, and selenium supplementation. *Biological Trace Element Research* 47: 279–285.

Theobald, S. 2006. Nutrition and prostate cancer—What is the scientific evidence? *Medizinische Monatsschrift für Pharmazeuten* 29(10): 371–377.

Thompson, I. M. 2007. Chemoprevention of prostate cancer: Agents and study designs. *Journal of Urology* 178(3) Supplement 1: S9–S13.

Thorpe, J. F., S. Jain, T. H. Marczylo, A. J. Gescher, W. P. Steward, and J. K. Mellon. 2007. A review of phase 111 clinical trials of prostate cancer chemoprevention. *Annals of the Royal College of Surgeons, England* 89(3): 207–211.

USGS, United States Geological Survey. 2005. http://www.minerals.usgs.gov/ds/2005/140/ (accessed October 24, 2007).

Vanderpas, J. B., B. Contempre, and N. L. Duale et al. 1990. Iodine and selenium deficiency associated with cretinism in northern Zaire. *American Journal of Clinical Nutrition* 52: 1087–1093.

Von Low, E. C., F. G. Perabo, R. Siener, and S. C. Muller. 2007. Review. Facts and fiction of phytotherapy for prostate cancer: A critical assessment of preclinical and clinical data. *In Vivo* 21(2): 189–204.

Welch, R. M. and R. D. Graham. 2004. Breeding for micronutrients in staple food crops from a human nutrition perspective. *Journal of Experimental Botany* 55: 353–364.

Whanger, P. D. 2004. Selenium and its relationship to cancer: An update. *British Journal of Nutrition* 91: 11–28.

WHO (World Health Organization). 2003. Joint WHO/FAO expert consultation on diet, nutrition and the prevention of chronic diseases (2002: Geneva, Switzerland). *World Health Technical Report Series,* 916, 1. Geneva, Switzerland: World Health Organisation.

Wong, A. H., I. Gottesman, and A. Petronas. 2005. Phenotypic differences in genetically identical organisms: The epigenetic perspective. *Human Molecular Genetics* 14 Spec No 1: R11–8.

Xu, J., F. Yang, X. An, and Q. Hu. 2007. Anticarcinogenic activity of selenium-enriched green tea extracts in vivo. *Journal of Agricultural and Food Chemistry* 55(13): 5349–5353.

Ylaranta, T. 1983. Effect of added selenite and selenate on the selenium content of Italian rye grass (*Lolium multiflorum*) in different soils. *Annales Agriculturae Fenniae* 22: 139–151.

3 Uptake and Accumulation of Selenium in Plants in Relation to Chemical Speciation and Biotransformation

Zhi-Qing Lin

CONTENTS

3.1 INTRODUCTION

The scientific evidence observed in the 1990s showed that selenium (Se) plays an important role in cancer prevention (Clark et al. 1996). Many studies have been carried out during the last decade on accumulation, speciation, and biological functions of Se in relation to human health. Indeed, the concept of Se biofortification is highly dependent on the level of Se uptake and the chemical composition of Se accumulated in dietary materials. The accumulation of Se in an organism is determined by the mass of total Se, or a specific chemical form of Se, retained in the organism (Lin 2005). Selenium speciation, however, identifies and quantifies the composition and distribution of different chemical forms or compounds of Se. For example, previous speciation studies (Whanger 2002) show that cucumber (*Cucumis sativus* L.) primarily accumulates selenate (SeO_4^{2-}); wheat grain and rice accumulate selenomethionine (SeMet); garlic, onion, and broccoli florets mainly accumulate Se-methylselenocysteine (MeSeCys). Clearly, the chemical composition of Se in plant tissues used as dietary supplements of Se varies significantly. Because the chemical form of Se determines its bioavailability, metabolic pathways, and biofortification functions (Ip et al. 2000; Whanger 2002), Se speciation analysis in dietary materials is an important research subject. Although several critical reviews have been published in recent years with different focuses of research (Whanger 2002; Rayman 2004; Abdulah et al. 2005; Dumont et al. 2006), the recent research on the biological transformation process that alternates the chemical species of Se needs to be addressed further.

The aim of this chapter is to discuss the importance of biotransformation and speciation of Se in soil–plant systems, which will promote better understanding of the effect of different Se compounds on their bioavailability or effectiveness by implementing the biofortification strategy for sustaining human health. Particularly, impacts of soil Se speciation on plant Se accumulation and subsequent biotransformation of Se in plant tissues are deliberated, and the analytical techniques currently available for Se speciation analysis are discussed.

3.2 CHEMICAL BEHAVIORS OF SELENOCOMPOUNDS IN SOIL

Selenium is a metalloid and chemical analog to sulfur. Under naturally occurring environmental conditions Se exists primarily in four different oxidation states of VI (selenate), IV (selenite), 0 (elemental Se), and II (selenide). Elemental Se has two amorphous forms, including powder in red color and vitreous in grey.

Various chemical species of Se have different physical, chemical, and biological properties, and therefore, have distinctive chemical behavior in water, soil, and air. For example, selenate is water soluble, but elemental Se is not water soluble and not bioavailable. The dominant chemical forms of Se vary significantly in the environment, which is primarily affected by pH, Eh (redox-potential) and the presence of metal oxides. In general, selenate (SeO_4^{2-}) is the dominant chemical form in alkaline soils under aerobic conditions, where selenite predominates in neutral and well-drained soils, and selenide is mainly observed in acidic and organic rich soils (McNeal and Balisteri 1989; Kabata-Pendias 2001). Selenite tends to be adsorbed on mineral particles and organic matter, and therefore, has a reduced bioavailability (Christensen et al. 1989). In anaerobic soils, selenate and selenite can be reduced to elemental Se (Adriano 2001). Such a chemical reduction of Se is the control mechanism of removing water soluble Se from water column to soil or sediment.

Selenium in water is dominated by selenate. In the San Joaquin Valley (SJV) of California, Gao et al. (2000) indicated that up to 92% of the total Se in agricultural drainage water was in the form of selenate, along with 5% selenite and 3% selenide. In Se-polluted industrial wastewater from the oil refinery process, however, selenite is the predominant chemical form (Hansen et al. 1998). Selenium compounds that can dissolve in water are biologically available for uptake and accumulation by plants. In the United States, the maximum Se contaminant level in drinking water is 50 μg L^{-1} (U.S. EPA 2008a), while the current national recommended water quality criteria for Se in surface water is 5 μg L^{-1} (U.S. EPA 2008b). In the SJV, the biomagnification of Se through the food chains has resulted in detrimental effects on local aquatic ecosystems (Schuler et al. 1990; Ohlendorf 2002).

In human and animal bodies, Se is required for the function of the glutathione peroxidase enzyme that protects intracellular structures against oxidative damage. With low levels of Se in soil and diets, Se deficiency results in human cardiomyopathy disease observed in the Keshan area of China. Selenium has a very narrow margin between the nutritionally required and toxic concentrations. Selenium toxicity to humans is rarely observed under naturally occurring environmental conditions, except for people living in the Enshi region of Hubei Province in China (Zhu and Zheng 2001). Patching and Gardiner (1999) indicated that the major biofunctions of Se can be attributed to its antioxidative properties and its role in the regulation of thyroid hormone metabolism, cell growth, and eicosanoid biosynthesis.

3.3 UPTAKE AND ACCUMULATION OF SELENIUM IN PLANTS

Selenium is an essential trace element for animals and humans, but not for higher plants (Terry et al. 2000). There are studies suggesting that Se is a beneficial element for Se hyperaccumulator species such as *Astragalus bisulcatus* (Broyer et al. 1972). Two recent excellent reviews on Se in higher plants were given by Terry et al. (2000) and Sors et al. (2005), both with a special emphasis on physiological and biochemical aspects of Se metabolism. In these reviews, they clearly illustrated that plant Se metabolism pathways are different in hyperaccumulators versus non-hyperaccumulators. Earlier studies demonstrated that high-affinity sulfate transporters are utilized by selenate. Like sulfate, selenate uptake from soil solution is mediated by the sulfate transporter that is localized in root epidermal cells (Sors et al. 2005). Selenite uptake differs from that of selenate. Although Arvy (1989, 1993) suggested that the uptake of selenite be partially metabolically linked (or partially as an active process), Sors et al. (2005) and Terry et al. (2000) concluded that there was no evidence that the uptake of selenite is mediated by a membrane transporter. In a recent study, Li et al. (2008) indicated that selenite uptake is an active process partially mediated by the phosphate transporter. Clearly, further plant molecular studies are needed to elucidate the mechanism involving selenite uptake across the root membrane.

The uptake of Se from soil or water and its transport and distribution in plant highly depend on the chemical form of Se supplied to the plant. Zayed et al. (1998) reported that plant shoots and roots accumulated Se differently with the different chemical forms of Se in higher plant species, such as broccoli (*Brassica oleracea* var. *botrytis* L.), Indian mustard (*Brassica juncea* L.), sugar-beet (*Beta vulgaris* L.) and rice (*Oryza sativa* L.). The highest Se concentrations in shoots were observed with the selenate-treatment, followed by SeMet, and then the selenite-treatment, while the highest Se concentrations in roots were attained when SeMet was supplied, followed by selenite, then selenate treatments. In a different study, Anakor et al. (2007) found that concentrations of Se in Enokitake mushroom (*Flammulina populicola*) treated with SeMet, Se(IV) and Se(VI) were 18.9, 11.8, and 3.7 mg Se kg^{-1}, respectively. The accumulation of Se in the mushroom tissues increased from 20.6 to 65.2 mg kg^{-1} with increasing Se concentrations in the substrate from 2 to 8 mg kg^{-1}. Lin and Duthie (2007) also indicated that when pickleweed (*Salicornia bigelovii* Torr.) seedlings were supplied with 10 mg kg^{-1} of Se(VI), Se(IV), or SeMet, concentrations of total Se from selenate (142 mg kg^{-1}) was about three to five times greater than the concentrations of Se from selenite (40 mg kg^{-1}) or SeMet (26 mg kg^{-1}) in pickleweed shoots 5 days after the inception of the Se treatments. These earlier studies show that effects of Se speciation on plant Se uptake and accumulation varied among different plant species.

To enhance the level of Se accumulated in specific plant tissues, fertilizing the plant with selenate or selenite compounds would be practically effective, as shown in Finland (Pyykko 1988). Ip et al. (1992) and Ip and Lisk (1995) reported that concentrations of Se in garlic tissues could be substantially increased from < 0.05 to 100 mg kg^{-1} dry weight via Se fertilization. Larsen et al. (2006) studied the uptake of Se in garlic cultivated in soil amended with selenate and/or *mycorrhiza*. They found that the addition of *mycorrhiza* to the natural soil increased the uptake of Se by garlic from 1.5 to 15 mg kg^{-1} (dry weight). However, fertilization of the natural soil with selenate along with addition of *mycorrhiza* increased Se accumulation in garlic to nearly 1000 mg kg^{-1}. Additionally, manipulating environmental conditions may also enhance the uptake of Se. For example, generating wetting and drying cycles in soil can increase Se-bioavailability by transforming more Se to water soluble selenate in soil (Shrestha et al. 2006). Indeed, Se speciation determines Se solubility and therefore its bioavailability for plant uptake and potential for transport in the environment (Ryser et al. 2006).

3.4 SPECIATION AND BIOTRANSFORMATION OF SELENIUM

3.4.1 Speciation of Selenium in Plants

Specific dominant chemical species of Se have been observed in different plant tissues and plant species. Freeman et al. (2006) found that Se in trichomes of *Astragalus bisulcatus* was present in the organic forms methylselenocysteine (MeSeCys, 53%) and γ-glutamyl-MeSeCys (47%), whereas in young leaves of *A. bisulcatus*, inorganic selenate and selenite accounted for 30% of the total Se and MeSeCys for 70%. In comparison with *A. bisulcatus*, the other Se hyperaccumulator *Stanleya pinnata* contained 88% of MeSeCys and 12% of selenocystathionine in young leaves (Freeman et al. 2006). Zhang and Frankenberger (2001) reported that 73–85.5% of the total water soluble Se was in the form of selenoamino acids in *Stanleya pinnata* extract, 7.5–19.5% selenate, and < 7% nonamino acid organic Se. Selenite was not detectable in the plant extract.

Kannamkumarath et al. (2002) reported that in Brazil nut (*Berthollectia excelsa*), walnut, cashew, and peanut tissues, nearly 12% of the total Se was weakly bound to proteins. SeMet was the dominant species, ranging from 19 to 25% of the total Se accumulated in the plant tissues. Selenium-enriched vegetables (such as mushroom, garlic, onion, and leek) can be important dietary supplements of Se for people living in Se-deficient areas. Among these species, Se-enriched garlic is known to accumulate Se as selenoamino acid derivatives, such as Se-methyl-l-selenocysteine (MeSeCys) and γ-glutamylmethylselenocysteine (GluMeSeCys) (Dong et al. 2001). For instance, Kotrebai et al. (1999) reported that Se-methylselenocysteine is the dominant chemical form of Se in Se-enriched garlic tissues, while Ogra et al. (2005) found that the aqueous extracts of the Se-enriched odorless garlic (*Allium sativum* L. Shiro) contained GluMeSeCys. Larsen et al. (2006) recently reported that γ-glutamyl-Se-methylselenocysteine accounted for nearly 67% of the total Se extracted from the garlic tissues, whereas methylselenocysteine, SeMet, and SeO_4^{2-} each accounted for a few percent. Whanger et al. (2000) quantified the predominant chemical forms of Se in bulbs of ramp (or wild leek), including Se-methylselenocysteine (~42%), along with selenate (~25%), Se-cystathionine (~2.5%) and glutamyl-Se-methylselenocysteine (1.5%). In shiitake mushroom (*Lentinula edodes*) supplied with selenite, the main Se compound was identified as SeMet and was likely bound to the water-extractable high molecular mass protein(s) (Ogra et al. 2004).

3.4.2 Biotransformation of Selenium

Several earlier studies have demonstrated that the chemical forms of Se can be transformed in soil and plant. When selenate is supplied to broccoli, most of the Se accumulated in broccoli leaf remained as SeO_4^{2-} with a small fraction being converted into SeMet-like organoselenium compounds (Zayed et al. 1998). When broccoli was supplied with SeO_3^{2-} Se is present as SeMet-like organoselenium compounds in the leaves. Similar results have been observed for roots of broccoli treated with SeO_4^{2-}, SeO_3^{2-} or SeMet, respectively, except that more Se was transformed from selenate to organoselenium in roots than in leaves (Zayed et al. 1998). A similar speciation study was conducted recently by Pedrero et al. (2007), showing the Se transformation process in broccoli: when broccoli was supplied with selenite for 40 days, SeMet was the major chemical form in roots, but the Se-methylselenocysteine compound predominated in florets. Because concentrations of Se accumulated in broccoli florets grown in the Westside of the SJV soil are 3–4 mg kg^{-1} (Bañuelos, personal communication, 2007), the broccoli production in the Central Valley would provide consumers an important source of Se-methylselenocysteine. Clearly, these studies demonstrate that biological transformation of Se in plant is a critical process for developing Se-biofortified agricultural products.

Different plant species have different abilities to transform inorganic Se to organoselenium. Yathavakilla et al. (2005) reported that the dominant chemical forms of Se in leaves of Indian mustard treated with SeO_3^{2-} contained MeSeMet and methylselenocysteine (MeSeCys), along with a small proportion of SeMet. When Indian mustard was supplied with SeO_4^{2-}, the selenate remained unchanged in the leaves. Further, Zhang and Frankenberger (2001) pointed out that considerable amounts of the selenate accumulated in *Stanleya pinnata,* a Se hyperaccumulator, would be transformed to selenoamino acids during plant growth. But, as already mentioned, although the addition of selenate and *mycorrhiza* to soil substantially increased the uptake and accumulation of Se in plant tissues, the chemical species composition (or biotransformation of Se) in the treated garlic was not significantly alternated (Larsen et al. 2006). Lee et al. (2001) examined the role of microbes in biotransformation of Se in pickleweed and found that the shoots and roots of the plants supplied with SeO_4^{2-} accumulated Se predominantly (65–75%) in organic Se forms, regardless of whether the plants were grown axenically or in the presence of microbes. The findings from this study are different from those observed with Indian mustard and broccoli species. Pickleweed has an unusual capacity to reduce selenate to organic forms that is independent of the presence of microbes (Zayed et al. 1998; Orser et al. 1999; Lee et al. 2001).

3.4.3 Speciation Analysis Methods and Technical Limitations

Different speciation methods have been developed and applied to determine chemical species of Se in various environmental samples and organisms. Some techniques are quite dependent on both expensive instruments and sequential chemical extraction procedures. For example, Zhang et al. (1999) developed a sequential extraction method to determine chemical species of selenate, selenite and organic Se compounds in agricultural drainage waters or soil/sediment extracts using hydride generation atomic absorption spectrophotometry. Sequential extraction is one of the most widely used methods to speciate Se in soil and other environmental samples. However, speciation results can be potentially affected by the extraction processes.

Infante et al. (2005) have recently reviewed the advancement in mass spectrometry for Se speciation and demonstrated the advantages of using the mass spectrometric method as the conventional approach. Several combined instrument systems have been used to separate and quantify the amount of each selenocompound separated from samples, including high-performance liquid chromatography with inductively coupled plasma mass spectrometry (HPLC-ICP-MS) (Cabanero et al. 2005), HPLC equipped with electrospray mass spectrometry (HPLC-ESI-MS) (Kotrebai et al. 1999, 2000), chromatography with atomic emission detection (GC-AED) (Uden 2002), multidimensional chromatography coupled to ICP-MS (Gergely et al. 2006), and LC (liquid chromatography)-ICP-MS technique (Pedrero et al. 2007). Moreover, Zhang and Frankenberger (2001) used ion exchange resin (Dowex 1-10X) to separate Se into four groups of nonamino acid organic Se, Se-amino acids, SeO_3^{2-} and SeO_4^{2-} in water extracts of princes plume (*Stanleya pinnata*) tissues, and determine Se concentrations by atomic absorption spectrometry with hydride generation.

Different extractants have been selected to extract selenocompounds from plant tissues. Montes-Bayón et al. (2002) compared the extraction effectiveness of HCl, Tris-HCl buffer, and enzymatic hydrolysis (i.e., using proteinase K and protease XIV). Their study revealed that proteinase K was the best, with nearly 75% of the total Se in tissues of Indian mustard, but where enzymatic hydrolysis extracts mostly SeMet in leaf. With a water extraction efficiency of 68%, Ogra et al. (2004) reported that Se in the water extract of the shiitake mushroom (*Lentinula edodes*) is mainly SeMet that is likely bound to the water extractable high molecular mass protein(s). Further, Huerta et al. (2006) found significant differences in Se speciation between natural and Se-enriched *Agaricus* mushrooms: SeMet was free in natural mushrooms but incorporated into proteins in the selenized mushrooms. The pretreatment procedure of extraction can be difficult to achieve high extraction efficiency. For example, according to Kannamkumarath et al. (2002), nut samples contain high proportions of lipids (50–70%) and lipids interfere with the extraction of Se compounds. In addition, with the use of ICP-MS for Se analysis, the presence of volatile Se metabolites (e.g., dimethylselenide [DMSe] and dimethyldiselenide [DMDSe]) in liquid or extraction samples imposes severe restrictions on the quantification of those Se metabolites because those volatile Se compounds might compromise quantitative Se analysis by ICP-MS (Juresa et al. 2006). Therefore, there is a need for a method that can avoid extracting and digesting samples (or destructive) for Se speciation analysis.

X-ray absorption near-edge spectroscopy (or XANES) technique for Se speciation was developed approximately a decade ago (Pickering et al. 1995). Although the conventional Se speciation methods require sample pretreatments such as digestion or extraction that may alter the chemical form of Se (Andrahennadi et al. 2007), the XANES technique offers an excellent technical option for elucidating the chemical forms of an element in biological tissues or intact organisms in a nondestructive manner (i.e., no chemical extraction and separation involved). In brief, this technique uses synchrotron-generated x-rays to characterize the local electronic environment and the structural parameters of the element (Pickering et al. 1995). Selenium K-edge XANES is sensitive to the chemical form and has previously been used for the analysis of organic and inorganic Se in different environment samples (Lin et al. 2002; Thavarajah et al. 2007). To obtain additional insights into the micromorphology and chemical speciation of heterogeneous environmental samples, the micro-XANES technique has also been used in several recent studies (Freeman et al. 2006; Ryser et al. 2006). For instance, Se speciation in *Astragalus bisulcatus* and *Stanleya pinnata* shows that Se in trichomes was present in the organic forms MeSeCys (53%) and γ-glutamyl-MeSeCys (47%) (Freeman et al. 2006). Although XANES has become a popular Se speciation technique, its availability to researchers is still limited and associated with high cost. In addition, the detection limit for Se speciation in many environmental samples is still relatively high in comparison to other mass spectroscopy methods (such as LC-ICP-MS). If the concentration of Se in the sample is low (e.g., < 5 mg kg^{-1}), a long scanning time for data collection will be required (e.g., several hours per sample).

3.5 BIOFORTIFICATION OF SELENIUM IN RELATION TO SELENIUM SPECIATION

The concept of biofortification provides us with a hope to search for naturally produced agricultural products that accumulate high contents of Se in a desirable chemical form to inhibit cancer development (Clark et al. 1996). As indicated by Abdulah et al. (2005), this is a challenging process but represents a valuable and promising direction to the human health in many developing countries. Previous studies have clearly indicated that the total Se mass accumulated in dietary materials is not an effective measurement of biofortification value of Se. Monomethylated Se compounds, such as Se-methylselenocysteine, are the most biologically effective chemical forms for cancer prevention (Ip et al. 2000; Abdulah et al. 2005). The SeMet-enriched yeast is less effective in rat tumor prevention compared with MeSeCys-enriched garlic, showing that various chemical forms of Se in dietary materials have different effectiveness of antitumors (Whanger et al. 2000). That is due to the fact that the Se speciation affects the level of Se bioavailability, uptake, accumulation and distribution in human and animal bodies. Thus, it is essential that Se speciation information (i.e., the composition of different chemical forms of Se) must be provided so that the effective and functional sources of Se can be evaluated. The importance of identifying health-effective Se compounds in dietary materials has been demonstrated by many earlier studies (Clark et al. 1996; Whanger et al. 2000).

Levels of Se in the diet and the chemical forms have been the attractive topic of much research in recent years because of the relation to human health (Whanger and Butler, 1988; Wilburn et al. 2004). As an example, Se-methylselenocysteine and other methylated selenocompounds are good sources of Se for combating some forms of cancer. These methylated Se compounds are capable of being metabolized in human and animal bodies to methylselenol and methylselenol-related compounds, and as such have been identified as inducers of cellular apoptosis according to Ip et al. (2002).

3.6 CONCLUSIONS

Different chemical forms of Se show various physical, chemical, and biological properties in the environment. Speciation of Se controls the transport pathways and fate of Se in soil and plant systems. Due to biological metabolisms, Se can be transformed from one chemical form to another. Therefore, Se biotransformation processes play an important role in Se accumulation in plant tissues, not only the amount of Se accumulated but also the specific chemical forms of Se being accumulated. To implement the concept of Se biofortification, crops, vegetables and other agricultural products must be developed to maximize the uptake and accumulation of the functional chemical form(s) of Se in the plant tissues. Future studies should include the potential of the manipulation of environmental factors to enhance biotransformation of Se in soil and plant systems.

ACKNOWLEDGMENTS

The author acknowledged the financial support from the Office of Science (BER), U.S. Department of Energy (DE-FG02-03ER63621) for the Se speciation and biotransformation studies at Southern Illinois University–Edwardsville.

REFERENCES

Abdulah, R., K. Miyazaki, M. Nakazawa, and H. Koyama. 2005. Chemical forms of selenium for cancer prevention. *Journal of Trace Elements in Medicine and Biology* 19(2–3): 141–150.

Adriano, D. C. 2001. *Trace Elements in Terrestrial Environments,* 2nd ed. New York: Springer.

Anakor, I., M. Swieciciki, C. Mclntyre, and Z.-Q. Lin. 2007. Decomposition of selenium-laden phytomaterials: Accumulation and volatilization of selenium by fungi. In *Proceedings of 4th International Conference on Phytotechnologies,* 24–26 September 2007, 95. Denver, CO: International Phytotechnology Society, p. 95.

Andrahennadi, R., M. Wayland, and I. J. Pickering. 2007. Speciation of selenium in stream insects using x-ray absorption spectroscopy. *Environmental Science and Technology* 41(22): 7683–7687.

Arvy, M. P. 1989. Some factors influencing the uptake and distribution of selenite in the bean plant (*Phaseolus vulgaris*). *Plant and Soil* 117(1): 129–133.

Arvy, M. P. 1993. Selenate and selenite uptake and translocation in bean plants (*Phaseolus vulgaris*). *Journal Experimental Botany* 44: 1083–1087.

Broyer, T. C., C. M. Johnson, and R. P. Huston. 1972. Selenium and nutrition of *Astragalus*: I. Effects of selenite or selenate supply on growth and selenium content. *Plant and Soil* 36(1–3): 635–649.

Cabanero, A., C. L. M. Carvalho, Y. Madrid, M. C. C. Batoréu, and C. Camara. 2005. Quantification and speciation of mercury and selenium in fish samples of high consumption in Spain and Portugal. *Biological Trace Element Research* 113: 17–36.

Christensen, B. T., F. Bertelsen, and G. Gissel-Nielsen. 1989. Selenite fixation by soil particle-size separates. *European Journal of Soil Science* 40(3): 641–647.

Clark, L. C., G. F. Combs, Jr., B. W. Turnbull, et al. 1996. Effects of selenium supplementation for cancer prevention in patients with carcinoma of the skin. *Journal of the American Medical Association* 276: 1957–1963.

Dong, Y., D. Lisk, E. Block, and C. Ip. 2001. Characterization of the biological activity of γ-glutamyl-Se-methylselenocysteine: A novel, naturally occurring anticancer agent from garlic. *Cancer Research* 61: 2923–2928.

Dumont, E., F. Vanhaecke, and R. Cornelis. 2006. Selenium speciation from food source to metabolites: A critical review. *Analytical and Bioanalytical Chemistry* 385(7): 1304–1323.

Freeman, J. L., L. H. Zhang, M. A. Marcus, S. Fakra, S. P. McGrath, and E. A. H. Pilon-Smits. 2006. Spatial imaging, speciation, and quantification of selenium in the hyperaccumulator plants *Astragalus bisulcatus* and *Stanleya pinnata*. *Plant Physiology* 142: 124–134.

Gao, S., K. K. Tanji, D. W. Peters, and M. J. Herbel. 2000. Water selenium speciation and sediment fractionation in a California flow-through wetland system. *Journal of Environmental Quality* 29(4): 1275–1283.

Gergely, V., K. M. Kubachka, S. Mounicou, P. Fodor, and J. A. Caruso. 2006. Selenium speciation in *Agaricus bisporus* and *Lentinula edodes* mushroom proteins using multidimensional chromatography coupled to inductively coupled plasma mass spectrometry. *Journal of Chromatography A* 1101: 94–102.

Hansen, D., P. J. Duda, A. Zayed, and N. Terry. 1998. Selenium removal by constructed wetlands: Role of biological volatilization. *Environmental Science and Technology* 32: 591–597.

Huerta, V. D., M. Luisa, F. Sánchez, and A. Sanz-Medel. 2006. An attempt to differentiate HPLC-ICP-MS selenium speciation in natural and selenised *Agaricus* mushrooms using different species extraction procedures. *Analytical and Bioanalytical Chemistry* 384(4): 902–907.

Infante, H. G., R. Hearn, and T. Catterick. 2005. Current mass spectrometry strategies for selenium speciation in dietary sources of high selenium. *Analytical and Bioanalytical Chemistry* 382: 957–967.

Ip, C., and D. J. Lisk. 1995. Efficacy of cancer prevention by high-selenium garlic is primarily dependent on the action of selenium. *Carcinogenesis* 16: 2649–2652.

Ip, C., D. J. Lisk, and G. S. Stoewsand. 1992. Mammary cancer prevention by regular garlic and selenium enriched garlic. *Nutrition and Cancer* 17: 279–286.

Ip, C., M. Birringer, E. Block, et al. 2000. Chemical speciation influences comparative activity of selenium-enriched garlic and yeast in mammary cancer. *Journal of Agriculture and Food Chemistry* 48: 2062–2070.

Ip, C., Y. Dong, and H. E. Ganther. 2002. New concepts in selenium chemoprevention. *Cancer and Metastasis Reviews* 21: 281–289.

Juresa, D., D. Kuehnelt, and K. A. Francesconi. 2006. Consequences of vapor enhancement on selenium speciation analysis by HPLC/ICPMS. *Analytical Chemistry* 78: 8569–8574.

Kabata-Pendias, A. 2001. *Trace Elements in Soil and Plants,* 3rd ed. Boca Raton, FL: CRC Press.

Kannamkumarath, S. S., K. Wrobel, K. Wrobel, A. Vonderheide, and J. A. Caruso. 2002. HPLC–ICP-MS determination of selenium distribution and speciation in different types of nuts. *Analytical and Bioanalytical Chemistry* 373: 454–460.

Kotrebai, M., M. Birringer, J. F. Tyson, E. Block, and P. C. Uden. 1999. Identification of the principal selenium compounds in selenium-enriched natural sample extracts by ion-pair liquid chromatography with inductively coupled plasma- and electrospray ionization-mass spectrometric detection. *Analytical Communications* 36: 249–252.

Kotrebai, M., J. F. Tyson, P. C. Uden, M. Birringer, and E. Block. 2000. Selenium speciation in enriched and natural samples by HPLC-ICP-MS and HPLC-ESI-MS with perfluorinated carboxylic acid ion-pairing agents. *Analyst* 125: 71–78.

Larsen, E., R. Lobinski, K. Burger-Meÿer, et al. 2006. Uptake and speciation of selenium in garlic cultivated in soil amended with symbiotic fungi (*mycorrhiza*) and selenate. *Analytical and Bioanalytical Chemistry* 385: 1098–1108.

Lee, A., Z.-Q. Lin, I. J. Pickering, and N. Terry. 2001. X-ray absorption spectroscopy study shows that the rapid selenium volatilizer, pickleweed (*Salicornia bigelovii* Torr.), reduces selenate to organic forms without the aid of microbes. *Planta* 213: 977–980.

Li, H. F., S. P. McGrath, and F. J. Zhao. 2008. Selenium uptake, translocation and speciation in wheat supplied with selenate or selenite. *New Phytologist* Doi:10.1111/j.1469-8137.2007.02343.x.

Lin, Z.-Q. 2005. Bioaccumulation. In *Water Encyclopaedia: Surface and Agricultural Water,* Lehr, J. H., and J. Keeley (Eds.). Hoboken, NJ: John Wiley, pp. 34–36.

Lin, Z.-Q., and A. B. Duthie. 2007. Bioavailability of trace elements in plants in relation to chemical speciation: A case study with pickleweed (*Salicornia bigelovii*). In *Biogeochemistry of Trace Elements: Environmental Protection, Remediation and Human Health,* Y. Zhu, N. Lepp, and R. Naidu (Eds.). Beijing, China: Tsinghua University Press, pp. 275–276.

Lin, Z.-Q., V. Cervinka, I. J. Pickering, A. Zayed, and N. Terry. 2002. Managing selenium-contaminated agricultural drainage water by the integrated on-farm drainage management system: Role of selenium volatilization. *Water Research* 12: 3150–3160.

McNeal, J. M., and L. S. Balisteri. 1989. Geochemistry and occurrence of selenium: An overview. In *Selenium in Agriculture and the Environment,* L. W. Jacobs (Ed.). Madison, WI: American Soil Science Society of America. Special publication No. 23, pp. 1–13.

Montes-Bayón, M., E. G. Yanes, C. P. de León, K. Jayasimhulu, A. Stalcup, J. Shann, and J. A. Caruso. 2002. Initial studies of selenium speciation in *Brassica juncea* by LC with ICPMS and ES-MS detection: An approach for phytoremediation studies. *Analytical Chemistry* 74(1): 107–113.

Ogra, Y., K. Ishiwata, J. R. Encinar, R. Lobinski, and K. T. Suzuki. 2004. Speciation of selenium in selenium-enriched shiitake mushroom, *Lentinule edodes. Analytical and Bioanalytical Chemistry* 379: 861–866.

Ogra, Y., K. Ishiwata, Y. Iwashita, and K. T. Suzuki. 2005. Simultaneous speciation of selenium and sulfur species in selenized odorless garlic (*Allium sativum* L. Shiro) and shallot (*Allium ascalonicum*) by HPLC-inductively coupled plasma-(octopole reaction system)-mass spectrometry and electrospray ionization-tandem mass spectrometry. *Journal of Chromatography* 1093: 118–125.

Ohlendorf, H. M. 2002. The birds of Kesterson Reservoir: A historical perspective. *Aquatic Toxicology* 57(1–2): 1–10.

Orser, C. S., D. E. Salt, I. J. Pickering, R. C. Prince, A. Epstein, and B. D. Ensley. 1999. Brassica plants to provide enhanced human mineral nutrition: Selenium phytoenrichment and metabolic transformation. *Journal of Medical Food* 1(4): 253–261.

Patching, S. G., and P. H. E. Gardiner. 1999. Recent developments in selenium metabolism and chemical speciation: A review. *Journal of Trace Elements in Medicine and Biology* 13(4): 193–214.

Pedrero, Z., D. Elvira, C. Cámara, and Y. Madrid. 2007. Selenium transformation studies during broccoli (*Brassica oleracea*) growing process by liquid chromatography-inductively coupled plasma mass spectrometry (LC–ICP-MS). *Analytica Chimica Acta*. 596(2): 251–256.

Pickering, I. J., G. E. Brown Jr., and T. K. Tokunaga. 1995. Quantitative speciation of selenium in soils using x-ray absorption spectroscopy. *Environmental Science and Technology* 29: 2429–2436.

Pyykko, K., R. Tuimala, R. Kroneld, M. Roos, and R. Huuska. 1988. Effect of selenium supplementation to fertilizers on the selenium status of the population in different parts of Finland. *European Journal of Clinic Nutrition* 42(7): 571–579.

Rayman, M. P. 2004. The use of high-selenium yeast to raise selenium status: How does it measure up? *British Journal of Nutrition* 92(4): 557–573.

Ryser, A. L., D. G. Strawn, M. A. Marcus, S. Fakra, J. L. Johnson-Maynard, and G. Moller. 2006. Microscopically focused synchrotron x-ray investigation of selenium speciation in soils developing on reclaimed mine lands. *Environmental Science and Technology* 40: 462–467.

Schuler, C. A., R. G. Anthony, and H. M. Ohlendorf. 1990. Selenium in wetlands and waterfowl foods at Kesterson Reservoir, California, 1984. *Archives of Environmental Contamination and Toxicology* 19(6): 845–853.

Shrestha, B., S. Lipe, K. A. Johnson, T. Q. Zhang, W. Retzlaff, and Z.-Q. Lin. 2006. Soil hydraulic manipulation and organic amendment for the enhancement of selenium volatilization in a soil-pickleweed system. *Plant and Soil* 288: 189–196.

Sors, T. G., D. R. Ellis, and D. E. Salt. 2005. Selenium uptake, translocation, assimilation and metabolic fate in plants. *Photosynthesis Research* 86(3): 373–389.

Terry, N., A. M. Zayed, M. P. de Souza, and A. S. Tarun. 2000. Selenium in higher plants. *Annual Review of Plant Physiology and Plant Molecular Biology* 51: 401–432.

Thavarajah, D., A. Vandenberg, G. N. George, and I. J. Pickering. 2007. Chemical form of selenium in naturally selenium-rich lentils (*Lens culinaris* L.) from Saskatchewan. *Journal of Agriculture and Food Chemistry* 55: 7337–7341.

Uden, P. C. 2002. Modern trends in the speciation of selenium by hyphenated techniques. *Analytical and Bioanalytical Chemistry* 373(6): 422–431.

U.S.EPA. 2008a. *Current National Recommended Water Quality Criteria*. Washington, DC: U.S.EPA. http://www.epa.gov/waterscience/criteria/wqcriteria.html.

U.S.EPA. 2008b. *Consumer Factsheet on: Selenium*. Washington, DC: U.S.EPA. http://www.epa.gov/safewater/contaminants/dw_contamfs/selenium.html.

Whanger, P. D. 2002. Selenocompounds in plants and animals and their biological significance. *Journal of the American College of Nutrition* 21(3): 223–232.

Whanger, P. D., and J. A. Butler. 1988. Effects of various dietary levels of selenium as selenite or selenomethionine on tissue selenium levels and glutathione peroxidase activity in rats. *Journal of Nutrition* 118: 846–852.

Whanger, P. D., C. Ip, C. E. Polan, P. C. Uden, and G. Welbaum. 2000. Tumorigenesis, metabolism, speciation, bioavailability, and tissue deposition of selenium in selenium-enriched ramps (*Allium tricoccum*). *Journal of Agriculture and Food Chemistry* 48: 5723–5730.

Wilburn, R. T., A. P. Vonderheide, R. S. Soman, and J. A. Caruso. 2004. Speciation of selenium in the mushroom *Boletus edulis* by high-performance liquid chromatography coupled to inductively coupled plasma-mass spectrometry with a collision cell. *Applied Spectroscopy* 58(10): 1251–1255.

Yathavakilla, S. V. K., M. Shah, S. Mounicou, and J. A. Caruso. 2005. Speciation of cationic selenium compounds in *Brassica juncea* leaves by strong cation-exchange chromatography with inductively coupled plasma mass spectrometry. *Journal of Chromatography* A 1100: 153–159.

Zayed, A. M., C. M. Lytle, and N. Terry. 1998. Accumulation and volatilization of different chemical species of selenium by plants. *Planta* 206: 284–292.

Zhang, Y., and W. T. Frankenberger, Jr. 2001. Speciation of selenium in plant water extracts by ion exchange chromatography-hydride generation atomic absorption spectrometry. *The Science of the Total Environment* 269(1–3): 39–47.

Zhang, Y., J. N. Moore, and W. T. Frankenberger, Jr. 1999. Speciation of soluble selenium in agricultural drainage waters and aqueous soil-sediment extracts using hydride generation atomic absorption spectrometry. *Environmental Science and Technology* 33: 1652–1656.

Zhu, J., and B. Zheng. 2001. Distribution of selenium in a mini-landscape of Yutangba, Enshi, Hubei Province, China. *Applied Geochemistry* 16(11–12): 1333–1344.

4 Phytoremediation of Selenium-Contaminated Soil and Water Produces Biofortified Products and New Agricultural Byproducts

Gary S. Bañuelos

CONTENTS

4.1 INTRODUCTION

The use of green plants and associated microbes for environmental remediation has been called "phytoremediation" (Chaney et al. 2000). This green technology is being developed for the cleanup/management of metal(loid)-contaminated soils and waters via the processes of phytoextraction, phytovolatilization, and phytostabilization. Based upon these processes, a plant management remediation strategy for selenium (Se) has been developed for the ecological area designated as "westside" of central California (Frankenberger and Karlson 1995; Zayed et al. 2000; Wu et al.

2000; Bañuelos et al. 2002; Lin et al. 2002; Bañuelos and Lin 2006). Developing a sustainable Se phytoremediation strategy under field conditions is, however, still in the nascent stage. Using crops successfully for field phytoremediation will require a wide range of knowledge regarding plant selection, and soil and crop management practices. In addition, the production of viable plant products and the development of economically feasible remediation systems (Schnoor et al. 1995; Chaney et al. 2000; Bañuelos 2000) will encourage more widespread implementation of an integrated phytoremediation technology.

Bañuelos (2002) has reported that potential crops that can be used for the phytoextraction of Se in central California, include moderate Se accumulators in the *Brassica* family, for example, broccoli (*B. oleracea*), canola (*B. napus*), and Indian mustard (*B. juncea*). He hypothesized that by using these types of plants for managing soluble Se from soil, high value products such as Se-enriched animal feed or Se-enriched vegetables can also be produced (Bañuelos 2002). In addition, other commodity byproducts can be generated, such as using extracted oil from canola and mustard seeds for the production of biofuel. Oleaginous crops, for example, *Brassica*, are considered to be potential alternative fuel sources for diesel-powered engines, which could perform well as a biofuel blend with chemical and physical properties similar to those of commercial-grade diesel fuel (Dorado et al. 2004). In addition, recent research by Bañuelos (2006) indicated that after the oil has been extracted from *Brassica* seeds, residual seed by-products are also of value as Se-enriched feed meal, organic fertilizer, and even as a biological herbicide. The objective of this paper is to discuss the derivation and novel production of Se-fortified products and other potential byproducts that were developed primarily from *Brassica* species and alternative crops grown for the management of natural-occurring Se in soils and waters in the westside of central California.

4.2 METHODS/MATERIALS FOR EXPERIMENTAL SITES

Multiyear field phytoremediation studies were conducted between 2002 and 2007 on sites located at Red Rock Ranch (RRR) in California (courtesy of John Diener, Five Points, California) and at the USDA-ARS research facility in Parlier, California. The soil in the experimental area at RRR was classified as an oxalis silty clay loam (fine montmorillonitic, thermic Pachic haploxeral with a well-developed salinity profile). Two water sources were available for irrigation; canal water from the California aqueduct, and Se-laden drainage water produced on RRR. The canal water had soluble Se concentrations of < 0.001 mg L^{-1}, soluble B concentrations of < 1 mg L^{-1}, and electrical conductivity (EC) for salinity EC of < 1 dS m^{-1}. The drainage water collected using subsurface tile drainage systems was pumped from a drainage sump, and had a range of soluble Se from 0.100 to 0.150 mg L^{-1}, soluble B from 4 to 7 mg L^{-1}, and a sodium sulfate-dominated salinity from 5 to 8 dS m^{-1}. Field experiments initially received "canal water" until plant species were deemed established (between true second or third leaves). Thereafter, irrigation scheduling with drainage water was based upon weather data reported from California Irrigation Management Information System (CIMIS) located at the University of California Westside Research Station situated less than 5 km from RRR field sites.

The USDA-ARS Parlier field site contained Se-laden drainage sediment collected at 0–25-cm depth from the San Luis Drain, Mendota, California. The sediment had a range of water soluble Se from 0.8 to 1.4 mg L^{-1}, soluble B from 8 to 14 mg L^{-1}, and an EC from 15 to 19 dS m^{-1}. Drainage sediment was spread to a depth of 25 cm in a previously excavated field. About 10 cm of good quality soil (Hanford sandy loam with pH 7.2 and EC of 0.9 dS m^{-1}) was applied as the topsoil to enhance plant survival and to encourage biological activity. The topsoil and drainage sediment were mechanically mixed by a tractor and fifteen 33-m long and 1-m wide beds were created. A surface drip irrigation system was installed consisting of in-line turbulent flow emitters with a 4 L h^{-1} emitter discharge rate and an emitter spacing of 0.45 m. All plants (described later) were surface-drip-irrigated with good quality water in part based upon weather data, for example, precipitation, evapotranspiration (ET), collected from a CIMIS weather station located less than 1 km away.

4.2.1 Plant Selection

Many common crops in the San Joaquin Valley (SJV) in California—for example, cotton, tomatoes, and alfalfa—are generally unable to remove great amounts of Se via phytoextraction from Se-rich soils under high sulfate-salinity conditions because high sulfate levels reduce the plant uptake of soluble Se (Zayed and Terry 1994). To circumvent sulfate's competition on the uptake of Se, one option may be to select a Se-accumulating plant species such as *Astragalus* that is a Se-hyperaccumulator capable of accumulating Se up to 10,000 mg Se kg^{-1} (Rosenfeld and Beath 1964). Using an *Astragalus* species for phytoremediation of Se may, however, not be advisable due to its slow-growing rate, unfamiliarity to growers, and lastly, because of the uncertainty associated with utilization of the plant products at harvest containing excessively high tissue Se concentrations. Moreover, because salt, B, and Se all coexist at high levels in waters and soils of the westside of central California, only salt- and B-tolerant plant species can be sustainably used under such poor growing conditions. For this reason, Bañuelos and colleagues have screened and selected plants initially for their salt and B tolerance, and moderate ability to accumulate Se—for example, Indian mustard, canola, broccoli, Salado grass/alfalfa—and secondarily for their ability to produce Se-biofortified products or other valuable commodities. The ability to produce cash-value agricultural products will encourage growers to accept and use phytoremediation as a green strategy for managing Se in problematic soils and waters in a sustainable manner.

4.3 RESULTS OF THE PRODUCTION OF NEW BIOFORTIFIED PRODUCTS

4.3.1 Selenium-Enriched Broccoli

Broccoli can also be used for the phytomanagement of Se via extraction (Bañuelos 2002) and volatilization (Zayed et al. 2000) under moderate adverse conditions. The accumulation of Se in the edible florets produces a high cash-value Se-biofortified commodity for the grower. Se-enriched broccoli can certainly be

TABLE 4.1
Fresh and Dry Weights and Tissue Se Concentrations for Broccoli Irrigated with Drainage Water during Two Growing Seasons[a]

Broccoli	Fresh Weight ($Mg\ ha^{-1}$)		Dry Weight ($Mg\ ha^{-1}$)		Se Concentration ($mg\ kg^{-1}$)	
	1st Year	2nd Year	1st Year	2nd Year	1st Year	2nd Year
Stalks	34.7(1.2)a[b]	36.5(1.6)a	3.4(.36)a	2.5(.12)a	3.7(.34)a	2.9(.09)a
Leaves	62.7(2.4)a	68.3(1.1)b	6.0(.29)a	6.5(.29)a	3.7(.15)a	3.2(.14)a
Florets	10.2(.42)a	12.5(.51)a	1.3(.21)a	2.0(.26)b	4.5(.12)a	4.0(.13)a
Roots	10.1(.89)a	11.8(.78)b	2.1(.33)a	2.6(.30)a	2.6(.11)a	2.9(.12)a

[a] Drainage water had a mean EC of 6.2 dS m^{-1}, soluble Se and B of 150 μg L^{-1} and 5 mg L^{-1}, respectively.

[b] Values represent the means and standard deviation in parenthesis. Means followed by the same letter are not significantly different within each row for each respective fresh weight yield, dry weight yield, and Se concentration at the $P < 0.05$ level by Duncan's multiple range test.

considered as a potential source of supplemental dietary Se for human nutrition. Concentrations of Se (less than 5 mg kg^{-1}) in the floret (see Table 4.1) are well below potentially toxic concentrations for human consumption (knowledge of individual selenocompounds is necessary to evaluate selenium's potential bioavailability; see Finley 2007, Thavarajah et al. 2007). In this regard, Se is an essential trace element for humans, and a recommended dietary allowance (RDA) has been established for 55 μg Se on a daily basis by USDA. Finley (2007) has presented an extensive review on the benefits from increased intakes of Se-enriched foods on human health. Because Se is known as an antioxidant, it may be capable of depressing some cancer cell activity (Whanger 2004). Finley et al. (2005) has conducted research on metabolism and potential health benefits of Se-enriched broccoli, whereas Erdrman et al. (personal communication) are presently investigating the potential anticarcinogenic properties of Se-enriched broccoli on prostate cancer cells at the University of Illinois at Urbana–Champaign in Illinois. Importantly, the glucosinolate species and the type of Se-amino acids contained within Cruciferous food products like broccoli may be so important because Se-methylselenocysteine allegedly may reduce growth of some cancer cells (Finley 2007; Robbins et al. 2005). For Se-phytoremediation sites using *Brassica* species (e.g., broccoli), the plants uniquely contain this species of Se when irrigated with Se-laden water. Hence, Se-biofortified broccoli may have the potential to be a unique health product in aiding mankind's struggle with cancer prevention (Kristal and Lampe 2002). More cancer-related studies using Se-enriched food products are in the planning for determining the possible link between Se nutrition and human health. Additional information can be found in Chapter 2 by Lyons et al. in this book.

4.3.2 Selenium-Enriched Animal Feed

Selenium, although not required by plants, is an essential trace element at concentrations of 0.1–0.3 mg Se kg^{-1} DM for normal nutrition and health of animals (Mayland 1994; NAS 1980). Selenium deficiencies are generally a far greater problem than Se toxicities in animals in the United States. In the SJV dairies, cow producers must use Se supplements for over 2.5 million head of cows. To ensure an adequate level of Se they have a variety of techniques at their disposal, which include giving Se by injection, Se-metal ballasts, Se-salt blocks, or by incorporating Se as an additional organic supplement into their feed ration (Echevarria et al. 1988). In this regard, Bañuelos et al. (2003) suggested that Se-enriched perennial crops (e.g., salado grass and alfalfa) that had been used for long-term management of Se in the western SJV could be harvested and safely used as sources of organic Se for preserving Se levels in animal nutrition. Moreover, these crops with Se concentrations of 2–4 mg Se kg^{-1} were low maintenance, and as perennials they could be planted on large-scale areas, and repeatedly cut, and used as a source of Se biofortified animal feed (Table 4.2).

Although bioavailablity of organic Se supplied as Se-enriched plant material and its subsequent absorption by animal tissues may be greater than inorganic forms of Se (Agbossamey et al. 1998; Bañuelos and Mayland 2000), Se absorption will always depend on type of animal (e.g., cow or sheep) and major chemical species of Se (e.g., selenomethionine contained within the organic matrix of Se). It does appear, in any

TABLE 4.2
Dry Matter Yield and Se Concentration in Different Forage Species Planted Both in Field Soils Incorporated with Se-Enriched *Brassica* Plant Material and in Se-Enriched Drainage Sediment

Species	Yield (Mg ha^{-1})	Mean Leaf Se Concentration[a] (mg Se kg^{-1})
Brassica-enriched soil[b]		
Tall fescue (*Festuca arundincea*)	1.8 (0.28)[c]	1.4 (0.28)
Alfalfa (*Medicago sativa*)	4.4 (0.47)	7.7 (0.82)
Birdsfoot trefoil (*Lotus corniculatus*)	4.7 (0.43)	3.9 (0.46)
Sediment-enriched soil[d]		
Salado grass (*Sporobulus airoides*)	2.3 (0.31)	11.2 (1.6)
Salado alfalfa (*Medicago sativa* "Salado")	1.4 (0.21)	8.4 (0.91)
Cordgrass (*Spartina patens* "Flageo")	1.6 (0.22)	10.1 (1.0)

[a] Selenium leaf concentrations did not exceed 0.2 mg Se kg^{-1} DM under normal growing conditions.

[b] Mustard leaves containing 250 mg Se kg^{-1} DM were incorporated into the soil at a rate of 325 g Se ha^{-1}.

[c] Mean values with standard deviation in parenthesis.

[d] Total Se concentrations were 1.5 mg kg^{-1} DM in soil enriched with Se-laden drainage sediment.

case, that either providing animals with Se-enriched plant material at safe levels or growing known forage crops in soils amended with Se-laden plant material are both practical strategies for disposing of plants used for managing Se on farm sites.

4.3.3 Selenium-Enriched Seed Meal

Canola grown for large-scale management of Se in the SJV produces total biomass yields over 12–13 metric tons (Mg) ha^{-1} (Table 4.3) and conservatively 2 Mg ha^{-1} of seed under high salt and B growing conditions. Because the oil content of canola seed can range from 35 to 40%, its oil is a potential commodity that has many uses for human nutrition and even for the production of alternative biofuels or biodiesel extenders (Meher et al. 2006), to be discussed later in this chapter. After pressing and extracting oil from seed, the residual seed meal is usable as an Se-enriched feed meal. In this regard, canola seed meal is one of the most widely used protein supplement in the animal feed trade, and generally competes as a second tier product to soy meal. Results in Table 4.4 show that Se, nutrient and vitamin concentrations in canola meal are adequate when compared to other reported values for grass and legume forage (NRC 1985). Importantly, the Se concentration in the canola meal produced from the SJV was less than 2 mg kg^{-1} DM (Bañuelos 2006), which allows its safe use as part of the daily feed ration for one Holstein lactating cow, at a mixing ratio of 2.5 kg Se-biofortified canola meal per 50 kg animal feed. Other vital animal forage components within the Se-biofortified meal such as acid detergent fiber (ADF), neutral detergent fiber (NDF), nonfibrous carbohydrates (NFC), were slightly lower for canola meal than typical values for good quality legume, grass and legume grass hays and silages (NRC 1985); however, the crude protein (CP), crude fat (CF), soluble protein (SP), and total digestible nutrients (TDN) values were slightly better than legume hays. The low glucosinolate content in the canola meal (21 μmol g^{-1}) is an important feed quality of the canola seed meal (Table 4.4), because at high concentrations glucosinolates break down into toxic aglucones and their bitter taste

TABLE 4.3

Fresh and Dry Weights and Tissue Se Concentrations for Canola Irrigated with Drainage Water during Two Growing Seasons[a]

Crop/Organ	Fresh Weight Yield		Dry Weight Yield		Se Concentration	
	1st Year	2nd Year	1st Year	2nd Year	1st Year	2nd Year
Canola						
Stems	44.7(1.6)a[b]	52.2(1.8)b	5.1(.32)a	5.1(.29)a	4.3(.13)a	3.8(.11)a
Leaves	70.1(2.7)a	78.2(2.9)b	7.6(.39)a	8.2(.31)a	5.8(.18)a	6.2(.15)a
Roots	8.5(.71)a	10.1(.83)a	2.3(.36)a	9.0(.35)a	2.9(.10)a	3.1(.13)a

[a] Drainage water had a mean EC of 6.2 dS m^{-1}, soluble Se, and B of 150 μg L^{-1} and 5 mg L^{-1}, respectively.

[b] Values represent the means and standard deviation in parentheses. Means followed by the same letter are not significantly different within each row for each respective fresh weight yield, dry weight yield, and Se concentration at the $P < 0.05$ level by Duncan's multiple range test.

TABLE 4.4
Examples of Nutritional Content of Canola Seed Meal after Oil Extraction

Elemental Analysis	Se	B	Ca	Cr	Cu	Fe	K	Mg	Mn	Mo	Na	Ni	P	S	Zn
	mg kg^{-1} DM														
	1.85	41	4455	0.1	5	184	12601	3647	39	1	958	0.55	6661	6743	64

Feed Quality Analysis[a]	ADF	Ash	CF	CP	SP	Fiber	Lignin	Moisture	NFE	NDF	Nitrogen	NFC	TDN
	(%)												
	20.6	3.5	35.7	40	51	12.2	7.2	6.8	15.6	39.1	4.2	17.4	62

Vitamin Analysis	Vitamin A	Vitamin D	Vitamin E	Niacin	Riboflavin
	(μg kg^{-1})			(mg kg^{-1})	
	192	75	57	115	13

Glucosinolates (21μ mol g^{-1})

[a] ADF: acid detergent fiber; CF: crude fat; CP: crude protein; SP: soluble protein; NFE: neutral fiber efficiency; NDF: neutral detergent fiber; NFC: nonfibrous carbohydrates; TDN: total digestible nutrients

results in reduced feed intake. Moreover, some forms of glucosinolates, especially at higher concentrations, are considered natural toxicants and are associated with goiter and liver damage when consumed in large quantities. At the current market price (in 2007) typical unselenized canola seed meal cost dairy producers about \$200 per ton. With our phytoremediation system we have produced Se-biofortified canola meal that has a higher cash value (> \$200/ton) than most non-seleniferous canola meal because of the Se enrichment. Moreover, this Se-enriched meal typically has Se concentrations ranging from 1.85 to 2.50 mg kg^{-1} when grown on the SJV; therefore, it can be of special importance for providing Se to central California's livestock and dairy industries.

4.3.4 Selenium-Enriched Fertilizer

Another product from seed meal is its use as an organic source of Se-enriched fertilizer. The addition of canola or mustard seed meal to soil supporting food crops can lead to increased concentrations of Se and other macro- and micronutrients in the edible tissue of the crop. The gradual breakdown of the meal and the slow release of Se, N, and other essential nutrients make seed meal an ideal source of green manure for growing crops under organic production. For example, we have produced Se-enriched strawberry after incorporating *Brassica* seed meal into the soil (Table 4.5). This observation shows the greatest potential for increasing Se levels in food crops that are produced in Se-deficient soils. In countries, such as China, India, or South Africa, where Se deficiencies are observed in the population consuming rice, grains, or maize low in Se, Se-enriched fertilizers could be very useful. Similarly, Se-enriched *Brassica* leaf material has been successfully tried for increasing Se levels in alfalfa (Bañuelos et al. 1991, 1992; Table 4.2) and other animal fodders (Dhillon et al. 2007) for animal-related uses. Incorporating Se-enriched meals into Se-deficient soils not only has the potential to increase Se levels in food crops and in animal forages, but also develop a beneficial and practical way of disposing of Se-enriched plant or residual seed meal material.

4.3.5 Byproducts from Biofortification

4.3.5.1 Biofumigant

Indian mustard, another oil-producing *Brassica* crop, is also used for managing Se in poor quality soil and water conditions. At harvest the mustard seed can be processed for its oil, which can be used similarly to canola oil in the production of biofuel (to be discussed later in this chapter). In contrast to canola meal, the residual mustard seed meal also possesses bioactive properties, which make them useful for biofumigation (Stapleton et al. 2000). Much of the interest revolves around a constituent class of nitrogens and sulfur compounds called glucosinolates. The high concentrations of glucosinolates (up to 345 μmol g^{-1}) are found in the meal from seeds harvested from plants grown at RRR with Se-tainted waters. They can hydrolyze into isothiocyanates (ITC) with potentially broad-spectrum biocidal activity. Isothiocyanates can be used for biofumigation and are similar to those produced by the synthetic commercial fumigants, Vapam and Basamid; albeit at much lower concentrations. Although

TABLE 4.5
Preliminary Data on Selected Nutrients in Strawberry Fruit Grown with Either Se-Enriched Canola or Mustard Seed Meal under Greenhouse and Field Conditions

Treatment[a]	Ca	Mg	K	P	Fe	Mn	Zn	Cu	Se
	mg kg^{-1}								μg kg^{-1}
				Greenhouse					
Control (no meal)	1605 (486)c[b]	1014 (242)c	11579 (1770)c	1801 (269)b	105 (37)d	41 (12)c	19 (13)b	5 (1)b	150 (78)b
Canola Meal	2700 (316)a	1531 (607)a	14664 (3790)a	2544 (630)a	16 (101)b	69 (20)b	32 (16)a	9 (2)b	183 (2)b
Mustard Meal	2234 (177)b	1297 (517)b	13065 (3704)b	2090 (504)b	125 (48)c	59 (24)b	36 (18)a	7 (2)b	197 (57)a
				Field[c]					
Control (no meal)	1594 (310)c	1239 (126)b	12785 (1678)bc	2347 (161)a	181 (4)c	41 (14)d	16 (11)b	6 (3)	29 (14)d
Canola Meal	2104 (339)b	1501 (168)a	13435 (1706)b	2417 (349)a	159 (65)b	80 (25)b	19 (5)b	17 (7)a	58 (25)c
Mustard Meal	1454 (224)c	1327 (265)ab	13372 (1965)b	2470 (304)a	167 (58)ab	166 (59)a	21 (5)b	14 (9)a	59 (32)c

[a] Rates of application were equivalent to 4.3 Mg ha^{-1}.
[b] Values represent the means and standard deviation in parenthesis from fruit harvested over a minimum of 4 weeks. Within each column standard deviations followed by the same letter are not significantly different at the $P < 0.05$ level by Duncan's multiple range test.
[c] Visual observations showed that there were almost no spittlebugs, mites, lygus, beetles, slugs, thrips, deer, or birds found in the canola and mustard microplots in contrast to the observed but uncounted numbers observed in the control plots.

high levels of glucosinolates are not desirable in plant material used for animal feed, their presence in mustard seed meals at high concentrations has prompted research by Hanson and Bañuelos to evaluate the potential herbicidal effects of mustard seed meal when applied to soil (Hanson and Bañuelos, unpublished data). In this regard, they have successfully demonstrated an effective herbicidal effect (e.g., observed as emergence inhibition) on ten different weed species after incorporating the mustard seed meal into soils at rates equivalent to 4.3–13.2 Mg ha^{-1} (Table 4.6) under field conditions. Creating an alternative biofumigant product from mustard seed meal with its high glucosinolate content may have more use as a plant-derived herbicide in organic agriculture. In this regard, management of soil-born pests can be exceedingly difficult in organic production, and the combination of weeds, nematodes, and other diseases contribute to reduced productivity over time compared to conventional fields. Moreover, factors such as soil type and texture, chemical composition,

TABLE 4.6

Weed Emergence Counts 21 Days after Planting in Biofumigation Preliminary Strawberry Field Trial

Meal[a]	Rate gm^{-2}	BYG[b]	Bluegr[b]	Chick[b]	Crabgr[b]	Horse[b]	Mallow	Pigweed	Purslane	Ryegrass	YNS[b]
		Number per plot									
None	—	9.3	37.3	19.0	10.3	54.0	11.5	17.8	20.5	38.8	2.8
Mustard	3000	5.8	16.8	0	6.8	0	3.3	0	0.8	1.3	2.3
Canola	3000	16.3	61.5	5.8	6.5	6.8	5.3	3.5	2.0	31.3	4.3
LSD (0.05)[c]	—	12.7	73.9	8.4	17.2	37.6	NS	9.6	5.6	26.2	3.9

[a] Application of respective seed meal by incorporation into soil.

[b] BYG is barnyardgrass, bluegr is annual bluegrass, chick is common chickweed, crabgr is crabgrass, horse is horseweed, and YNS is yellow nutsedge.

[c] LSD values often were fairly large due to problems in planting uniform numbers of very small seed.

temperature and moisture content, composition of native soil microflora, cropping history, will affect the specific performance of an organic amendment like mustard seed meal used for soil disinfestations (Stapleton 2006). Bañuelos is currently investigating the impact of the Se-enriched mustard meal on controlling specific insects, for example, cabbage looper and worm in broccoli and spittlebugs, mites, lygus, beetles, slugs, and even deer in strawberry production under field conditions. Similarly, other studies have shown that *Brassica* species that absorbed Se have also deterred insect infestation in crop species listed in the review by Boyd (2007).

4.3.5.2 Biofuel

In central California's SJV, growing canola and mustard plants for Se management produces an array of biofortified products already discussed above, may include the byproduct biofuel, whereby vegetable oils and their mono-fatty acid low alkyl esters are used for biodiesel or as a diesel fuel lubricity additive (Kulkarni et al. 2007). Biodiesel derived from vegetable oils is rapidly gaining market share as a diesel fuel extender worldwide (Meher et al. 2006) for environmental benefits and for a new renewable resource of biodiesel (Kulkarni and Dalai 2006). With a "horizontal press" and "extruder" Bañuelos and Diener (Bañuelos 2006) were able to extract oil at a conservative rate of 2.7–4.5 Mg of canola and mustard seeds per day at RRR. The seeds were put through screw presses or expellers (i.e., horizontal press), which mechanically removed the liquid oil from the seeds. The resultant "press cake" was further processed by the "extruder," which utilized friction as the sole source of heat accompanied by pressure and attrition, and resulted in the additional recovery of oil. The expelled oil is remarkably stable due to the high levels of natural antioxidants, and low levels of phospholipids and peroxides. The resulting "extrudate" was then pressed one more time for the remaining oil with the "horizontal press." All recovered oil was stored in airtight tanks, allowing for natural separation of canola oil and any remaining residual particles. At this stage, the resulting cake meals from pressing and extruding canola and mustard seeds from the "press" and the "extruder" were now available for other uses (e.g., animal feed, fertilizer, and herbicide), as described earlier.

Currently, the expected seed yield from 40 ha of drainage- and Se-impacted soil planted to canola is 136 Mg (under ideal harvesting conditions), which when processed, will yield approximately 30% by weight of recoverable vegetable oil (48.1 Mg) or a volume of 53,200 L, and approximately 90.7 Mg of canola seed meal. The vegetable oil can be converted to 47,880 L of American Society for Testing and Materials (ASTM) standard D-6751 biodiesel or BD 100 (100% biodiesel when transesterfied). The 47,880 L of BD 100 produced from 40 ha can provide 239,400 L of BD 20 (20% canola oil and 80% diesel) that is immediately available and a typical blend for use in diesel-powered engines. The example of growing canola on 40 ha, therefore, will provide 25% (239,400 L of BD 20) of RRR's annual requirements for 950,000 L petroleum diesel fuel that can be directly used for powering diesel engines on either irrigation booster pumps and on-site tractors. Preliminary analysis on using BD 20 biofuels in irrigation booster pumps at RRR showed positive effects on lowering air pollutant emissions of particulate matter, CO, and sometimes NOx in the exhaust (Bañuelos, unpublished data). Reduced combustion and consumption of diesel, in addition to lowering emissions of specific pollutants, is a positive secondary effect for air quality upon using BD 20 fuels in the SJV, which has one of the most contaminated valley basins in the United States. The United States plans on increasing first-generation biofuel (fuels produced from farm crops) production to 1.3×10^{11} L by 2017, whereas Europe proposes that 5.75% of all petrol and diesel for transportation purposes (calculated on the basis of energy content) to be placed on the market by December 31, 2010 (McDermott 2006). Producing oil for biofuel is a significant byproduct resulting from the biofortification process that contributes overall to the sustainability of field phytoremediation of Se.

4.4 CONCLUSIONS

Coupling phytomanagement of Se with the creation of new biofortified and byproducts (e.g., Se-enriched vegetables, animal feed, fertilizers, biofumigants), and even a byproduct like biofuel, may provide California growers unique opportunities to increase the environmental and economic sustainability of a plant Se-management system, while supporting the agriculture community in central California. Successful production of canola and mustard crops may offer Westside region growers an economic incentive to reuse Se-laden agricultural drainage water on a sustained basis, and hence, reduce the volume of drainage water requiring treatment or discharge. If this multifaceted phytoremediation system proves to be sustainable and is accepted by growers, there could be long-term economic benefits, as well as improved water and air quality associated with agricultural production in the SJV. Moreover, sustained operation of this agronomic based system demonstrates the effectiveness of attempting to develop a holistic phytoremediation system for managing contaminants (e.g., Se), and transforming them into valuable biofortified products and byproducts.

REFERENCES

Agbossamey, Y. R., H. V. Petit, J. R. Seoane, and G. J. St.-Laurent. 1998. Performance of lambs fed either hay or silage supplemented with canola or fish meals. *Canadian Journal of Animal Science* 78: 135–141.

Bañuelos, G. S. 2006. Phyto-products may be essential for sustainability and implementation of phytoremediation. *Journal of Environmental Pollution* 144: 19–23.

Bañuelos, G. S. 2002. Irrigation of broccoli and canola with boron and selenium-laden effluent. *Journal Environmental Quality* 31: 1802–1808.

Bañuelos, G. S. 2000. Factors influencing field phytoremediation of selenium-laden soils. In *Phytoremediaiton of Contaminated Soils and Waters*, N. Terry and G. S. Bañuelos, eds. Boca Raton, FL: CRC Press, pp. 41–59.

Bañuelos, G. S., Z. Q. Lin, L. Wu, and N. Terry. 2002. Phytoremediation of selenium-contaminated soils and waters: Fundamentals and future prospects. *Reviews on Environmental Health* 17: 291–306.

Bañuelos, G. S., and H. Mayland. 2000. Adsorption and distribution of selenium in animals consuming canola grown for selenium phytoremediation. *Journal of Exotoxicology and Environmental Safety* 46: 322–328.

Bañuelos, G. S., R. Mead, and S. Akohoue. 1991. Adding selenium-enriched plant tissue to soil causes the accumulation of selenium in alfalfa. *Journal of Plant Nutrition* 14: 701–713.

Bañuelos, G. S., R. Mead, L. Wu, P. Beuselinck, and S. Akohoue. 1992. Differential selenium accumulation among forage plant species grown in soils amended with selenium-enriched plant tissue. *Journal of Soil and Water Conservation* 47(4): 338–342.

Bañuelos, G. S., S. Sharmasarkar, D. Cone, and G. Stuhr. 2003. Vegetative approach for improving the quality of water provided from soils in the Westside of central California. *Plant and Soil* 249: 229–236.

Bañuelos, G. S., and Z.-Q. Lin. 2006. Reuse of agricultural drainage water in central California: Phytosustainability in high levels of salinity and toxic trace elements. In *Function of Soils for Human Societies and the Environment,* Frossard E. Blum, W. E. H., and B. P. Warkentin, Eds. *Special Publication of the Geological Society,* London, England, Vol. 266, pp. 78–88.

Boyd, R. S. 2007. The defense hypothesis of elemental hyperaccumulation status, challenges and new direction. *Plant and Soil* 293: 153–176.

Chaney, R. L., Y. M. Li, S. L. Brown, et al. 2000. Improving metal hyperaccumulator wild plants to develop commercial phytoextraction systems: Approaches and progress. In *Phytoremediation of Contaminated Soil and Water,* N. Terry and G.S. Bañuelos, eds. Boca Raton, FL: CRC Press, pp. 129–159.

Dhillon, S. K., B. K. Hendol, and K. S. Dhillon. 2007. Bioavailability of Se to forage crops in a sandy loam soil amended with Se-rich plant materials. *Chemosphere* 66: 1734–1743.

Dorado, M. P., E. Ballesteros, F. J. Lopez, and M. Mittelbach. 2004. Optimization of alkali-catalyzed transesterification of *Brassica carinata* oil for biodiesel production. *Energy and Fuels* 18: 77–83.

Echevarria, M. G., P. R. Henry, C. B. Ammerman, and P. V. Rao. 1988. Effects of time and dietary Se concentrations as sodium selenite on tissue Se uptake by sheep. *Journal of Animal Science* 66: 2299–2305.

Finley, J. W. 2007. Increased intakes of Se-enriched foods may benefit human health. *Journal of Science, Food, and Agriculture* 87: 1620–1629.

Finley, J. W., A. Sigrid-Keck, R. J. Robbins, and K. J. Hintze. 2005. Selenium enrichment of broccoli: Interaction between selenium and secondary plant compounds. *Journal of Nutrition* 1236–1238.

Frankenberger, W. T. Jr., and U. Karlson. 1995. Volatilization of selenium from a watered seleniferous sediment: A field study. *Journal of Industrial Microbiology* 14: 226–237.

Kristal, A. R., and J. W. Lampe. 2002. Brassica vegetables and prostate cancer risk: A review of the epidemiological evidence. *Nutrition Cancer* 42: 1–9.

Kulkarni, M. G., and A. K. Dalai. 2006. Waste cooking oil an economical source for biodiesel: A review. *Industrial and Engineering Chemical Research* 45: 2901–2913.

Kulkarni, M. G., Dalai, A. K., and N. N. Bakhshi. 2007. Transesterificaiton of canola oil in mixed methanol/ethanol system and use of esters as lubricity additive. *Bioresource Technology* 98: 2027–2033.

Lin, Z. Q., V. Cervinka, I. J. Pickering, A. Zayed, and N. Terry. 2002. Managing selenium contaminated agricultural drainage water by the integrated on-farm drainage management system: Role of selenium volatilization. *Water Research* 36: 3150–3160.

Mayland, H. F. 1994. Selenium in plant and animal nutrition. In *Selenium in the Environment,* W.T. Frankenberger, Jr. and S. Bensen, eds. New York: Marcel Dekker, pp. 29–46.

McDermott, W. E. 2006. Biofuels in Europe. *McDermott Newsletter* Oct. 17, 2006.

Meher, L. C., D. Vidya Sagar, and S. N. Naik. 2006. Techinical aspects of biodiesel production by transesterification-a review. *Renewable and Sustainable Energy Reviews* 10: 248–268.

National Academy of Sciences (NAS). 1980. Selenium. In *Mineral Tolerance of Domestic Animals.* Subcommittee on Mineral Toxicity in Animals. Washington, DC: National Academic Press.

National Research Council (NRC). 1985. *Mineral Tolerances of Domestic Animals.* Washington, DC: National Academic Press.

Robbins R. J., A. Sigrid-Keck, G. S. Bañuelos, and J. W. Finley. 2005. Cultivation conditions and selenium fertilization alter the phenolic profile, glucosinolate, and sulforaphane content of broccoli. *Journal of Medicinal Food* 8: 204–214.

Rosenfeld, I., and O. A. Beath. 1964. *Selenium, Geobotany, Biochemisty, Toxicity, and Nutrition.* New York: Academic Press.

Schnoor, J. L., L. A. Licht, S. C. McCutcheon, N. L. Wolfe, and L. H. Carreira. 1995. Phytoremediation of organic and nutrient contaminants. *Environmental Science Technology* 29: 318–323.

Stapleton, J. J. 2006. Biocidal and allelopathic properties of gramineaus crop residue amendments as influenced by soil temperature. In *Proceedings, California Conference on Biological Control* V, Riverside, CA, July 25–27, 2006: 179–181. *http://www.cnr.berkeley.edu/biocon/complete%20Proceedings%20for%20CCBC%20V.pdf.* (Accessed November 1, 2007.)

Stapleton, J. J., C. L. Elmore, and J. E. DeVay. 2000. Solarization and biofumigation help disinfest soil. *California Agriculture* 52: 42–45.

Thavarajah, D., A. Vandenberg, G. N. Graham, and I. J. Pickering. 2007. Chemical form of selenium in naturally selenium-rich lentils (*Lens culinaris L.*) from *Saskatchewan Journal of Agricultural and Food Chemistry* 10.1021/jf070681i.

Whanger, P. D. 2004. Selenium and its relationship to cancer: An update. *British Journal of Nutrition* 91: 11–28.

Wu, L., G. S. Bañuelos, and X. Guo. 2000. Changes of soil and plant tissues selenium status in an upland grassland contaminated by selenium-rich agricultural drainage sediment after ten years transformed from a wetland habitat. *Ecotoxicology Environmental Safety* 47: 201–209.

Zayed, A., E. Pilon-Smits, M. de Souza, Z. Q. Lin, and N. Terry. 2000. Remediation of selenium polluted soils and waters by phytovolatilization. In *Phytoremediation of Contaminated Soil and Water*, N. Terry and G. S. Bañuelos, eds. Boca Raton, FL: CRC Press, pp. 61–83.

Zayed, A. M. and N. Terry. 1994. Selenium volatilization in roots and shoots: Effects of shoot removal and sulfate level. *Journal of Plant Physiology* 143: 8–14.

5 Soil and Crop Management for Improving Iron and Zinc Nutrition of Crops

Fusuo Zhang, Mingsheng Fan, Xiaopeng Gao, Chunqin Zou, and Yuanmei Zuo

CONTENTS

5.1 INTRODUCTION

The diets of populations subsisting on cereals are often lacking in micronutrients such as iron (Fe) and zinc (Zn). This can be ascribed partly to crop production failing to deliver proper nutrition. For instance, the concentrations of Fe and Zn in wheat grain show a decreasing trend with the use of high-yield varieties. Although the traditional strategies, such as supplementation or food fortification, cannot always be universally successful, biofortification can be achieved by utilizing crop and soil management with plant breeding for increasing micronutrient concentrations in the edible parts of crops. The concept of biofortification is attractive not only for improving growing conditions for crops but also for the exploiting a plant's potential for micronutrient mobilization and utilization.

The concentrations of Fe and Zn in crop grains can be increased through soil and foliar applications of Fe and Zn fertilizers. The choices of fertilizer forms, application methods, and timing of fertilization are discussed in this chapter. As the adsorption and bioavailability of Fe and Zn can be inhibited by high concentrations of phytic acid, we also briefly report a recent field study on the molar ratio of phytic acid to Zn in aerobic rice cultivations in China. Nutrient availability in the rhizosphere can be increased by either a higher spatial exploration area (i.e., root surface and mycorrhizas) or root-induced chemical changes (i.e., pH changes and root exudates). In this chapter, we evaluate the potential to increase nutrient uptake and accumulation into plants by manipulating both spatial and chemical changes in the rhizosphere. Managing cropping systems, such as rotation and intercropping strategies, may have numerous advantages in terms of increasing micronutrient availability. With several recent studies, including peanut–maize intercropping, introduction of aerobic rice, and ground cover rice production system, our recent research progress in soil and crop management has provided the means to resolve complex plant Fe and Zn nutritional problems through manipulating the rhizosphere, managing cropping systems, screening for Fe and Zn efficient species, and applying fertilizers. Substantial efforts need to be made in increasing not only crop yield but also micronutrient density through a combination of rhizosphere management, genetic and molecular approaches, consequently improving the nutritional value of food crops and ultimately improving human nutrition and health.

5.2 IRON (Fe) AND ZINC (Zn) IN CROPS

5.2.1 Variation in Iron and Zinc in the Edible Portions of Plants

There appears to be considerable genetic variation in the Fe and Zn concentrations present in the edible portions of most crop species (see Table 5.1), which can be used in breeding programs to increase Fe and Zn concentrations in crops. Generally, grains of cereal crops such as wheat and rice have lower Zn and Fe concentrations than legume seeds (Frossard et al. 2000). As a result, many people in developing countries subsisting on cereal-based diets are deficient in Fe and Zn. Such countries include those in Asia whose staples are based on rice or wheat, or based on maize-like crops in sub-Saharan Africa and Latin America (White and Broadley 2005). Within cereal crops, rice grain contains the lowest Fe and Zn concentration (Welch

TABLE 5.1
The Variation in Fe and Zn in the Edible Portions of Plants

Plant Species	Accessions	Fe (mg kg^{-1})	Zn (mg kg^{-1})	References
Rice grain	1138	6–24	14–58	Gregorio et al. (2000)
Wheat grain	324	25–73	25–92	Monasterio and Graham (2000)
Maize grain	1814	10–63	13–58	Bänziger and Long (2000)
Soybean	322	48–110	29–67	Wolnik et al. (1983)
Common bean	Over 1000	34–89	21–54	Beebe et al. (2000)
Cassava storage roots	162	4–49	4–18	Maziya-Dixon et al. (2000)

and Graham 1999). Gregorio et al. (2000) reported a fourfold difference in Fe and Zn concentrations, which suggests a genetic potential to increase the concentrations of the micronutrients in rice grain. In wheat lines, a high correlation between grain Fe and Zn concentrations was reported, indicating that it is possible to simultaneously improve Fe and Zn levels in wheat grain.

5.2.2 Impact of Agricultural Practices on Fe and Zn in Crops

5.2.2.1 Historical Changes of Fe and Zn Concentration in Wheat Grain

Wheat is the most widely grown food crop with a global production of about 600 million metric tons annually (FAO 2006). Wheat and wheat products are important sources of Fe and Zn in the human diet (Pomeranz 1988). However, modern plant breeding has been historically oriented toward higher agronomic yield rather than the nutritional quality (Welch and Graham 1999; Morris and Sands 2006). For example, wheat yields have been increased dramatically by more than threefold compared to the yield of 860 kg per hectare in the early 1900s (Percival 1921; Agricultural Statistics 2004). The increased yield may have resulted in lower concentrations of minerals in the grain. Garvin et al. (2006) showed that grain Fe and Zn concentrations decreased significantly with the date of cultivar release in a set of 14 U.S. wheat cultivars from production eras spanning more than a century. McGrath et al. (2007) found a decrease in the concentrations of Fe and Zn in wheat grain with time, based on archived samples (1845–2005) from the Broadbalk Wheat Experiment at Rothamsted, England. Further analysis revealed that the yield and harvest index accounted for the decreasing Zn and Fe concentrations in wheat grain in the Broadbalk experiment. Significant negative relationships between yield and grain Fe and Zn concentrations were also observed in the study of Garvin et al. (2006). Thus, it appears that Fe and Zn concentrations in wheat grain have decreased with time, partly due to a "dilution" effect resulted from increased yields.

5.2.2.2 Changes in Zn Concentration in Response to the Shift of Flooded Rice to Aerobic Rice Crops

Rice production in China is now undergoing an important change brought about by water constraints (see Chapter 11 by Slingerland et al. in this book). Traditional high water-consuming flooded rice cultivation is gradually being replaced by the new aerobic rice production system in which rice is grown as a dry field crop in irrigated but nonflooded and nonpuddled fertile soils (Bouman et al. 2005). The newly bred aerobic rice varieties are currently being developed by crossing high-yield lowland rice cultivars with drought-resistance upland rice. In China, breeders have produced aerobic rice varieties with an estimated yield potential of 6–7 t ha^{-1} that are now being pioneered by farmers on some 190,000 ha in northern China where increasing water scarcity makes lowland rice uneconomic (Wang et al. 2002). The shift from flooded to aerobic conditions results in the problem of micronutrient deficiency, especially for Zn in rice.

Many factors that determine Zn bioavailability are expected to change after the shift to aerobic cultivation. Bulk soil pH will change from pH 7 to the original soil pH (Liu 1996). Redox potential will increase (Gao et al. 2002), causing the formation of Fe and Mn oxides where Zn might be adsorbed. Meanwhile, Zn solubility may increase due to the oxidation of sulfide and consequent dissolution of ZnS (Castro 1977). The content of organic matter may decrease because of oxidation. The consequence of all these changes for Zn bioavailability is difficult to predict. In field experiments, Gao et al. (2006) demonstrated that the cultivation shift from flooded to aerobic condition may increase Zn deficiency problems. The uptake of Zn in rice plants was less in aerobic fields than those in flooded fields. In a Zn deficient soil in central China, the shift to aerobic rice decreased not only the quantity (grain yield) but also the quality (grain Zn concentration) of the grain (Gao 2007). Zinc harvest index significantly decreased as a consequence of the shift to aerobic conditions, and there was also a large genotypic difference in Zn harvest index among different rice genotypes (see Table 5.2; Gao et al. 2006). These results demonstrate that the introduction of aerobic rice systems may increase Zn deficiency problems. In practice, Zn fertilization should be recommended to local farmers in areas where lowland rice production is going to be replaced by aerobic rice production. Also, the genotypic variation in tolerance to Zn deficiency and Zn harvest index can be used as an alternative way to overcome the Zn deficiency in aerobic rice production.

5.2.2.3 Improvement of Fe and Zn Nutrition of Dicot Plants by Intercropping with Graminaceous Species

Cropping systems, such as rotation and intercropping, may have numerous advantages in terms of increasing bioavailability of Fe and Zn. In an intercropping system containing Chinese peanut (*Arachis hypogaea* L.) and maize (*Zea mays* L.), Zuo et al. (2000) found that the excretion of phytosiderophores by maize in the rhizosphere played an important role in improving Fe nutrition of peanut crops. Recently, some research groups have reported that interspecific root interactions and rhizosphere effects from intercropping could be linked to improved nutrient uptake in plants (Li et al. 2007; Gunes et al. 2007). For example, enhanced release of phytosiderophores

TABLE 5.2
Effects of Cultivation and Genotype on Zn Harvest Index [Grain Zn Content/(Shoot + Grain Zn Content)] in Rice

	Zn Harvest Index (%)	
Genotypes	**Flooded**	**Aerobic**
Qiuguang	46.9	24.6
K150	48.3	37.2
Han72	56.4	23.5
89B-271-17(hun)	53.1	46.1
Han277	41.7	34.5
Han297	42.6	44.3
Mean	48.2	35.0

Source: Adapted from Gao et al. 2006. *Plant and Soil* 280: 41–47.

by plants may mobilize Zn in the soil and enhance plant Zn uptake (Zhang et al. 1991). The mugineic acid family phytosiderophores (MAs) play a major role in Fe acquisition, and may also contribute to the acquisition of Zn or other metal nutrients by graminaceous plants (Welch 1995; Von Wirén et al. 1996). In a maize–peanut intercropping system, the maize not only acquired adequate Fe to meet the growth needs, but also improved the Fe status of peanut. Notably, intercropping has increased the Zn content in the peanut shoot (see Table 5.3), which means that agronomic intercropping helps mobilize and uptake limiting nutrient elements, as well as provides benefits through effects on plant growth, development and adaptability to adverse environments. The possible reason for such different effects on Zn concentrations of peanut plants caused by intercropping could be root exudates from gramineous species. Specifically, the production and release of phytosiderophores from gramineous species may improve the solubility of Fe and Zn by chelation, which helps plants obtain these essential elements from the soil (Rengel 2002; Schmidt 2003; Inal et al. 2007). Further investigations are needed to quantify the role of rhizosphere processes in intercropping systems.

5.3 MICRONUTRIENT BIOFORTIFICATION STRATEGIES

Iron and Zn are essential nutrients that are often lacking in human diets, either due to insufficient intake or poor absorption of the elements from food. The focus has been placed on Fe and Zn because their deficiencies are common not only in developing countries, where the diet is essentially composed of cereal crops, but also in some industrialized countries (Welch and Graham 1999). It is estimated that there are over 4–5 billion Fe-deficient individuals and about 2 billion people who are or are at risk of developing Zn efficiency disorders (Ghandilyan et al. 2006). Recommended daily Fe and Zn intakes between the age of 25 and 50 years were 10 mg Fe and 15 mg Zn for male, and 15 mg Fe and 12 mg Zn for female, respectively (FAO/WHO 2000). Traditional strategies to deliver these minerals to susceptible populations have relied

TABLE 5.3

The Effects of Intercropping Peanut with Maize on Fe and Zn Concentration in Peanut (mg kg^{-1} DW) Grown on a Calcareous Soil

Plant Tissue	Monocropping	Intercropping	References
	Fe		Zuo et al. 2000
Shoots	28.0 ± 7.0b	65.5 ± 8.9a	
Roots	159.5 ± 13.1b	203.1 ± 16.8a	
Seeds	22.2 ± 2.9b	31. 8 ± 3.9a	
	Zn		Inal et al. 2007
Shoot	10.37	26.23	
F values:	14.01**		

Note: All data were analyzed using the SAS software, expressed as means of three replicates with standard deviation, and the means were subject to another test by using the least significant difference (LSD) method at 5% probability level (Zuo et al. 2000). Statistical significance of difference was determined by analysis of variance (ANOVA) and the LSD test at $P \leq 0.05^{**}$ for multiple comparisons (Inal et al. 2007). Standard deviation values followed by the same letter in each row are not significant.

on supplementation or food fortification programs. Unfortunately, these interventions have not always been universally successful because of various reasons (White and Broadley 2005). An alternative solution is to increase mineral concentrations in edible crops by the concept of biofortification, that is, increasing the bioavailable concentrations of essential elements in the edible portions of crop plants through agronomic intervention or genetic selection (Graham et al. 2001).

Ultimately, micronutrients such as Fe and Zn enter the food chain through plants that in turn take up Fe and Zn from the soil. However, it is estimated that about 50% of the soils used for cereal production worldwide contain low levels of plant-available Zn (Alloway 2004) and 40% of the world's soils are Fe-deficient (Chen and Barak 1982). Traditional agricultural practices always focus on maximizing production while minimizing costs. Only recently, more researchers are becoming aware that increasing bioavailability of micronutrients in the edible parts of staple crops through agricultural management is a cost-effective and sustainable way to alleviate micronutrient malnutrition in vulnerable groups. Some global joint projects such as HarvestPlus (http://www.harvestplus.org/) are being conducted with the aim of improving micronutrient status of human nutrition. It is recognized that the available scientific knowledge could greatly improve the micronutrients output of farming systems and the bioavailability of micronutrients in edible crops (Welch and Graham 2002). The following discussion briefly presents some agronomic practices that could be used to enhance the bioavailability of Fe and Zn in crop grains and hence minimize the deficiency of these micronutrients in plants.

5.4 SOIL AND CROP MANAGEMENT FOR IMPROVING Fe AND Zn NUTRITION OF CROPS

5.4.1 Fertilization

Both macro- and micronutrient fertilizers can have significant effects on the accumulation of micronutrients in edible plant products. However, fertilizer application by either soil or foliage may be successful in increasing grain yield and grain micronutrient density only in well-defined circumstances. Different forms of fertilizer vary considerably in physical and chemical structure and are influenced by biological properties associated with plants. Methods of fertilization also differ depending on the soil, crop, and the farming system. The extent of micronutrient fertilization is largely dependent on the dose and timing. In general, care must be taken in selecting the suitable form, dose, and method of application at the proper time.

5.4.1.1 Iron Fertilization

Researchers are facing many challenges when working with Fe fertilization. Soil application of inorganic Fe fertilizers to Fe-deficient soils is usually ineffective because the applied Fe is quickly converted into plant unavailable Fe (III) forms. For example, application of 100 kg Fe ha^{-1} to the soil may be required to correct the Fe deficiency in sorghum (Loué 1988). Olson (1950) found that 1120 kg ha^{-1} of Fe ($FeSO_4{\cdot}7H_2O$) mixed into an Fe-deficient soil did not significantly increase the aboveground dry matter of sorghum grain in a greenhouse study. The main reason is that $FeSO_4$ added to calcareous soils quickly reacts with $CaCO_3$ to form Fe oxides that are less available for plant uptake. Although soil application of Fe chelates, such as Fe-EDTA or Fe-EDDHA, has shown better results than inorganic Fe salts such as $FeSO_4$, these compounds are expensive and the outcome is rarely successful. Moreover, even though some treatments (e.g., localized placement strategies) have been identified to help correct Fe deficiency, few have been found feasible as a whole-field treatment for crop production.

Foliar application of $FeSO_4$ or Fe chelates was shown to be more efficient than soil application because of the direct uptake of Fe by plant through cuticular pores on the leaf surface. Foliar application can correct Fe chlorosis in bean, sorghum, peanut, and some other fruit crops. Spraying of 2500 mg L^{-1} $FeSO_4$ solution for two or three times during the flowering period can increase the yield of soybean by 17.5–22.9% (Han et al. 1994). The most widely used Fe sources for spraying are inorganic Fe forms and chelates mixed with inorganic Fe forms (Vempati and Loeppert 1988). Chelated forms of Fe are usually more effective in reducing Fe chlorosis than inorganic forms. However, chelates are rarely economically feasible for crops, such as corn and sorghum, because the spray treatment often has to be repeated for several times during a growing season.

Iron deficiency in China is a serious nutritional problem in the production of fruits, including apple, peach, and orange (Liu et al. 2002). The traditional practices, such as foliar sprays and soil fertilization, have shown little effect on correcting the Fe deficiency. Some simple and effective practices, however, have been put forward and widely used by local farmers, such as trunk injection (injecting Fe^{2+}), root suc-

FIGURE 5.1 Examples of special practices to treat iron deficiency of fruit trees. *Left*: bag fertilization; *right*: root suction (Lu and Xue, personal communication).

tion (putting the cut roots in Fe fertilizer solution), and bag fertilization (putting the cut braches in a bag with Fe fertilizer solution) as shown in Figure 5.1.

There are a few studies that aimed to increase yield and Fe concentration in grains by different Fe fertilizers. Although foliar fertilization increases crop yield to a greater extent than it increases the Fe density in grain, it might be the only available fertilization practice that can slightly increase the Fe density in grain (Frossard et al. 2000; Rengel et al. 1999).

5.4.1.2 Zinc Fertilization

Compared to Fe, soil application of inorganic Zn ($ZnSO_4$) on Zn deficient soils is usually effective, for increasing both yield and Zn density in grains (Cakmak et al. 1999). In a Zn-deficient calcareous soil in Turkey, Yilmaz et al. (1997) reported that Zn application as soil or foliar or a combination of both could increase yield and seed Zn concentrations of wheat by a factor of 3 or more (Table 5.4). Long-term fertilization with $ZnSO_4$ on an alkaline soil did not increase Zn concentration in maize grain (Payne et al. 1988). In most cases, concentrations of Zn in crop grains does not increase to the desired levels to meet human requirements even by using high rates of fertilizer application (Rengel et al. 1999). Srivastava et al. (1999) showed that soil Zn application at transplanting only increased grain Zn mass fraction of lowland rice from 18 to 25 mg kg^{-1}. Similarly, Gao et al. (2005) recently found little increase in Zn concentration in rice grains with soil application of $ZnSO_4$ at a rate of 23 kg ha^{-1} on a calcareous soil. On a clay soil in a different year, soil application of Zn at sowing showed a minor influence on Zn concentration in brown rice, even though the effect was statistically significant (Gao et al. 2007; see Figure 5.2). Our results, in combination with similar results from Srivastava et al. (1999), clearly showed that it was more difficult to improve Zn density in the edible grain of rice than in other crops. More investigations are required to understand these differences.

Zinc sulfate is the most commonly used source of Zn fertilizer. In most cases, inorganic and chelated Zn fertilizer sources are equally effective in improving crop yield. Hergert et al. (1984) compared the grain yields of maize with five Zn fertilizer sources and found no systematic difference between different Zn sources (Table 5.5). Similar results were also reported in rice (Slaton et al. 2005). The chelated or inorganic $ZnSO_4$ fertilizer sources are equally effective in Zn solutions applied both

TABLE 5.4
The Effects of Different Zn Application Methods on Grain Yield and Zn Concentration in Grain of Two Bread Wheat Cultivars Grown on a Zn-Deficient Calcareous Soil

	Yield (t ha^{-1})		Grain Zn concentration (mg kg^{-1})	
Treatments[a]	**Gerek-79**	**Bezostaja**	**Gerek-79**	**Bezostaja**
Control	0.74	0.81	9	10
Soil	2.70	2.34	17	17
Seed	2.05	1.96	11	8
Leaf	1.47	1.55	30	28
Soil + leaf	2.71	2.33	34	38
Seed + leaf	2.77	2.38	34	25
LSD (0.05)	0.45	0.74	6	9

[a] Control, no Zn application; soil application, 23 kg Zn ha^{-1} as $ZnSO_4 \cdot 7H_2O$; seed application, 1.0 L of 30% $ZnSO_4 \cdot 7H_2O$ for 10 kg of seed; leaf application, 200 g Zn ha^{-1} as $ZnSO_4 \cdot 7H_2O$ in 450 L during tillering and stem elongation stages.

Source: Adapted from Yilmaz et al. 1997. *J. Plant Nutr.* 20: 461–471.

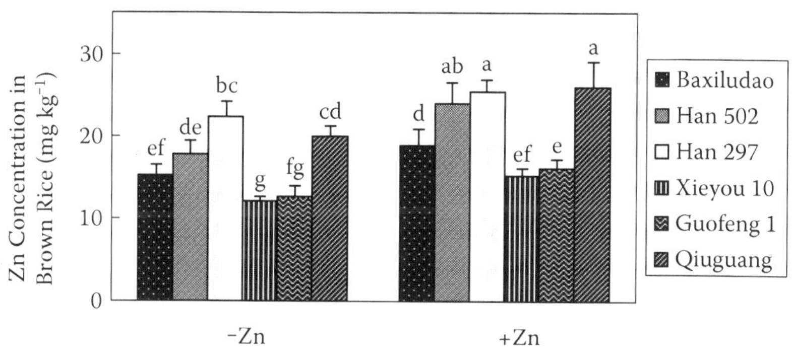

FIGURE 5.2 Effects of Zn application on Zn mass fraction in brown rice of six rice genotypes (means ± SD). Data are from a field experiment with rice plants grown aerobically on a low Zn soil. The experiment was designed as a randomized complete block with Zn fertilizer rates (0 and 10 kg ha^{-1} Zn as $ZnSO_4$) as main plots and six different rice genotypes as subplots. At harvest, rice grains were dehulled and Zn mass fractions in brown rice were then determined after digestion in an acid mixture (HNO_3 + $HClO_4$).

pre- and post-emergence. Haslett et al. (2001) reported that Zn from $ZnSO_4$ and Zn-EDTA solutions applied to wheat foliage at the same rates was absorbed and translocated with equivalent efficiency. Foliar application with $ZnSO_4$, Zn-EDTA, or other chelates has also been used in fruit, vegetables, and field crops. Zinc sulfate is generally effective as a synthetic chelate, so it is mostly used because of lower costs

TABLE 5.5
Yield Responses of Maize (in t ha^{-1}) to Zn Sources Band Applied with an Ammonium Polyphosphate Starter Fertilizer

	Zn Applied Rate (kg ha^{-1})		
Zn sources	0	0.11	1.12
Control	3.9	—	—
ZnO	—	8.2	8.6
$ZnSO_4$	—	8.3	8.7
Zn-EDTA	—	8.6	9.5
$ZnSO_4$-NH_3 complex	—	7.8	8.5
$Zn(NO_3)_2$-UAN	—	8.2	8.9

Source: Adapted from Hergert et al. 1984. *Soil Sci. Soc. Am. J.* 48: 1190–1193.

(Mortvedt and Gilkes 1993). The need for chelates for Zn fertilization (in sprays) is not as critical as for Fe.

The timing of Zn fertilization is crucial because early Zn deficiency may predispose a crop to a grain yield loss (Rengel et al. 1999). Soil application at sowing is a preferred option for wheat (Brennan 1991), and foliar application of $ZnSO_4$ to barley. Applying Zn in the late growing season of wheat can lead to a high seed Zn concentration (I. Cakmak, personal communication). For farmers, selecting the appropriate Zn-fertilization strategy that includes Zn fertilizer source and application rate and time will depend on a variety of economic and agronomic factors, as well as the chemical and physical properties of the selected fertilizer. In sorghum production, it is suggested that Zn fertilizer combined with organic amendment could significantly increase grain yield and grain Zn concentrations (Traore 2006).

Existence of high concentrations of phytic acid in cereal-based foods is a major cause for the occurrence of Zn deficiency to human beings (Welch 1993). As the main storage form of P in cereal seeds, phytic acid usually accounts for 70–80% of the total P in seeds (Raboy et al. 1991). Phytic acid impairs bioavailability of Zn by binding it in the grain. In most cases, phytic acid-to-Zn molar ratio in foods is considered as a predictor of Zn bioavailability. The ratio of > 20 has been reported to reduce Zn absorption and growth in animals (Solomons 1982). A survey was conducted on 41 aerobic rice fields in five provinces of China (Chen et al. 2005). Results showed that brown rice phytic acid-to-Zn molar ratios ranged from 29 to 100, indicating the low bioavailability of Zn in rice grain. There were weak correlations between soil P-Olsen-to-DTPA-Zn molar ratios and brown rice phytate-to-Zn molar ratios. This observation indicated that soil P/Zn bioavailability was a very poor predictor for the grain phytate-to-Zn ratio. It was hypothesized that excessive application of P fertilizer might contribute to the high phytic acid-to-Zn molar ratios and consequently the low bioavailability of Zn. From a collection of 54 winter and summer wheat cultivars, phytic acid-to-Zn molar ratios ranged between 39 in the grains with the highest Zn concentration (31 mg kg^{-1}) and 137 in the grain with the lowest Zn concentration (8 mg kg^{-1}) suggesting low bioavailability in wheat (Cakmak et al.

1999). On the other hand, this finding also suggested that Zn fertilization could be used as an effective way to significantly decrease phytic acid-to-Zn ratio. With 20 wheat cultivars grown on a Zn-deficient soil in Turkey, Erdal et al. (2002) reported substantial variation in seed phytic acid concentration, indicating this variation can be exploited for breeding genotypes with low concentrations of phytic acid and high bioavailability of Zn. By fertilizing Zn at the rate of 23 kg ha^{-1}, the average grain Zn concentration increased from 8.8 to 16.8 mg kg^{-1}, whereas phytic acid-to-Zn ratio decreased dramatically from 127 to 56 indicating the beneficial role of Zn fertilization on bioavailability of Zn in grains (Table 5.6).

5.4.2 Rhizosphere Management

5.4.2.1 Fe and Zn Availability in Soils

For optimal plant growth, an Fe concentration of 10^{-6} to 10^{-5} mol L^{-1} is required (Marschner 1995). In most soils with a pH ranging between 7 and 9, however, the concentration of free Fe^{2+} and Fe^{3+} in the solution is extremely low, not exceeding

TABLE 5.6
The Effect of Zn Fertilization (23 kg Zn ha^{-1}) on Concentrations of Grain Zn and Phytic Acid to Zn Molar Ratios in 20 Wheat Cultivars Grown in Zn-Deficient Calcareous Soils in Turkey

	Grain Zn Concentration (mg kg^{-1} DW)		Phytic-Acid-to-Zn Ratio	
Cultivars	-Zn	+Zn	-Zn	+Zn
Kırac 66	9.8	22.8	126	45
Bolal 2973	7.3	13.8	167	55
Gerek 79	7.3	15.3	129	61
Atay-85	10.3	17.0	107	61
ES-14	9.0	18.0	133	54
Gun-91	9.8	17.8	130	68
Kırkpınar	10.3	18.5	124	45
Partizanka	9.8	14.5	118	68
P. Niska	10.0	20.0	123	56
Katea	7.5	16.8	139	55
Bezostaja	9.0	16.0	107	55
Dagdas	8.0	17.8	135	62
BDME-10	7.0	14.3	135	59
Zitarka	9.8	16.5	118	50
ES 90-3	7.8	13.8	130	60
ES 90-11	7.5	16.5	144	58
ES 90-14	8.5	15.8	137	55
ES 90-8-1	7.0	13.8	116	51
Kunduru*	10.0	17.3	116	49
Kızıltan*	10.5	19.8	106	53
Mean	8.8	16.8	127	56

* Durum wheat; others are bread wheat

Source: Adapted from Erdal et al. 2002. *J. Plant Nutr.* 25: 113–127.

10^{-10} mol L^{-1}. This is why Fe deficiency in plants occurs mostly on calcareous and alkaline soils, especially under arid and semiarid conditions.

Zinc in soil may be found (1) in the soil solution, as the free ion Zn^{2+}, and associated with organic and inorganic ligands; (2) on exchange sites of reactive soil minerals; (3) bound by organic matter; (4) occluded in oxides and hydroxides of Al, Fe, and Mn; and (5) entrapped in primary and secondary minerals (Kiekens 1995). The relative amount of each of these forms of Zn is affected by various soil physical and chemical properties. Only the mentioned soluble Zn (present either as the free Zn^{2+} or Zn-ligand complexes) is commonly supposed to be directly available to plants. In most soils, the concentration of Zn^{2+} ranges from 10^{-7} to10^{-5} mol L^{-1}. Similar to Fe, Zn deficiency in plants was also mostly found in calcareous and alkaline soils.

5.4.2.2 Fe and Zn Uptake by Plant Roots

Two major plant-involved mechanisms that improve nutrient uptake by roots are: (1) increased surface area for nutrient acquisition including root and mycorrhizal hyphal surface and (2) root-induced chemical changes in rhizosphere (Marschner 1998). In general, a large root surface area (fine roots and long root hairs) is of key importance for nutrient acquisition by roots (Marschner 1998). An increase in root surface area can be either an inherent property or deficiency induced, such as N or P deficiency. The nutrient-absorbing surface area can also be enhanced by mycorrhizal colonization of roots. A beneficial effect of mycorrhizal colonization on Zn uptake has been shown for maize (Faber et al. 1990; Sharma et al. 1992; Liu et al. 2000), pigeneo pea (Wellings et al. 1991), wheat (Khare et al. 1998; Ryan and Angus 2003), lowland rice (Purakayastha and Chhonkar 2001) and aerobic rice (Gao et al. 2007). Upon colonization, the mycelium of the AMF increases the nutrient absorbing surface area of the symbiosis, enhances exploration of a larger soil volume, and thereby increases uptake of immobile nutrients such as Zn (Smith and Read 1997). Caris et al. (1998) also reported that AMF-colonized peanut and sorghum can mobilize and take up more Fe from a calcareous soil than nonmycorrhizal control. These studies highlight the vital importance of mycorrhizas in promoting crop productivity and nutritional value of food products, as well as increase the feasibility of employing mycorrhizas in biofortification programs (He and Nara 2007).

Conditions in the rhizosphere differ from those in the bulk soil, such as pH, redox potential, microbial activity, and concentrations of low and high molecular weight exudates. Many studies show that root-induced changes in rhizosphere are of particular relevance to the dynamics and mobilization of Fe and Zn. The question is how we can use the available knowledge to manipulate root growth and rhizosphere chemistry to increase availability of both Fe and Zn and other nutrients. In the following tests, we briefly discuss the physiological mechanisms of the acquisition of Fe and Zn by plant roots, and the possibility of extending the knowledge and technology in agricultural management schemes.

5.4.2.3 Role of Root Growth

The efficiency by which acquire Fe and Zn plant roots depends on the root architecture, the presence of mycorrhizal fungi, and root physiology (Marschner 1998).

Generally, a healthy root growth is the basis for maintaining a high efficiency of nutrient acquisition, shoot growth and yield. Root morphology and more specifically the distribution of root hairs determine the volume of soil from which nutrients can be extracted. By studying root morphology of three wheat genotypes differing in tolerance to Zn deficiency, Dong et al. (1995) found the efficient genotype could grow longer and had a greater proportion of thinner roots in the early growth period. In a pot experiment with 23 rice genotypes, Gao et al. (2005) reported a significantly positive correlation between Zn uptake and root surface area, indicating the importance of root growth for Zn acquisition in rice.

The regulation of root growth has been thoroughly reviewed by Forde and Lorenzo (2001). The root growth can be regulated or adapted by internal signal substances, such as carbohydrates (Van den Driessche 1987), nitrate (Forde and Lorenzo 2001), auxin and cytokinins (Neumann et al. 2000). In addition, several environmental factors, such as nutritional status, light density, water availability, and temperature, can also influence root growth (Marschner 1995). We can develop improved management strategies in agricultural production with a better knowledge of root growth regulation. A localized supply of macro- or micronutrients may affect root development. For example, supplying NO_3^- stimulated lateral root elongation by two- to threefold in the NO_3-rich zone (Zhang et al. 1999). Similar results were also found on the Zn hyperaccumulator species *Thlaspi caerulescens*, where in a split-root system, roots exposed to soil enriched for Zn produced higher biomass and length (Whiting et al. 2000).

5.4.2.4 Manipulation of Rhizosphere pH by NO_3^-/NH_4^+ Supply

Among soil factors, pH has the most significant influence on plant availability of Fe and Zn. In the pH range of 5.5–7.0, the concentration of Zn in the plant may decrease 30–45 times for each unit increase in soil pH (Moraghan and Mascagni 1991). Rhizosphere pH may differ from that of the bulk soil by up to two units, depending on plant and soil factors. Changes in pH are caused by the excretion of protons (H^+), hydroxyl (OH^-) or bicarbonate (HCO_3^-) ions due to cation–anion imbalance in the plant, the evolution of CO_2 by respiration and the excretion of low molecular weight organic acids. The form of N supply has a major role in determining cation–anion uptake ratio and its subsequent effect on rhizosphere pH (Marschner 1993). In neutral and alkaline soils, plants provided N as NH_4^+-N enhance the mobilization of P, and the micronutrients of Fe and Zn. Silber et al. (2004) showed the rhizosphere pH of rice flower (a new crop in Israel) supplied with 3:1 NH_4^+ to NO_3^- ratio was 2 units lower than plants supplied with 1:3 NH_4^+ to NO_3^- ratio. Moreover, the leaf concentrations of Fe and Zn in plants grown in a 3:1 NH_4^+ to NO_3^- ratio were plants 32% and 38%, respectively, higher than that plants grown in 1:3 NH_4^+ to NO_3^- ratio.

5.4.2.5 Role of Root Exudates

In the last two decades, efforts have been attempted to manage root response to nutrient deficiency. The importance of root response to Fe acquisition has been well documented (Marschner and Römheld 1994), and the mechanisms involved (strategy I and strategy II) are relatively well understood. Strategy I, used by all plants except graminaceous plants, involves the release of protons into the rhizosphere to decrease the rhizosphere pH. As a result, the induction of Fe^{3+}-chelate reductase expression reduces Fe^{3+} to the more soluble Fe^{2+} form, and the resulting Fe^{2+} is taken up via an Fe^{2+} transporter. Strategy II is limited to graminaceous species, with enhanced release of phytosiderophores (PS), a group of hydroxy- and amino-substituted iminocarboxylic acids. Besides Fe, PS are also highly efficient in complexing other micronutrients, such as Zn, Mn, and Cu (Takagi et al. 1988; Treeby et al. 1989), which indicates that the release of PS from Fe-deficient plants can increase micronutrient availability in rhizosphere. Plants that are capable of releasing high amounts of PS are generally more resistant to Fe deficiency than the plants without releasing high amount of PS (Römheld and Marschner 1986; Jolley and Brown 1989). The existence of a large difference in tolerance to Fe deficiency between various cereal species correlated well with the release rate of PS from roots (Marschner et al. 1986; Kawai et al. 1988; Römheld and Marschner 1990).

Root-induced changes in the rhizosphere also occur in response to Zn deficiency. At Zn deficiency, the rhizosphere of cotton and sunflowers can be acidified when N is supplied in the form of NO_3^-, which can occur by an increase in the cation–anion uptake ratio (Cakmak and Marschner 1990). Zhang et al. (1989, 1991) demonstrated that the Zn-deficient graminaceous species can release the same PS as released under Fe deficiency. The beneficial effect of PS on plant tolerance to Zn deficiency has been shown in several studies for graminaceous plant species, including wheat (Walter et al. 1994), barley (Erenoglu et al. 2000), sorghum and maize (Hopkins et al. 1998). The higher tolerance to Zn deficiency of bread wheat than durum wheat correlated well with their difference in release rate of PS (Rengel et al. 1998a). However, when comparing the genotypes of a given cereal species, tolerance to Fe or Zn deficiency was not always correlated with PS release, as shown in oats with Fe deficiency (Hansen and Jolley 1995) and wheat with Zn deficiency (Erenoglu et al. 1996).

Unlike most graminaceous crops, rice is well adapted for growth under submerged conditions, in which Fe^{2+} is normally abundant. As a strategy II plant, rice releases low levels of PS into the rhizosphere under Fe deficiency (Takagi 1976). To the best of our knowledge, there is no report on the PS exudation in response to Zn deficiency. However, rice plants have been found to exude low molecular weight organic anions (LMWOAs) (Aulakh et al. 2001; Bhattacharyya et al. 2003). When grown hydroponically or in soil at nutrient stress, rice roots were reported to exude malate or citrate, which are highly effective in the solubilization and mobilization of Zn. Malate exudation increased as a response to higher bicarbonate levels in the root growth medium (Yang et al. 2003; Hajiboland et al. 2005). Recently, with several lowland rice genotypes, Hoffland et al. (2006) found that rice plants responded similarly to P and Zn deficiency by increasing citrate exudation. Under both hydroponic and soil conditions, Gao (2007) found that genotypic variation of aerobic rice in Zn uptake was correlated with root exudation of malate.

5.4.2.6 Role of Microbial-Plant Interactions

Microbial activity in the rhizosphere of plants differs considerably from nonrhizosphere soil because in rhizosphere there is a constant supply of readily available carbon sources for the heterotrophic microflora (Bonkowski et al. 2000). Organic and inorganic compounds released from roots can stimulate microbial activity (Lynch and Whipps 1990), and bacterial directly supported by root exudates may have a biomass up to 36% of root dry weight in sand culture (Lynch and Whipps 1983).

There are several studies that showed the variation of Zn efficiency in wheat genotypes is positively correlated with the microbial populations in the rhizosphere (Rengel et al. 1996; Rengel 1997; Rengel et al. 1998b). At Zn deficiency, the numbers of fluorescent pseudomonads in rhizosphere of all wheat genotypes increased, but the effect was particularly obvious for Zn-efficient genotypes (Rengel 1997). Similarly, some Fe-efficient plants have been shown a high capacity to acquire Fe from various microbial siderophores (Crowley et al. 1991). The mechanisms by which microorganisms can positively increase plant nutrient uptake were suggested as: (1) biological control agents of root pathogens and pests, (2) nitrogen fixation, (3) mycorrhiza, (4) mineralization, (5) plant-growth-promoting bacteria, (6) phosphate solubilization, and (7) biochelation (Darrah 1993). Although there is no doubt that microorganisms play an important role in plant nutrition, the quantitative estimations of microbial-plant interactions are still unclear.

5.4.3 Molecular Regulation of Fe and Zn Homeostasis in Plants

Development of crop cultivars with elevated levels of micronutrients is being increasingly recognized as one of the effective approaches to provide sustainable solutions to various health problems associated with micronutrient malnutrition, especially in developing countries (Velu et al. 2007). Another alternative approach to overcome Fe and Zn deficiency in crops is to exploit the genetic variation in Fe and Zn efficiency within the plant genome. The observed genetic variation in Fe and Zn efficiency among various plant genotypes offers opportunities for breeding as a tool to resolve these nutrient deficiencies.

Significant progress of molecular aspects of Fe and Zn homeostasis in plants has been made in recent years in enhancing our understanding of how metals are obtained from the soil and distributed throughout the plant. For example, in strategy I plants, Fe is first reduced on the root surface from ferric to ferrous ion by a plasma membrane-bound Fe(III)-chelatereductase (FRO gene family), and subsequently transferred across the rhizodermal plasma membrane barrier by a high-affinity Fe(II) transporter, such as the IRT gene family into the root cell (Varotto et al. 2002; Vert et al. 2002; Connolly et al. 2003; Mukherjee et al. 2005). For other Fe transporter genes, such as AtNRAMP3, AtNRAMP4, and AtVIT1, are all expressed in the vasculature of the roots and shoots, and the proteins that these genes encode appear to play a role in vacuolar Fe homeostasis (Thomine et al. 2003; Lanquar et al. 2005). Vacuolar Fe storage is also critical for seedling development that will ultimately aid the development of nutrient-rich seed, benefiting both human health and agricultural productivity (Kim et al. 2006). Recent results confirmed that the FRD3-mediated efflux of citrate into the root vasculature is necessary for efficient Fe translocation, a process that is important for the translocation of Fe to leaves. The FRD3 transports a small, organic iron-chelator that is necessary for the correct localization of Fe throughout the plant into the xylem (Durrett et al. 2007).

For strategy II plants, the first YS1 gene was identified from maize roots that could transport PS-Fe(III) from rhizosphere (Curie et al. 2001). Recently, 18 YSL genes have been identified in rice, and many of them are expressed in both roots and the shoots (Koike et al. 2004). In fact, the YSL genes also play a role in strategy I plants. The YSL family, consisting of eight members in *Arabidopsis*, has been implicated in the intercellular transport of Fe chelates, specifically Fe complexed to nicotianamine (NA) that regulates Fe and Zn homeostasis and plays a role in Fe and NA seed loading (Le Jean et al. 2005; Schaaf et al. 2005). Recently, several lines of evidence have indicated a role of nicotianamine (NA) in metal homeostasis through the chelation and transport of NA–metal complexes in *Thlaspi caerulescens*. This molecular, histological and biochemical evidence supports a role for YSL transporters in the overall scheme of NA and NA–metal circulation (Gendre et al. 2007). A number of small organic molecules have been implicated in metal ion homeostasis as metal ion ligands to facilitate uptake and transport of metal ions with low solubility, and also as chelators that are implicated in sequestration for metal tolerance and efficient storage metals in fruits and seeds (Haydon and Cobbett 2007). The role of mugineic acid, nicotianamine, organic acids (citrate and malate), histidine, and phytate as ligands for Fe, Zn, copper (Cu), manganese (Mn), and nickel (Ni) in plants could make a significant contribution to our understanding of metal homeostasis in plants.

Although the molecular mechanisms for Zn efficiency are not well understood, it has been suggested that an increased secretion of PS by Zn-efficient plants might be involved. Furthermore, several ZIP proteins have been characterized in the strategy II plant rice, suggesting that this family plays a role in monocots, such as grasses (Ramesh et al. 2003). Genetic engineering approaches have been applied to increase plant tolerance to low Zn soils. At present, knowledge of the genes controlling specific steps of the Zn network in soil–plant systems is still rudimentary, but increasing rapidly. Transformation and overexpression of known Zn transporters from *Arabidopsis* to barley (*Hordeum vulgare* cv. Golden Promise) can increase plant Zn uptake and seed Zn content (Ramesh et al. 2004). Recently, a NAC gene was identified in wheat, which can accelerate senescence and increase Zn and Fe remobilization from leaves to developing grains (Uauy et al. 2006). These results clearly show the contribution of using molecular genetic tools for increasing Zn and Fe efficiency in crops and for enhancing Zn and Fe translocation into edible plant tissues. In addition, breeding strategies have been developed based on these genetic findings.

5.5 CONCLUSIONS

The recent progress in soil and crop management has provided the means to tackle complex plant Fe and Zn nutritional problems through manipulating the rhizosphere, managing cropping systems, screening Fe and Zn efficiency species, and applying novel approaches to fertilizer application. The fact that many of the molecular and biochemical changes in response to Fe and Zn deficiency occur in synchrony suggests that the genes involved could be coordinately expressed and share a common regulatory system. Possibly, some soil and crop management protocols might regulate the Fe and Zn efficiency gene network into improving Fe and Zn mobilization from soil to plant, translocating the nutrients within the plant, increasing Fe and Zn content and availability in grains of crops, and ultimately enhancing the nutritional value of food crops, and improving human nutrition and health. A combination of all these approaches, together with phenotypic analyses, will enable us to elucidate and address the effects and mechanisms of soil and crop management on Fe and Zn homeostasis in plants at the molecular, cellular, whole-plant, and population levels. Substantial efforts are being made in increasing plant Fe and Zn efficiency through rhizosphere management and through genetic and molecular approaches.

REFERENCES

Alloway, B. J. 2004. *Zinc in Soils and Crop Nutrition*. Brussels, Belgium: International Zinc Association.

Aulakh, M. S., R. Wassmann, C. Bueno, J. Kreuzwieser, and H. Rennenberg. 2001. Characterization of root exudates at different growth stages of ten rice (*Oryza sativa* L.) cultivars. *Plant Biology* 3: 139–148.

Bänziger, M., and J. Long. 2000. The potential for increasing the iron and zinc density of maize through plant-breeding *Nutrition Bulletin* 21: 397–400.

Beebe, S., A. V. Gonzalez, and J. Rengifo. 2000. Research on trace minerals in the common bean. *Nutrition Bulletin* 21: 387–391.

Bhattacharyya, P., S. C. Datta, and P. Dureja. 2003. Interrelationship of pH, organic acids, and phosphorus concentration in soil solution of rhizosphere and non-rhizosphere of wheat and rice crops. *Communications in Soil Science and Plant Analysis* 34: 231–245.

Bonkowski, M., W. Cheng, J. Alphei, B. S. Griffiths, and S. Dcheu. 2000. Microbial-fungal interactions in the rhizosphere and effects on plant growth. *European Journal of Soil Biology* 36: 135–147.

Bouman, B. A. M., S. Peng, A. R. Castañeda, and R. M. Visperas. 2005. Yield and water use of irrigated tropical aerobic rice systems. *Agricultural Water Management* 74: 87–105.

Brennan, R. F. 1991. Effectiveness of zinc sulfate and zinc chelate as foliar sprays in alleviating zinc deficiency of wheat grown on zinc-deficient soils in Western Australia. *Australian Journal of Experimental Agriculture* 31: 831–834.

Cakmak, I., and H. Marschner. 1990. Decrease in nitrate uptake and increase in proton release in zinc deficient cotton, sunflower and buckwheat plants. *Plant and Soil* 129: 261–268.

Cakmak, I., M. Kalayci, H. Ekiz, H. J. Braun, Y. Kilinç, and A.Yilmaz. 1999. Zinc deficiency as a practical problem in plant and human nutrition in Turkey: A NATO-science for stability project. *Field Crops Research* 60: 175–188.

Caris, C., W. Hoerdt, H. J. Hwkins, V. H. Roemheld, and E. George. 1998. Studies on the iron transport by arbuscular mycorrhizal hyphae from soil to peanut and sorghum plants. *Mycorrhiza* 8: 35–39.

Castro, R. U. 1977. Zinc deficiency in rice: A series research at the IRRI. *International Reliability Physics Symposium* 9: 1–18.

Chen, Y., and P. Barak. 1982. Iron nutrition of plants in calcareous soils. *Advances in Agronomy* 35: 217–240.

Chen, K., E. Hoffland, C. Q. Zou , X. P. Gao, O. Oenema, and F. S. Zhang. 2005. Is grain phytate/zinc ratio in rice dependent on soil phosphorus and zinc availability. In *Plant Nutrition for Food Security, Human Health and Environmental Protection*, C. J. Li, ed. Beijing: Tsinghua University Press, pp. 372–373.

Connolly, E. L., N. Campbell, N. H. Grotz, C. L. Prichard, and M. L. Guerinot. 2003. Overexpression of the FRO2 ferric chelate reductase confers tolerance to growth on low iron and uncovers posttranscriptional control. *Plant Physiology* 133: 1102–1110.

Crowley, D. E., Y. C. Wang, C. P. P. Reid, and P. J. Szansiszlo. 1991. Mechanism of iron acquisition from siderophores by microorganisms and plants. *Plant and Soil* 130: 179–198.

Curie, C., Z. Panaviene, C. Loulergue, S. L. Dellaporta, J. F. Briat, and E. L. Walker. 2001. Maize *yellow stripe1* encodes a membrane protein directly involved in Fe(III) uptake. *Nature* 409: 346–349.

Darrah, P. R. 1993. The rhizosphere and plant nutrition: A quantitative approach. *Plant and Soil* 155: 1–20.

Dong, B., Z. Rengel, and E. D.Graham. 1995. Root morphology of wheat genotypes differing in zinc efficiency. *Journal of Plant Nutrition* 18: 2761–2773.

Durrett, T. P., W. Gassmann, and E. E. Rogers. 2007. The FRD3-mediated efflux of citrate into the root vasculature is necessary for efficient iron translocation. *Plant Physiology* 144: 197–205.

Erdal, I., A. Yilmaz, S. Taban, S. Eker, B. Torun, and I. Cakmak. 2002. Phytic acid and phosphorus concentrations in seeds of wheat cultivars grown with and without zinc fertilization. *Journal of Plant Nutrition* 25: 113–127.

Erenoglu, B., I. Cakmak, H. Marschner, et al. 1996. Phytosiderophore release does not correlate well with zinc efficiency in different bread wheat genotypes. *Journal of Plant Nutrition* 19: 1569–1580.

Erenoglu, B., S. Eker, I. Cakmak, R. Derici, and V. Römheld. 2000. Effect of iron and zinc deficiency on release of phytosiderophores in barley cultivars differing in zinc efficiency. *Journal Plant Nutrition* 23: 1645–1656.

Faber, B. A., R. J. Zasoski, R. G. Burau, and K. Uriu. 1990. Zinc uptake by corn as affected by vesicular-arbuscular mycorrhizae. *Plant and Soil* 129: 121–130.

FAO. 2006. *Food and Agriculture (FAO) Statistical Yearbook 2005/06* (issue 1 and issue 2). Rome, Italy: Food and Agricultural Organization of the United Nations.

FAO/WHO, 2000. *Preliminary Report on Recommended Nutrient Intakes: Joint FAO/WHO Expert Consultation on Human Vitamin and Mineral Requirements*, 1998. Revised July 13, 2000. Rome, Italy: Food and Agricultural Organization of the United Nations and Geneva, Switzerland: World Health Organization.

Forde, B., and H. Lorenzo. 2001. The nutritional control of root development. *Plant and Soil* 232: 51–68.

Frossard, E., M. Bucher, F. Mächler, A. Mozafar, and R. Hurrell. 2000. Potential for increasing the content and bioavailability of Fe, Zn and Ca in plants for human nutrition. *Journal of Science Food Agriculture* 80: 861–879.

Gao, X. 2007. Bioavailability of zinc to aerobic rice. PhD dissertation. Wageningen University, Wageningen, The Netherlands.

Gao, S., K. K. Tanji, S. C. Scardaci, and A. T. Chow. 2002. Comparison of redox indicators in a paddy soil during rice-growing season. *Soil Science Society of America Journal* 66: 805–817.

Gao, X., C. Zou, F. Zhang, S. E. A. T. M. van der Zee, and E. Hoffland. 2005. Tolerance to zinc deficiency in rice correlates with zinc uptake and translocation. *Plant and Soil* 278: 253–261.

Gao, X., C. Zou, X. Fan, F. Zhang, and E. Hoffland. 2006. From flooded to aerobic conditions in rice cultivation: Consequences for zinc uptake. *Plant and Soil* 280: 41–47.

Gao, X., T. M. Kuyper, C. Zou, F. Zhang, and E. Hoffland. 2007. Mycorrhizal responsiveness of aerobic rice genotypes is negatively correlated with their zinc uptake when nonmycorrhizal. *Plant and Soil* 290: 283–291.

Garvin, D. F., R. M. Welch, and J. W. Finley. 2006. Historical shifts in the seed mineral micronutrient concentration of US hard red winter wheat germplasm. *Journal of Science Food Agriculture* 86: 2213–2220.

Gendre, D., P. Czernic, and G. Conejer, et al. 2007. TcYSL3, a member of the YSL gene family from the hyperTcYSL3, a member of the YSL gene family from the hyperaccumulator *Thlaspi caerulescens*, encodes a nicotianamine-Ni/Fe transporter. *Plant Journal* 49: 1–15.

Ghandilyan, A., D. Vreugdenhil, and M. G. M. Aarts. 2006. Progress in the genetic understanding of plant iron and zinc nutrition. *Physiology Plant* 126: 407–417.

Graham, R. D., R. M. Welch, and H. E. Bouis. 2001. Addressing micronutrient malnutrition through enhancing the nutritional quality of staple foods: Principles, perspectives, and knowledge gaps. *Advance Agronomy* 70: 77–142.

Gregorio, G. B., D. Senadhira, H. Htut, and R. D. Graham. 2000. Breeding for trace mineral density in rice. *Food Nutrition Bulletin* 21: 382–386.

Gunes, A., A. Inal, M. S. Adak, et al. 2007. Mineral nutrition of wheat, chickpea and lentil as affected by intercropped cropping and soil moisture. *Nutrient Cycling in Agroecosystem* 78: 83–96.

Hajiboland, R. X. E. Yang, V. Römheld, and G. Neumann. 2005. Effect of bicarbonate on elongation and distribution of organic acids in root and root zone of Zn-efficient and Zn-inefficient rice (*Oryza sativa* L.) genotypes. *Environmental Experimental Botany* 54: 163–173.

Han, C., C. Li, R. Li, Q. Wang, M. Wu, and C. Xiao. 1994. Technical approach to application of iron fertilization on soybean. *Oil Crops of China* 16: 43–45 (in Chinese).

Hansen, N. C., and V. D. Jolley. 1995. Phytosiderophore release as a criterion for genotypic evaluation of iron efficiency in oat. *Journal of Plant Nutrition* 18: 455–465.

Haslett, B. S., R. J. Reid, and Z. Rengel. 2001. Zinc mobility in wheat: Uptake and distribution of zinc applied to leaves or roots. *Annual Botany* 87: 379–386.

Haydon, M. J., and C. S. Cobbett. 2007. Transporters of ligands for essential metal ions in plants. *New Phytologist* 174(3): 499–506.

He, X., and K. Nara. 2007. Element biofortification: Can mycorrhizas potentially offer a more effective and sustainable pathway to curb human malnutrition? *Trends in Plant Science* 12: 331–335.

Hergert, G. W., G. W. Rehm, and R. A. Wiese. 1984. Field evaluation of zinc sources band applied in ammonium polyphosphate suspension. *Soil Science Society of America Journal* 48: 1190–1193.

Hoffland, E., C. Wei, and M. Wissuwa. 2006. Organic anion exudation by lowland rice (*Oryza sativa* L.) at zinc and phosphorus deficiency. *Plant and Soil* 283: 155–162.

Hopkins, B. G., D. A. Whiteney, and R. E. Lamond. 1998. Phytosiderophore release by sorghum, wheat and corn under zinc deficiency. *Journal of Plant Nutrition* 21: 2623–2637.

Inal, A., A. Gunes, and F. Zhang. 2007. Peanut/maize intercropping induced changes in rhizosphere and nutrient concentrations in shoots. *Plant Physiology and Biochemistry* 45: 350–356.

Jolley, V. D. and J. C. Brown. 1989. Iron efficient and inefficient wheat. I. Differences in phytosiderophore release. *Journal of Plant Nutrition* 12: 423–435.

Kawai, S., S. Takagi, and Y. Sato. 1988. Mugineic acid-family phytosiderophores in root secretions of barley, corn and sorghum varieties. *Journal of Plant Nutrition* 11: 633–642.

Khare, A. K., A. K. Rawat, and S. B. Dubey, et al. 1998. Role of native vesicular-arbuscular mycorrhizal fungi in wheat-based cropping sequence for efficient use of phosphorus and zinc in black soils of Madhya Pradesh. *Journal of Agricultural Science* 68: 247–250.

Kiekens, L. 1995. Zinc. In *Heavy Metals in Soils,* B. J. Alloway, ed. London: Blackie Academic and Professional, pp. 284–305.

Kim, S. A., T. Punshon, A. Lanzirotti, et al. 2006. Localization of iron in *Arabidopsis* seed requires the vacuolar membrane transporter VIT1. *Science* 314: 1295–1298.

Koike, S., H. Inoue, and D. Mizuno. 2004. OsYSL2 is a rice metal-nicotianamine transporter that is regulated by iron and expressed in the phloem. *Plant Journal* 39: 415–424.

Lanquar, V., F. Lelièvre, and S. Bolte. 2005. Mobilization of vacuolar iron by AtNRAMP3 and AtNRAMP4 is essential for seed germination on low iron, *EMBO Journal* 24: 4041–4051.

Le Jean, M., A. Schikora, and S. Mari. 2005. A loss-of-function mutation in *AtYSL1* reveals its role in iron and nicotianamine seed loading. *Plant Journal* 44: 769–782.

Li, L., S. M. Li, and J. H. Sun. 2007. Diversity enhances agricultural productivity via rhizosphere phosphorus facilitation on phosphorus-deficient soils. *Proceedings of National Academy Science of the United States of America* 104: 11192–11196.

Liu, A., C. Hamel, and R. I. Hamilton. 2000. Acquisition of Cu, Zn, Mn and Fe by mycorrhizal maize grown in soil at different P and micronutrient levels. *Mycorrhiza* 9: 331–336.

Liu, W., L. Du, and D. Liu. 2002. A review on diagnosis for iron nutrition status of fruit trees and fertilization correction for Fe-deficiency chlorosis. *Chinese Agriculture Science Bulletin* 18: 67–69 (in Chinese).

Liu, Z. 1996. *Microelements in Soils of China*. Nanjing, China: Jiangsu Science and Technology Publishing House.

Loué, A. 1988. *Les oligoéléments en agriculture*. SCPA, Agri-Nathan International Paris (in French).

Lynch, J. M., and J. M. Whipps. 1983. Substrate flow and utilization in the rhizosphere of cereals. *New Phytologist* 95: 605–623.

Lynch, J. M. and J. M. Whipps. 1990. Substrate flow in the rhizosphere. *Plant and Soil* 129:1–10.

Marschner, H. 1993. Zinc uptake from soils. In *Zinc in Soils and Plants,* A. D. Robson, ed. Dordrecht, The Netherlands: Kluwer Academic Publishers, pp. 59–78.

Marschner, H. 1995. *Mineral Nutrition of Higher Plants*. London: Academic Press.

Marschner, H. 1998. Role of root growth, arbuscular mycorrhza, and root exudates for the efficiency in nutrient acquisition. *Field Crops Research* 56: 203–207.

Marschner, H., and V. Römheld. 1994. Strategies of plants for acquisition of iron. *Plant and Soil* 165: 261–274.

Marschner, H., V. Römheld, and M. Kissel. 1986. Different strategies in higher plants in mobilization and uptake of iron. *Journal of Plant Nutrition* 9: 695–713.

Maziya-Dixon, B., J. G. Kling, and A. Menkir. 2000. Genetic variation in total carotene, iron, and zinc contents of maize and cassava genotypes. *Food Nutrition Bulletin* 21: 419–422.

McGrath, S. P., F. J. Zhao, and S. Fan. 2007. Decreasing mineral concentration in wheat grain over the last 160 years. In *Biogeochemistry of Trace Elements: Environmental Protection, Remediation and Human Health*, Y. Zhu, N. Lepp, and R. Naidu, eds. Beijing, China: Tsinghua University Press, pp. 281–282.

Monasterio, I., and R. D. Graham. 2000. Breeding for trace minerals in wheat. *Food Nutrition Bulletin* 21: 392–396.

Moraghan, J. T., and H. J. Mascagni. 1991. Environmental and soil factors affecting micronutrient deficiencies and toxities. In *Micronutrients in Agriculture*, J. J. Mortvedt, F. R. Cox, L. M. Shuman, and R. M. Welch, eds. SSSA Book Series No. 4, Madison: Soil Science Society of America, pp. 371–425.

Morris, C. E., and D. C. Sands. 2006. The breeder's dilemma-yield or nutrition? *National Biotechnology* 24: 1078–1080.

Mortvedt, J. J., and R. J. Gilkes. 1993. Zinc fertilizers. In *Zinc in Soils and Plants*, A. D. Robson, Ed. Dordrecht, The Netherlands: Kluwer Academic Publishers, pp. 33–44.

Mukherjee. I., N. H. Campbell, and J. S. Ash. 2005. Expression profiling of the *Arabidopsis* ferric chelate reductase (*FRO*) gene family reveals differential regulation by iron and copper. *Planta* 223: 1178–1190.

Neumann, G., A. Massonneau, and N. Langlade, et al. 2000. Physiological aspects of cluster root function and development in phosphorus-deficient white lupin (*Lupinus albus* L.). *Annual Botany* 85: 909–919.

Olson, R. V. 1950. Effects of acidification, iron oxide addition, and other soil treatments on sorghum chlorosis and iron absorption. *Soil Science Society of America Journal* 15: 97–101.

Payne, G. G., D. C. Martens, C. Winarko, and N. F. Perera. 1988. Form and availability of copper and zinc following long-term copper sulfate and zinc sulfate applications. *Journal of Environmental Quality* 17: 707–711.

Percival, J. 1921. *The Wheat Plant: A Monograph*. London: Duckworth.

Pomeranz, Y. 1988. Chemical composition of kernel structures. In *Wheat: Chemistry and Technology*, Vol. 1. St Paul, MN: American Association of Cereal Chemists.

Purakayastha, T. J., and P. K. Chhonkar. 2001. Influence of vesicular-arbuscular mycorrhizal fungi (*Glomus etunicatum* L.) on mobilization of zinc in wetland rice (*Oryza sativa* L.). *Biology and Fertility in Soils* 33: 323–327.

Raboy, V., M. Mather, and G. Noaman. 1991. Grain phytic acid and protein are highly correlated in winter wheat. *Crop Science* 31: 631–635.

Ramesh, S. A., R. Shin, D. J. Eide, and D. P. Schachtman. 2003. Differential metal selectivity and gene expression of two zinc transporters from rice. *Plant Physiology* 133: 126–134.

Ramesh, S. A., S. Choimes, and D. P. Schachtman. 2004. Over-expression of an *Arabidopsis* zinc transporter in *Hordeum vulgare* increases short-term zinc uptake after zinc deprivation and seed zinc content. *Plant Molecular Biology* 54: 373–385.

Rengel, Z. 1997. Capacity of the take-all fungus to zinc to oxidise manganousions influences take-all decline. *Journal of Plant Nutrition* 20: 455–460.

Rengel, Z. 2002. Genetic control of root exudation. *Plant and Soil* 245: 59–70.

Rengel, Z., R. Gutteridge, P. Hirsch, and D. Hornby. 1996. Plant genotype, micronutrient fertilization and take-all infection influence bacterial populations in the rhizosphere of wheat. *Plant and Soil* 183: 269–277.

Rengel, Z., V. Römheld, and H. Marschner. 1998a. Uptake of zinc and iron by wheat genotypes differing in tolerance to zinc deficiency. *Journal of Plant Physiology* 152: 433–438.

Rengel, Z., G. Ross, and P. Hirsch. 1998b. Colonisation of wheat roots by soil bacteria depends on the plant genotype and micronutrient nutrition. *Journal of Plant Nutrition* 21: 99–113.

Rengel, Z., G. D. Batten, and D. E. Crowley. 1999. Agronomic approaches for improving the micronutrient density in edible portions of field crops. *Field Crops Research* 60: 27–40.

Römheld, V., and H. Marschner. 1986. Evidence for a specific uptake system for iron phytosiderophores in roots of grasses. *Plant Physiology* 80: 175–180.

Römheld, V., and H. Marschner. 1990. Genotypical differences among graminaceous species in release of phytosiderophores and uptake of iron phytosiderophores. *Plant and Soil* 123: 147–153.

Ryan, M. H., and J. F. Angus. 2003. Arbuscular mycorrhizae in wheat and field pea crops on a low P soil: Increased Zn uptake but no increase in P uptake or yield. *Plant and Soil* 250: 225–239.

Schaaf, G., A. Schikora, J. Häberle. 2005. A putative function for the *Arabidopsis* Fe-Phytosiderophore transporter homolog AtYSL2 in Fe and Zn homeostasis. *Plant Cell Physiology* 46: 762–774.

Schmidt, W. 2003. Iron solutions: acquisition strategies and signaling pathways in plants. *Trends in Plant Science* 8:188–193.

Sharma, A. K., P. C. Srivastava, and B. N. Johri. 1992. Kinetics of zinc uptake by mycorrhizal and non-mycorrhizal corn (*Zea mays* L.) roots. *Biology Fertility in Soils* 13: 206–210.

Silber, A., L. Ben Yones, and I. Dori. 2004. Rhizosphere pH as a result of nitrogen levels and NH_4/NO_3 ratio and its effect on zinc availability and on growth of rice flower (*Ozothamnus diosmifolius*). *Plant and Soil* 262: 205–213.

Slaton, N. A., R. J. Norman, and C.E. Wilson Jr. 2005. Effect of zinc source and application time on zinc uptake and grain yield of flood-irrigated rice. *Agronomy Journal* 97: 272–278.

Smith, S. E., and D. J. Read. 1997. *Mycorrhizal Symbiosis*. London: Academic Press.

Solomons, N. W. 1982. Biological availability of zinc in humans. *American Journal of Clinical Nutrition* 35: 1048–1075.

Srivastava, P. C., D. Ghosh, and V. P. Singh. 1999. Evaluation of different zinc sources for lowland rice production. *Biology and Fertility in Soils* 30: 168–172.

Takagi, S. 1976. Naturally occurring iron-chelating compounds in oat and rice-root washings: Activity measurement and preliminary characterizations. *Soil Science Plant Nutrition* 22: 423–433.

Takagi, S., S. Kawai, and M. H. 1988. Efficiency of iron extraction from soil by mugineic acid family phytosiderophores. *Journal of Plant Nutrition* 11: 643–651.

Thomine, S., F. Lelièvre, E. Debarbieux, J. I. Schroeder, and H. Barbier-Bygoo. 2003. AtNRAMP3, a multispecific vacuolar metal transporter involved in plant responses to iron deficiency. *Plant Journal* 34: 685–695.

Traore, K. 2006. Effects of soil amendments and drought on zinc husbandry and grain quality in Sahelian sorghum. PhD dissertation. Wageningen University, Wageningen, The Netherlands.

Treeby, M., H. Marschner, and V. Römheld. 1989. Mobilization of iron and other micronutrient cations from a calcareous soil by plant-borne, microbial, and synthetic metal chelators. *Plant and Soil* 114: 217–226.

Uauy, C., A. Distelfeld, T. Fahima, A. Blechl, and J. Dubcovsky. 2006. A NAC gene regulating senescence improves grain protein, zinc, and iron content in wheat. *Science* 314: 1298–1301.

United States Department of Agriculture. National Agricultural Statistics. 2004. U.S. Govt. Printing Office, Washington, DC, pp. I1–I12.

Van den Driessche, R. 1987. Seasonal changes in root growth capacity and carbohydrates in red pine and white spruce nursery seedlings. In *Proceedings of the IUFRO Symposium on Root Physiology and Symbiosis*, A. Riedacker, J. Gagnaire-Michard, and J. Nancy, eds. France.

Varotto, C., D. Maiwald, P. Pesaresi, P. Jahns, F. Salamini, and D. Leister. 2002. The metal ion transporter IRT1 is necessary for iron homeostasis and efficient photosynthesis in *Arabidopsis thaliana*. *Plant Journal* 31: 589–599.

Velu, G., K. N. Rai, V. Muralidharan, V. N. Kulkarni, T. Longvahand, and T. S. Raveendran. 2007. Prospects of breeding biofortified pearl millet with high grain iron and zinc content. *Plant Breeding* 126: 182–185.

Vempati, R. K. and R. H. Loeppert. 1988. Chemistry and mineralogy of Fe-containing oxides and layer silicates in relation to plant available iron. *Journal of Plant Nutrition* 11: 1557–1574.

Vert, G., N. Grotz, and F. Dédaldéchamp, et al. 2002. IRT1, an *Arabidopsis* transporter essential for iron uptake from the soil and for plant growth. *Plant Cell* 14: 1223–1233.

Von Wirén, N., H. Marschner, and V. Römheld. 1996. Roots of iron-efficient maize also absorb phytosiderophore-chelated zinc. *Plant Physiology* 111: 1119–1125.

Walter, A., V. Römheld, H. Marschner, and S. Mori. 1994. Is the release of phytosiderophores in zinc-deficient wheat plants a respnse to impaired iron utilization? *Physiological Plantarum* 92: 493–500.

Wang, H. Q., B. A. M. Bouman, Z. Dule, C. Wang, and P. E. Moya. 2002. Aerobic rice in northern China: Opportunities and challenges. In *Water-Wise RiceProduction*, B. A. M. Bouman, H. Hengsdijk, B. Hardy, P. S. Bindraban, T. P. Tuong, and J. K. Ladha, eds. Proceedings of the International Workshop on Water-Wise Rice Production, 8–11 April 2002. Los Baños, Philippines: International Rice Research Institute, pp. 143–154.

Welch, R. M. 1993. Zinc concentrations and forms in plants for humans and animals. In *Zinc in Soils and Plants*, A. D. Robson, ed. Dordrecht, The Netherlands: Kluwer Academic Publishers, pp. 183–195.

Welch, R. M. 1995. Micronutrient nutrition of plants. *Critical Reviews in Plant Science* 14: 49–82.

Welch, R. M., and R. D. Graham. 1999. A new paradigm for world agriculture: Meeting human needs. Productive, sustainable, nutritious. *Field Crops Research* 60: 1–10.

Welch, R. M., and R. D. Graham. 2002. Breeding crops for enhanced micronutrient content. *Plant and Soil* 245: 205–214.

Wellings, N. P., A. H. Wearing, and J. P. Thompson. 1991. Vesicular-arbuscular mycorrhizae (VAM) improve phosphorus and zinc nutrition and growth of pigeonpea in a Vertisol. *Australian Journal of Agriculture Research* 42: 835–845.

White, P. J., and M. R. Broadley. 2005. Biofortifying crops with essential mineral elements. *Trends in Plant Science* 10: 586–593.

Whiting, S. N., J. R. Leake, S. P. McGrath, and A. J. M. Baker. 2000. Positive responses to Zn and Cd by roots of the Zn and Cd hyperaccumulator *Thlaspi caerulescens. New Phytologist* 145: 199–210.

Wolnik, K. A., F. I. Fricke, S. G. Capar, et al. 1983. Elements in major raw agricultural crops in the United States. 2. Other elements in lettuce, peanuts, potatoes, soybeans, sweet corn, and wheat. *Journal of Agriculture, Food and Chemistry* 31: 1244–1249.

Yang, X., R. Hajiboland, and V. Römheld. 2003. Bicarbonate had greater effects than high pH on inhibiting root growth of zinc-inefficient rice genotype. *Journal of Plant Nutrition* 26: 399–415.

Yilmaz, A., H. Ekiz, B.Torun, et al. 1997. Different zinc application methods on grain yield and zinc concentrations in wheat grown on zinc deficient calcareous soils in central Anatolia. *Journal of Plant Nutrition* 20: 461–471.

Zhang, F. S., V. Römheld, and H. Marschner. 1989. Effect of zinc deficiency in wheat on the release of zinc and iron mobilizing root exudates. *Zeitschrift der Pflanzenernaehrung und Bodenkunde* 152: 205–210.

Zhang, F. S., V. Römheld, and H. Marschner. 1991. Release of zinc mobilizing root exudates in different plant species as affected by zinc nutritional status. *Journal of Plant Nutrition* 14: 675–686.

Zhang, H., A. Jennings, W. B. Peter, and G. F. Brian. 1999. Dual pathways for regulation of root branching by nitrate. *Plant Biology* 96: 529–534.

Zuo, Y., F. Zhang, X. Li, and Y. Cao. 2000. Studies on the improvement in iron nutrition of peanut by intercropping with maize on a calcareous soil. *Plant and Soil* 220: 1325.

6 Effects of Soil Management on Zinc Uptake and Its Bioavailability in Plants

Rainer Schulin, Amir Khoshgoftarmanesh, Majid Afyuni, Bernd Nowack, and Emmanuel Frossard

CONTENTS

6.1 INTRODUCTION: THE ROLE OF ZINC AS AN ESSENTIAL MICRONUTRIENT

Zinc (Zn) deficiency is now recognized as one of the most widespread mineral deficiencies in global human nutrition, affecting up to one-third of the world population, although estimates are based primarily on circumstantial evidence (Combs et al. 1996; Welch 2002; White and Broadley 2005). Diagnosis of Zn deficiency is difficult because even under a severe shortage in Zn supply, its concentrations in the human body may not be low due to the reduction of growth and because symptoms of Zn deficiency are not specific (Hurrell 2001; Gibson 2006). Approximately 300 enzymes have been described requiring Zn for their activity (Vallee and Falchuk 1993), and Zn-dependent enzymes are found in all major enzyme classes (Broadley at al. 2007). Zinc-binding sites not only occur in metalloenzymes but also in other proteins, membrane lipids, and nucleic acids. Consequently, intracellular Zn is also found to be involved in the regulation of nonmetalloenzymes, in the maintenance of the structure and function of biomembranes, as well as in the interactions between proteins and nucleic acids (Cousins 1979; House 1999; Broadley et al. 2007). Of particular importance is the role of the so-called Zn-finger domain-containing proteins in regulating transcription, RNA metabolism, and chromatin structure (Broadley et al. 2007).

Due to zinc's manifold involvement with the human body, its deficiency affects a wide range of biochemical and physiological functions, such as growth, reproduction, immune responses, and neurobehavioral development; and it may lead to underweight, delayed skeletal and sexual maturity, dermatitis, persistent diarrhoea, chronic nonhealing leg ulcers, recurrent infections, as well as mental handicaps (Gibson 2006). The most obvious manifestation of Zn deficiency at a population level is stunted growth. Although dwarfism in young males from the Middle East had been related to Zn deficiency already in the early 1960s (Prasad et al. 1963), the full scale of the problem of the "hidden hunger" for Zn has only been very recently recognized (Welch 2002; Gibson 2006).

Zinc deficiency in humans results from diets low in bioavailable Zn. Cereals and legume seeds are particularly poor sources of Zn because they are high in phytic acid, which strongly reduces the availability of Zn for intestinal absorption by forming poorly soluble Zn–phytate complexes (House 1999; Frossard et al. 2000). In contrast, meat and fish are rich Zn sources, followed by fruits and vegetables, due to their low phytate concentrations. Consequently, Zn deficiency is primarily a problem in rural areas of developing countries in which diets are based on cereal and legume staple foods and where consumption of animal and fish products is often limited for economic or other reasons. The problem is exacerbated in areas with low soil Zn availability for plant uptake.

Given that foods differ widely in available Zn, dietary modification or diversification would seem a straightforward and sustainable way to combat Zn deficiency. However, changes in dietary habits require individual and societal acceptance, as well as the availability of alternative foods at affordable prices. Supplementation with pharmaceutical Zn preparations can be an effective measure to alleviate Zn deficiency on an individual basis. However, this strategy has generally failed on a population level in developing countries due to lack of adequate infrastructure and education (Graham et al. 2000). Food fortification is a strategy that can be applied rather rapidly at the national level without personal contact to and change of dietary habits by the recipients. Fortification of maize and wheat flour with Zn and other micronutrients has been implemented in Mexico, Indonesia, and South Africa, but these programs still need further evaluation (Gibson 2006).

A disadvantage of food fortification is that, similar to supplementation, its successful implementation into society requires safe delivery systems, stable policies, appropriate social infrastructures, and continued financial support (White and Broadley 2005; Gibson 2006). A new strategy for combating hidden hunger, that is, micronutrient malnutrition, is to fortify plant-based food through increasing the accumulation and bioavailability of micronutrients in edible parts of crop plants (Bouis 1996; Welch and Graham 1999; Frossard et al. 2000; Welch 2002). This strategy, for which the term *biofortification* was coined, has the potential to be sustainable as well as economical. Moreover, biofortified food products can reach larger numbers of the poorer segments of the population without changing particular infrastructure or dietary habits (Graham et al. 2000; Yang et al. 2007).

Biofortification can be achieved in two distinct ways: (1) increase the enrichment of bioavailable micronutrients in the plant parts to be consumed through breeding or genetic engineering (Graham et al. 2000; Welch and Graham 2004; White and Broadley 2005), and (2) enhance the total accumulation of the deficient micronutrients through agricultural methods of crop cultivation, in particular fertilization (Rengel et al. 1999; Frossard et al. 2000; Graham et al. 2001; Welch 2002). In this chapter we will only address and highlight the second approach, which is also called *agronomic biofortification*, and focus on Zn biofortification of cereal and legume seeds. A comprehensive discussion of the two approaches for the case of cereals has recently been given by Cakmak (2008). For a more comprehensive treatment of the agronomic biofortification approach, readers are also referred to the excellent review by Rengel et al. (1999).

6.2 ZINC FERTILIZATION

The most direct way to increase the density of Zn in food crops is by application of Zn fertilizer. Zinc deficiency is the most widespread micronutrient deficiency in crop production worldwide (White and Zasoski 1999). Low availability of soil Zn is particularly common on calcareous and alkaline soils in arid and semiarid climates, but also on highly weathered soils in the tropics and subtropics (Alloway 2006). Zinc deficiencies affect, in particular, the production of cereals. Large contiguous areas of known Zn deficiency occur in Central Anatolia and Western Australia (Cakmak et al. 1996; Brennan and Bolland 2002). In China, about 40% of the land is estimated to be Zn deficient according to plant-response studies. Around 50% of the total land area has a DTPA-extractable Zn concentration of less than 1 mg kg^{-1} (Liu 1994; Yang et al. 2007). Also in India, a large proportion of the soils are low in available Zn (White and Zasoski 1999). Application of Zn fertilizer is a common practice in these areas to increase yields, and many studies have demonstrated the effectiveness of this strategy. The majority of these studies have, however, focused on yield, whereas the effects of fertilization on the concentration of Zn and other nutrients in the crops have received much less attention.

Nonetheless, there are a variety of studies that also assessed the effects of Zn fertilization on the nutritional quality of crops. They reported that the application of Zn fertilizer to low-Zn soil not only increased yield but also Zn concentrations in the seeds of crops, such as pea (Peck et al. 1980; Fawzi et al. 1993), navy bean (Brouwer et al. 1981; Moraghan 1994), soybean (Raboy and Dickinson 1984), sorghum (Rashid et al. 1997), chickpea (Khan et al. 2003), wheat, barley, and rice (Paterson et al. 1991; Graham et al. 1992; Zhu et al. 1997; Yilmaz et al. 1998; Erdal et al. 2002; Hu et al. 2003; Genc et al. 2004). Increasing the Zn supply to plants was not only found to increase total Zn but also bioavailability of Zn in seeds of pea, navy bean, and wheat (Welch et al. 1974; Moraghan 1994; House and Welch 1989).

Generally, a grain yield response of cereal crops to Zn fertilization is associated with an increase in grain Zn concentration, whereas Zn concentrations in the vegetative parts may remain almost unaffected as long as Zn supply is growth limiting. This observation suggests that, in contrast to shoot Zn, grain Zn is a good index of soil fertility status under conditions of low soil Zn concentrations. However, at higher levels of Zn supply, an increase in grain Zn concentration is not necessarily associated with a positive yield response. Genc et al. (2004) grew two barley cultivars with different Zn efficiencies at 6 different levels of Zn fertilization (ranging from 0 to 12.8 mg Zn per kilogram dry soil) in pots containing Zn-deficient sandy soil. They found that grain Zn concentrations continued to increase over the entire range of the Zn application levels up to 120 mg kg^{-1} in the more Zn-efficient of the two cultivars (Figure 6.1), whereas a maximum yield was already reached at Zn application rates between 0.8 and 3.2 mg kg^{-1}. Shoot Zn concentrations did not, however, show similar and more monotonous complex behavior and the relationship between shoot Zn concentration and Zn supply depended on the stage of plant development. At a marginal level of Zn supply and above, shoot Zn concentrations decreased with plant age, whereas no decrease was observed under Zn-deficient conditions. A negative response of Zn concentrations in older plant tissue was also found for chickpea under low levels of Zn fertilization (Khan et al. 2003). This observation indicates that Zn can be more exhaustively relocated from leaves to grains under Zn-sufficient than Zn-deficient conditions. The results of Genc et al. (2004) and Khan et al. (2003) show that Zn levels in excess of those required for maximum yield are required to achieve high Zn densities in seeds. They concur with similar findings for pea seeds and maize grains (Payne et al. 1988).

In the study of Genc et al. (2004) Zn application rates did not reach phytotoxic levels, in contrast to a study by Shute and Macfie (2006) on growth and Zn uptake of soybean. Shute and Macfie (2006) found that seed Zn concentrations in soybean

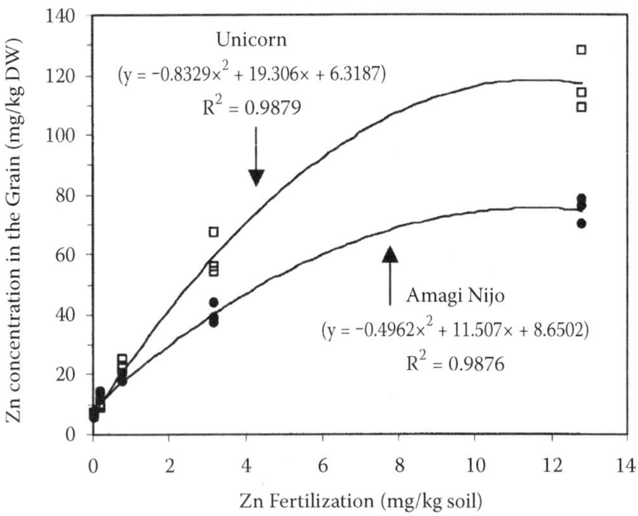

FIGURE 6.1 Grain Zn concentration in two barley genotypes with different Zn efficiency grown on an Zn-deficient siliceous sand at different levels of Zn fertilization. (From Genc et al. (2004). *Plant and Soil* 263: 273–287.)

were significantly higher after the addition of 2000 mg kg^{-1} Zn than after the addition of 50 mg kg^{-1} Zn in the form of Zn chloride to a clay loam soil, whereas growth became suppressed at total soil Zn concentrations above 100 mg kg^{-1}. Unfortunately, the authors did not report seed Zn concentrations at the Zn application levels between 50 and 2000 mg kg^{-1}. Increasing application of Zn beyond phytotoxic thresholds will not only result in reduced yield but may eventually also decrease grain Zn density (Nambiar and Motiramani 1983; Chaney 1993). Furthermore, it should be noted that high loads of Zn may have negative effects on soil organisms and impair the ecological functions of soil (Frossard et al. 2000).

Zinc fertilization has to be adapted not only to the requirements of the crop but also to the properties of the soil. The availability of soil Zn for uptake by crop plants varies widely among different types of soil. The solubility of Zn decreases with increasing soil pH. The binding of Zn to organic matter is less strong than that of other heavy metals such as Cu and Ni. Unlike other micronutrient elements (e.g., Fe, Mn, and Cu), Zn is also not subject to changes in its oxidation state under normal environmental conditions. Indirectly, soil redox conditions can, however, have a great influence on Zn availability, because surface complexation of Zn on iron oxides is a major retention mechanism in noncalcareous mineral soils. Moreover, the inhibition of aerobic respiration and also of abiotic oxidative processes under anaerobic conditions strongly affect the composition of solid and dissolved organic matter.

Important solid phases that can control Zn concentrations in soil include carbonates and hydroxides at high pH, sulfides at very low redox potential, and phosphates (Lindsay 1979; Srivastava and Gupta 1996). Even in a very poor soil, the pool of total Zn is still three magnitudes larger than the amount of Zn removed by a single crop. Thus, a deficiency in Zn very rarely results from a limitation in the total amount of soil Zn, but rather may be due to a shortage in readily plant-available Zn (i.e., the Zn in solution plus the Zn bound in an easily exchangeable form) combined with a kinetic limitation in the resupply of Zn to this pool from the larger pools of more tightly bound and less soluble Zn. Fertilization improves Zn supply to plants only as much as it increases the pool of readily available Zn. The addition of a soluble Zn fertilizer to the available pool in excess of the equilibrium concentration will cause a transfer (flux) of Zn from this pool into the less available pools by diffusion, sorption, and precipitation processes, which tends to reestablish thermodynamic equilibrium conditions. Usually, a fertilizer effect will disappear at a much faster rate than the rates at which Zn is lost from soil due to plant uptake, leaching, and other processes (usually of minor importance in comparison). The duration of fertilizer effectiveness, however, can vary considerably, depending on the type of fertilizer, application rate, and soil property.

Brennan and Bolland (2006) investigated the residual effectiveness of Zn oxide periodically applied at 0.5 and 1.0 kg Zn per ha to a lateritic soil cleared of native vegetation in southwestern Australia over a period of 17 years. The soil was a weakly acidic loamy sand (pH 5.3) with a low Zn concentration (< 0.2 mg kg^{-1} DTPA-extractable Zn) and a low phosphorus status (6 mg kg^{-1} bicarbonate-extractable P concentration). Wheat was grown on the average every 3 years (for a total of seven crops), and pasture was grown in the years between the wheat crops. The 0.5 kg ha^{-1} Zn treatment was found to be effective for four wheat crops, whereas no Zn deficiency was observed even after seven wheat crops in the 1.0 kg ha^{-1} treatment. Total amounts of 0.15 and 0.34 kg ha^{-1} Zn were removed with the grains in the 0.5 and 1.0 kg ha^{-1} treatments, respectively, over the duration of the experiment, which corresponded to about a third of the applied Zn in each treatment. The authors estimated that an additional 16–24% of the applied Zn was removed by sheep grazing on the plots in the pasture years. In addition, Brennan and Bolland (2006) performed pot trials with soil collected from the plots of the field experiment to determine how the availability of applied Zn for various crops had changed with time after fertilizer application. In order to characterize the phytoavailability of soil Zn, they measured Zn uptake in the "youngest mature growth" (YMG), that is, in the apex and the youngest fully emerged leaf after 50 days of growth. Zinc concentrations in YMG differed among crops and were always higher at the 1.0 kg ha^{-1} than in 0.5 kg ha^{-1} treatment. For all crops Zn concentrations decreased linearly with time for the two Zn treatments over the entire range of times after the initial fertilizer application.

In another recent study, Shaver et al. (2007) compared the residual fertilization effect of various Zn sources with different solubility on the Zn uptake by four consecutive corn crops grown in a loamy Ustollic Haplargid soil (pH 5.6, 0.4 mg kg^{-1} AB-DTPA-extractable Zn) with and without liming. The Zn fertilizers applied were 99.7% water-soluble Zn sulfate (33% total Zn), a 56% water-soluble industrial by-product with 23% total Zn, and a 6% water-soluble industrial by-product with 37.6% total Zn. The two industrial by-products resulted in similar Zn uptake, whereas Zn uptake was significantly higher from the Zn sulfate source in the first two crops. In the third crop, there were no longer differences observed in Zn uptake. Even at the highest application rate of 20 kg Zn per hectare, Zn uptake had decreased to that observed at control levels. Liming increased the soil pH to 7.5 and, as a result, there was a significant reduction in Zn uptake in the first crop. In the subsequent crops, the reduction in Zn availability became insignificant as the available Zn was increasingly exhausted or transformed to less-soluble phases in the nonlimed treatments. Averaging across 20 highly calcareous Iranian soils with carbonate contents ranging from 16 to 58% and soil pH ranging from 7.9 to 8.5, Yasrebi et al. (1994) found that 58 and 60% of Zn applied as Zn sulfate at rates of 10 and 20 mg kg^{-1}, respectively, was converted into the soil carbonate fraction after harvesting a greenhouse corn crop using sequential extraction analysis.

Compared to soluble Zn fertilizers (such as Zn sulphate) that become ineffective rather rapidly as the dissolved Zn reacts with soil minerals and organic matter, chelates have the advantage of keeping the applied Zn in solution in a less reactive form. This feature is particularly attractive for Zn applications to high-pH soils. A number of studies evaluated the fertilizer effectiveness of various synthetic and natural chelates as Zn carriers. Prasad and Sinha (1981) gave the following order of relative efficiency with respect to yield and Zn uptake by maize from a calcareous soil: Zn-DTPA > Zn-fulvate > Zn-EDTA > Zn-citrate > Zn-sulfate, which reflects the stability of these compounds. Maftoun and Karimian (1989) also observed that Zn-EDTA was generally more effective than Zn sulfate in increasing Zn uptake by maize on calcareous soils. Alvarez and Gonzalez (2006) found that Zn-EDTA was more efficient than Zn-EDDHA in enhancing Zn accumulation by maize on a neutral clay loam soil, and both were more effective than Zn-polyflavonoid, Zn-phenolate,

Zn-lignosulfonate, and Zn-glucoheptonate. Zn-EDDHA was, however, the only Zn carrier of the six that also produced a significant, though only slight yield increase. Similarly, Gonzalez et al. (2007) found that the application of Zn-EDTA to a calcareous soil caused the highest increase in Zn concentration of navy bean and had no effect on yield in comparison to five other Zn chelates, including Zn-S and S-EDDS. Unfortunately, the grain Zn concentrations were not determined in any of these studies. As pointed out by Gonzalez et al. (2007), care must be taken to control the leaching risk involved in the application of Zn chelates because the more mobile the chelate, or the less biodegradable the carrier, the greater the risk is for leaching.

Zinc fertilizers are mostly applied to the soil, but foliar application is also used, in particular in fruit and vegetable production (Rengel et al. 1999). The advantage of foliar fertilization is that the fertilizer is directly applied to the plant. Thus, the absorption of Zn is not interfered by soil–root transfer processes. The main barrier to be crossed is the leaf cuticle layer. However, foliar Zn fertilization is not effective if leaves are not sufficiently developed. Paterson et al. (1991) found that foliar application of Zn sulfate was more effective than soil application for increasing grain Zn density in barley. Similarly, Yilmaz et al. (1997) observed that foliar application of Zn sulfate to two wheat cultivars resulted in a larger increase of grain Zn density than soil application, but it had a smaller effect on yield. Brouwer et al. (1981) found that foliar application of Zn sulfate increased yield more (in proportion) than it increased Zn concentrations in navy bean seeds. Ozturk et al. (2006) showed that the Zn transfer from shoots into wheat grains is particularly high during early seed development, which suggests that the timing of foliar Zn application would be a critical factor for effective biofortification.

Another method to apply Zn to crop plants is the treatment of seeds with Zn fertilizers, a technique called *seed priming*. Applying Zn to seeds is much easier and less expensive than distributing Zn fertilizer evenly into the soil or spraying onto leaves, especially for resource-poor farmers who cannot afford the equipment and labor for repeated foliar applications. Pretreating (priming) seeds in fertilizer solutions can have high agronomic benefits, in particular for farming on soils deficient in micronutrients, by accelerating seedling emergence and improving seedling vigor. However, priming seeds with Zn alone was not found to increase the seed Zn density of the progeny. Yilmaz et al. (1997) reported that seed priming with Zn sulfate resulted in a stronger increase in the grain yield of wheat grown on Zn-deficient soil than foliar Zn application, but it did not significantly increase grain Zn concentration in contrast to foliar and soil application. Johnson et al. (2005) observed that Zn-primed grains produced larger seedlings in rice, but they also found no effect of seed priming on the Zn contents of progeny seeds in rice, lentil, chickpea, and wheat. These results agree with the findings on grain yield and grain Zn concentration in wheat grown from unprimed seeds with different Zn content. Rengel and Graham (1995) grew seeds with low and high Zn densities from two wheat cultivars with different Zn efficiencies on Zn-deficient sand. Plants grown from Zn-rich seeds produced more and larger grains than plants grown from Zn-poor seeds. Plants grown from Zn-rich seeds also utilized absorbed Zn more efficiently in terms of grain dry matter production per unit of Zn in the aboveground biomass. They also allocated a larger portion of absorbed Zn to the grains but did not produce seeds with an increased Zn concentration without fertilization. Yilmaz et al. (1998) investigated the grain yield and Zn density of wheat plants grown from seeds with different Zn content that had been produced by varying the number of foliar Zn sulfate applications to the previous crop. Again, the grain yield from the high-Zn seeds was significantly larger than from low-Zn seeds, whereas there was no significant effect on Zn accumulation in the plants. Although these results show that seed priming may not substitute soil or foliar fertilization in biofortifying staple food crops with Zn for human nutrition, they demonstrate that biofortified seeds are also very advantageous from the perspective of yield, and thus, food security.

6.3 NITROGEN AND PHOSPHORUS FERTILIZATION

Application of macronutrient fertilizers is another agricultural practice that can have significant effects on Zn accumulation in plants. Rengel et al. (1999) compiled data from various studies showing that concentrations of micronutrients are, in general, positively correlated with macronutrient and protein concentrations in whole grains of wheat, rice, and soybean. From these correlations they concluded that the addition of fertilizer to correct macronutrient deficiencies will also have a positive but weak influence on the micronutrient density of a crop due to increased uptake of micronutrients through the promotion of root and shoot development. The role of root traits in micronutrient acquisition was highlighted, for example, by Dong et al. (1995), and more recently by Gahoonia et al. (2006). Dong et al. (1995) investigated the effect of the herbicide chlorsulfuron on the uptake of Zn and Cu by three different wheat cultivars on an Zn-deficient sandy soil. The herbicide reduced shoot growth and root length, in particular the length of the fine roots, but increased the average root diameter. Therefore, as a net effect, the total root biomass remained unaffected. The herbicide treatment also decreased Zn accumulation in roots and shoots but increased Cu accumulation in the roots. The decreased concentrations in Zn and Cu were interpreted as a result of the reduced root surface area, because the rates of net uptake per unit root surface area were not affected by the chlorsulfuron treatment. Gahoonia et al. (2006) found that root length varied by more than a factor of two among ten different lentil genotypes, and that it was significantly correlated with the accumulation of Zn, Cu, and Mn in the shoots from a slightly alkaline soil (pH 7.7 in 0.01 M $CaCl_2$) with marginal Zn availability (1.7 mg kg^{-1} DTPA-extractable Zn). Nitrogen fertilization has been found to increase Zn accumulation in wheat and barley grains by Zebarth et al. (1992) and De Brouwer and Smolders (2006), whereas Brennan (1996) found that N fertilizer application decreased the Zn concentration of wheat grains grown on a range of soils in southwestern Australia in 20 out of 30 field trials.

Different effects on Zn uptake by crops have also been reported regarding P fertilization. In a 3-year field experiment performed at various locations in Alberta and Manitoba of Canada with different P sources, Grant et al. (2002) found that phosphate fertilization generally decreased grain Zn concentration but increased Cd concentration independently of the Cd content of the fertilizer. The P fertilization effect on grain P concentration was weak and inconsistent. The reduction in Zn density also occurred on plots where seed yield was not increased, and on those plots where seed yield increased, the reduction in Zn density was larger than what could have been attributed to dilution by the increased biomass. Verna and Minhas (1987) found that P fertilization increased the translocation of Zn into wheat and maize grains, whereas Bolland et al. (1993) observed no change in grain Zn density of wheat, lupin, and triticale upon phosphate application. Phosphorus-induced inhibition of Zn uptake can be due to antagonistic interactions between P and Zn in the soil as well as within the plant. For example, Agbenin (1998) found that incubation of a Nigerian savanna soil for three weeks with P fertilizer strongly increased Zn sorption by the soil, which the author attributed to the formation of Zn-phosphate surface complexes, and also to Zn phosphate precipitations at high Zn and P concentrations. On the other hand, Bogdanovic et al. (1999) reported results of a field study showing that P fertilization did not decrease the available Zn concentration in the soil over a period of 30 years but nonetheless significantly reduced Zn uptake by corn.

6.4 MYCORRHIZAL EFFECTS AND PLANT–PLANT INTERACTIONS

Mycorrhizal fungi may play an important role in determining macronutrient fertilization effects on Zn uptake by plants. Tarkalson et al. (1998a) studied mycorrhizal root colonization, yield, and nutrient uptake by wheat and corn grown on field plots with an eroded and leveled silt loam soil after treatment either with dairy manure or mineral fertilizer. For both plant species, shoot Zn and Mn concentrations generally increased with root colonization, whereas P, Ca, Mg, Fe, and Cu were not consistently or significantly influenced by the mycorrhizae. In another study, the same authors found that mycorrhizal root colonization also paralleled increased Zn uptake during the early growth of bean, whereas the uptake of other nutrients was not or even negatively related to mycorrhizal root colonization (Tarkalson et al. 1998b). Such mycorrhizal effects on crop nutrition also appear to explain the differences between the Zn density of wheat grains produced on acidic red loam soils (Kandosols) by organic and conventional farming in southeastern Australia (Ryan et al. 2004). At this location, grains from organic farms were significantly richer in Zn and Cu than from conventional farms, but in general somewhat poorer in P, whereas there were no significant differences in other analyzed macro- and micronutrients. The higher Zn concentrations obtained on organic farms were related to a higher degree of mycorrhizal root colonization, as evidenced by a very close positive correlation (r^2 = 0.94) between these two parameters in soil samples collected from fertilized and unfertilized field plots of neighboring organic and conventional farms (Figure 6.2). The lower mycorrhizal colonization of wheat roots in conventional farming was attributed to the application of soluble P fertilizer, which is known to suppress mycorrhizal infection. In a fertilizer trial on the same field plots, superphosphate application in fact strongly reduced mycorrhizal root colonization, while at the same time the treatment decreased Zn and increased P concentrations in grains from organic farms to the same levels as in grains produced from conventional farms.

In addition to beneficial effects of soil microorganisms on plant Zn nutrition, complementary interactions among different plant species, including plant residue effects in crop rotation schemes, may be utilized to enhance biofortification. Gunes et al. (2007) investigated the effect of intercropping wheat and chickpea on the

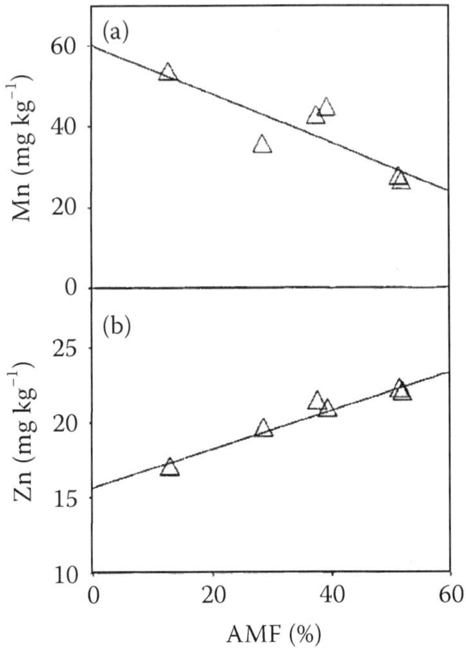

FIGURE 6.2 Manganese (a) and zinc (b) concentrations of wheat grains produced on various fields with different phosphorus fertilizer applications (no fertilizer, reactive phosphate rock, superphosphate) on neighboring organic and conventional farms in southeastern Australia in relation to the percentage of root length colonized by arbuscular mycorrhizal fungi (AMF). (From Ryan et al. 2004. *Journal of the Science of Food and Agriculture* 84: 207–216. With permission.)

mineral nutrition of the two crops and the mineral nutritive value of their seed produced under field conditions. The two crops were grown on a calcareous clay loam soil (pH 7.9 in water extract, 0.96% organic matter content) with very low DTPA-extractable Zn and Fe concentrations (0.11 and 0.91 mg kg^{-1}, respectively). All plots planted with chickpea were inoculated with a *Rhizobium* implant at sowing. In comparison to separate cultivation of the two crops in monoculture, intercropping increased the combined biomass of wheat and chickpea, as well as the total grain/seed yield per unit land area by 18 and 23%, respectively. This increase in total biomass and yield was primarily due to an increase in the growth and also in the harvest index of wheat. The harvest index is the percentage of grain in relation to the total aboveground biomass produced. The index increased from 24.1% in monoculture to 29.3% under intercropping conditions. As a result, the total wheat grain yield per unit area on the intercropped plots was not significantly less than on the monocultured wheat plots.

Intercropping also increased the concentrations of the micronutrients Zn, Fe, and Mn in shoots and seeds of both species, although the increased accumulation of

Zn and Mn in wheat grains and the increased accumulation of Fe in chickpea shoots were not statistically significant. Compared to the micronutrients, a different pattern of effects was observed for the macronutrients. Although N, P, and K concentrations were enriched in the seeds of both species under intercropping as compared to monocropping, shoot P and K concentrations were increased only in wheat, whereas shoot N, P, and K concentrations were decreased in chickpea. These intercropping effects on nutrient concentrations indicate that the grain yield increase in wheat was primarily due to an enhanced supply of soil nutrients in the root zone made available by the chickpea plants. Such interactions have been shown previously in regard to P nutrition by Li et al. (2003) and Li et al. (2004). Although chickpea growth suffered from the competition with wheat for macronutrients in the study of Gunes et al. (2007), the nutritional quality of chickpea seeds benefited from the interaction as the density of micronutrients became enhanced. This latter effect may be attributed to the well-known exudation of metal-complexing compounds by graminaceous roots, in particular phytosiderophores that increase the availability of the complexed metals for root uptake (Marschner 1995).

6.5 SOIL WATER REGIME

Very little information is available about the potential role that physical soil conditions play on the Zn biofortification of staple food crops. Drought stress applied under greenhouse conditions reduced the accumulation of micronutrients, including Zn, in shoots of wheat and chickpea, but enhanced Zn accumulation by lentil, both under mono- and intercropping conditions (Gunes et al. 2007). Unfortunately, no data on seed Zn concentrations were given by Gunes et al. (2007) for this drought experiment. Khan et al. (2003) compared the effects of imposed drought conditions on growth and Zn uptake of two drought-tolerant chickpea cultivars with different Zn-efficiencies grown under conditions of severe to marginal Zn deficiency. Water stress reduced seed Zn only in the Zn-efficient cultivar under marginal Zn deficiency, whereas the stress slightly increased Zn density in the seeds of the other cultivar under severe Zn-deficiency conditions. In all other treatment combinations, there was no significant effect on seed Zn concentration.

Increased Zn deficiency due to changes in water regime is of particular concern in rice cultivation. In order to conserve precious irrigation water, rice production is increasingly shifted from paddy to dry-field cultivation, in which the soil is irrigated but not flooded and puddled. As a result, soil conditions become aerobic. As the redox potential increases, N, Mn, Fe, S, and organic carbon are oxidized. These processes increase acidity, which in turn leads to the solubilization of sorbed Zn, and also decrease Zn retention through the decomposition of Zn-binding organic matter and the dissolution of Zn-bearing mineral phases such as Zn sulfide and franklinite. In addition, there may also be processes that reduce dissolved Zn concentrations. For example, the precipitation of Fe and Mn oxides and reduced concentrations of dissolved Zn-complexing organic ligands will lead to increased Zn sorption. Lower soil water content may also limit Zn mobility. As Zn deficiency is widespread in soils planted to rice, there is concern that the net effect of these changes is an aggravation of the Zn problem. Gao et al. (2006) compared several rice genotypes for growth and

Zn uptake under flooded and aerobic cultivation with and without Zn fertilization (0 versus 23 kg Zn ha^{-1}, applied as Zn sulfate). In a calcareous Cambisol soil with 2.0 mg kg^{-1} DTPA-extractable Zn, shoot biomass and shoot Zn concentration increased with Zn fertilization, whereas the grain yield did not change significantly. The grain yield and shoot Zn concentrations were lower under aerobic than under flooded conditions, whereas there was no significant flooding effect on the final shoot biomass. Neither cultivation nor fertilization had a significant effect on grain Zn density in this experiment. Much more research is still needed for a better understanding of these rather surprising effects.

6.6 ANTINUTRIENTS AND PROMOTERS OF FOOD ZINC BIOAVAILABILITY

The fraction of Zn content in food that is available for absorption by the human gastrointestinal tract depends on the presence of other food ingredients. There are nutrient promoters (or enhancers, i.e., substances increasing absorption) and antinutrients (i.e., substances inhibiting absorption). Table 6.1 gives a list of known or suspected promoters and antinutrients of human Zn nutrition. Research on the potential role of a variety of substances to act as promoters or antinutrients of Zn bioavailability to humans has been reviewed by House (1999). There is evidence for a promoter function of certain chelating low-molecular-weight organic acids (e.g., in particular, ascorbic and picolinic acid) as well as amino acids (e.g., methionine, cysteine, and

TABLE 6.1
Components in Plant Food that May Enhance (Promoters) or Reduce (Antinutrients) Dietary Zinc Absorption by Humans or Experimental Animals

Substance	Major Source in Human Diets
Promoters	
Ascorbic acid	Fresh fruits and vegetables
Amino acids: methionine, cysteine, histidine	Animal meats
Long-chain fatty acids (e.g., palmitate)	Breast milk
β-carotene	Green and orange vegetables
Antinutrients	
Phytic acid	Legume seeds and cereal grains
Fiber (e.g., cellulose, hemicellulose, lignin, cutin)	Whole cereal grains
Tannins and other polyphenols	Tea, coffee, beans, sorghum
Oxalic acid	Spinach leaves, rhubarb
Hemagglutinins (e.g., lectins)	Most legumes and wheat
Goitrogens	*Brassica and Allium*
Heavy metals (e.g., cadmium)	Contaminated leaf and root vegetables

Source: Modified from Graham et al. 2001. *Advances in Agronomy* 70: 77–142.

histidine). In contrast, there are also studies in which Zn complexes with amino acids (including methionine), picolinate, or ascorbate did not affect Zn absorption by experimental animals (House 1999). The list of proposed Zn antinutrients includes phytic acid, tannins and other polyphenols, oxalic acid, and some heavy metals, in particular cadmium (Cd) (Graham et al. 2001). Firm evidence for a strong antinutrient effect has only been given for phytate (House 1999). The antinutrient effect of phytate depends on the molar phytate:Zn ratio. Zinc absorption in the intestine is reduced at ratios above a value of around 20 (Frossard et al. 2000).

Around 80% of the P stored in cereal and legume seeds is present as phytate. Thus, it is not surprising that seed phytate concentrations strongly depend on the availability of P in the soil to plants. Because of the well-known antagonism between P and Zn uptake by Zn-deficient plants (Cakmak and Marschner 1986; Marschner 1995; Torun et al. 2001), seed phytate concentrations are not only enhanced by P fertilization but also decreased by Zn fertilization under Zn-deficient conditions (Raboy and Dickinson 1984). This effect has been confirmed for wheat by Erdal et al. (2002) in a field study with 20 different wheat cultivars grown on a Zn-deficient calcareous soil (0.10 mg kg^{-1} DTPA-extractable Zn) in central Anatolia. Zinc fertilization (23 kg Zn ha^{-1}, applied in the form of Zn sulfate) increased seed Zn concentration in all cultivars and reduced seed phytate and P concentrations in the majority of the cultivars. The average increase in seed Zn was from 8.8 to 16.7 mg kg^{-1}, whereas phytate decreased in average from 10.7 to 10.0 mg g^{-1} and P from 3.9 to 3.5 mg g^{-1} grain dry weight. Although the decrease in seed phytate was small, it added to the effect on the phytate-to-Zn molar ratio, which decreased in average from 126 to 56.

Although increasing the bioavailability of Zn in food is a desirable effect in fighting Zn deficiency in human nutrition, reducing the concentration of phytate in food may also create health and other problems that compromise the positive effect. There is evidence that dietary phytate has anticarcinogenic and antioxidant effects, and that it may enhance the absorption of dietary Cu (Lee et al. 1988; Harland and Morris 1995). Furthermore, a decrease in phytate content may reduce the vitality of seeds and thus lower seedling survival and growth, especially on low-P soils (Frossard et al. 2000). Also, tinkering with other antinutrients in food plants may have undesired side effects. For example, tannins and other polyphenols play important beneficial roles in the defense of plants against microorganisms and their toxins (Frossard et al. 2000). For these reasons, extreme care should to be taken in reducing the densities of these compounds in seeds and other edible parts of crop plants.

6.7 INTERACTIONS BETWEEN ZINC AND CADMIUM

Regarding the effects of other metals on the accumulation of Zn in plants, the interactions with Cd deserve special attention as they are particularly strong due to the chemical similarity between these two elements and because of the potential risks of Cd toxicity to humans that may arise from Cd accumulation in food plants grown on Cd-polluted agricultural soils. Such pollution is a widespread problem due to the atmospheric deposition of industrial and other emissions, as well as the intensive application of low-quality mineral fertilizers, sewage sludge, and other biowaste products. Although such Cd inputs are now much better controlled than in the past in many countries, existing soil pollution is going to remain for a long time at many contaminated sites because the mobility of metals is generally very low in agricultural soils and as long as soil remediation is not economically and technically feasible.

The interactions between Zn and Cd accumulation in plants have received particular attention after Zn deficiency was found to increase Cd accumulation in rice (Honma and Hirata 1978). Since then, Zn applications were not only found to decrease Cd accumulation in rice but also in other plants, for example in lettuce (McKenna et al. 1993), wheat shoots and grains (Oliver et al. 1994; Khoshgoftar et al. 2004), and potato tubers (McLaughlin et al. 1994). These effects were attributed to the competition between Cd and Zn for plant uptake through common transporter proteins in the cell membranes of root and phloem cells (Cakmak et al. 2000; Hart et al. 2002). Such a competition can also explain that an increase in soil Cd can reduce Zn accumulation as observed in all parts of soybean plants (in particular in the seeds) grown on an Zn-sufficient clay loam soil (Shute and Macfie 2006). In contrast to this, it was also found that Cd can increase Zn uptake and that the effect of Cd on Zn accumulation can strongly differ not only among plant species but also among different parts of the same plant. In the study of McKenna et al. (1993), Cd promoted Zn accumulation in young lettuce but not in spinach. Chaoui et al. (1997) observed that more Zn was accumulated in bean roots when Cd was added to solution cultures, whereas Zn accumulation in the shoots was reduced. Moreover, Cd effects on Zn uptake by plants may completely change with the level of Zn supply. Köleli et al. (2004) found that adding Cd at concentrations of 10 and 25 mg kg^{-1} to a Zn-deficient calcareous soil tended to decrease shoot Zn concentrations in durum and bread wheat. When the soil was fertilized with Zn sulfate at a rate of 10 mg Zn kg^{-1}, Zn accumulation increased from less than 10 mg kg^{-1} in the shoots up to a factor of 10 and was further enhanced by the Cd treatments. In contrast, the accumulation of Cu and Fe was strongly and significantly decreased by the Cd treatments under Zn deficiency and remained unaffected in combination with Zn fertilization.

Cadmium accumulation in shoots was reduced by the application of Zn at the lower rate (10 mg kg^{-1}), but not at the higher rate (25 mg kg^{-1}) of Cd application. Also, in soybean the interactions between Cd and Zn varied with the level of soil Cd and Zn concentrations, but not in the same way as reported by Köleli et al. (2004) for wheat. Shute and Macfie (2006) found that the addition of Cd substantially reduced Zn accumulation in soybean when a high rate of Cd (100 mg kg^{-1}) was applied to a soil in addition to a high rate of Zn (2000 mg kg^{-1}). This effect became insignificant when a low rate of Cd (2 mg kg^{-1}) was applied to the soil in addition to a low Zn concentration (50 mg kg^{-1}). The effect of Zn on the uptake of Cd was opposite to the effect of Cd on Zn uptake. Cadmium accumulation was decreased by Zn addition only at the low application rates of the two metals, but enhanced at the high rates. This asymmetrical behavior indicates that the competition for a common transporter protein is not sufficient to account for the observed interaction effects between Zn and Cd in plants. Shute and Macfie (2006) explained the observed interactions in part as the result of the mutual effect of one metal on the bioavailability of the other metal. In fact, they found that addition of Cd did not increase but rather reduced the solubility of Zn (as determined by an organic acid extraction procedure that mimicked plant exudates), whereas Zn addition strongly increased Cd availability. This result is in line with the generally higher affinity of soil minerals for Zn than for Cd.

The solubility and speciation of the two metals in soil solution was also found to be of particular relevance for their uptake by wheat on Zn-deficient saline soil, where high chloride concentrations enhanced Cd solubility through the formation of Cd–chloride complexes (Khoshgoftar et al. 2004). The complexity and diversity of Zn–Cd interactions demonstrates the difficulty in predicting how specific agronomic practices will affect the accumulation of these two metals by plants. Particular care has to be taken before new practices are proposed and applied on soils rich in Cd for the purpose of Zn biofortification.

6.8 CONCLUSIONS

The available evidence shows that agronomic biofortification of staple crops with Zn through the development of suitable cultivation methods is an achievable objective. This approach to biofortification is all the more realistic if it is not pursued as a stand-alone strategy but rather as a complementary approach to the improvement of genotypes by breeding and genetic engineering, as well as to improvements in the processing of crops into food products. We were unable to elaborate on these latter aspects for reasons of space and focus. An excellent discussion of the various approaches to biofortification has recently been presented by Cakmak (2008). Although much has been learned about the mechanisms and pathways of Zn acquisition by plants and the translocation of Zn into shoots and seeds, the processes and factors that limit Zn transfer from soil to the roots and shoots are far from being sufficiently understood. A better understanding is needed in particular on the movement of Zn from shoots into seeds. Research is also lacking on the roles that soil biota and soil organic matter play in the accumulation of Zn by plants. Published research related to the effects of manure, compost, and sewage sludge application on Zn uptake generally does not allow us to separate the effects of organic matter and macronutrient inputs from concomitant metal inputs. Much more work is also needed on the effects of cultivation methods on antinutrient (in particular phytate) and promoter concentrations in food crops, on the potential implications that changes in these components may have on human health, and on the agronomic performance of crops (such as yield, seed survival, and disease resistance, etc.). Apart from clarifying such issues, the reliability of new cultivation methods has to be tested sufficiently for a wide range of sites and cultivars, before they can be recommended for routine application in practice.

REFERENCES

Agbenin, J. O. 1998. Phosphate-induced zinc retention in a tropical semi-arid soil. *European Journal of Soil Science* 49: 693–700.

Alloway, B. J. 2006. *Zinc in Soils and Crop Nutrition.* Online book published by the International Zinc Association, Brussels, Belgium, http://www.zinc-crops.org/, pp. 116.

Alvarez, J. M. and D. Gonzalez. 2006. Zinc transformations in neutral soil and zinc efficiency in maize fertilization. *Journal of Agricultural and Food Chemistry* 54: 9488–9495.

Bogdanovic, D., M. Ubavic, and M. Cuvardic. 1999. Effect of phosphorus fertilization on Zn and Cd contents in soil and corn plants. *Nutrient Cycling in Agroecosystems* 54: 49–56.

Bolland, M.D.A., R.J. Jarvis, P. Coates, and D. J. Harris. 1993. Effect of phosphate fertilizers on the elemental composition of seed of wheat, lupin, and triticale. *Communications in Soil Science and Plant Analysis* 24: 1991–2014.

Bouis, H. 1996. Enrichment of food staples through plant breeding: A new strategy for fighting micronutrient malnutrition. *Nutrition Reviews* 54: 131–137.

Brennan, R.F. 1996. Availability of previous and current applications of zinc fertilizer using single superphosphate for the grain production of wheat on soils of South Western Australia. *Journal of Plant Nutrition* 19: 1099–1115.

Brennan, R.F. and M.D.A. Bolland. 2002. Relative effectiveness of soil-applied zinc for four crop species. *Australian Journal of Experimental Agriculture* 42: 985–993.

Brennan, R.F. and M.D.A. Bolland. 2006. Residual values of soil-applied zinc fertiliser for early vegetative growth of six crop species. *Australian Journal of Experimental Agriculture* 46: 1341–1347.

Broadley, M. R., P. J. White, J. P. Hammond, I. Zelko, and A. Lux. 2007. Zinc in plants. *New Phytologist* 173: 677–702.

Brouwer, H. M., G. R. Stevens, and J. G. Fletcher. 1981. Differential varietal response to zinc foliar sprays in navy beans (*Phaseolus vulgaris*). *Queensland Journal of Agriculture and Animal Sciences* 38: 181–185.

Cakmak, I. 2008. Enrichment of cereal grains with zinc: Agronomic or genetic biofortification? *Plant and Soil* 302: 1–17.

Cakmak, I. and H. Marschner. 1986. Mechanism of phosphate induced zinc deficiency in cotton. I. Zinc deficiency-enhanced uptake rate of phosphorus. *Physiologia Plantarum* 68: 483–490.

Cakmak, I., A. Yilmaz, M. Kalayci et al. 1996. Zinc deficiency as a critical problem in wheat production in Central Anatolia. *Plant and Soil* 180: 165–172.

Cakmak, I., R. M. Welch, B. Erenoglu, V. Römheld, W. A. Norvell, and L. V. Kochian. 2000. Influence of varied zinc supply on re-translocation of cadmium (^{109}Cd) and rubidium (^{86}Rb) applied on mature leaf of durum wheat seedlings. *Plant and Soil* 219: 279–284.

Chaney, R. L. 1993. Zinc phytotoxicity. In *Zinc in Soil and Plants*, A.D. Robson (Ed.), 135–150, Dordrecht: Kluwer Academic Press, pp. 135–150.

Chaoui, A., M. H. Ghorbal, and E. El Ferhani. 1997. Effects of cadmium-zinc interactions on hydroponically grown bean (*Phaseolus vulgaris* L.). *Plant Science* 126: 21–28.

Combs, G. F. Jr., R. M. Welch, J. M. Duxbury, N. T. Uphoff, and M. C. Nesheim. 1996. *Food-based Approaches to Preventing Micronutrient Malnutrition: an International Research Agenda*. 68. Ithaca, NY: Cornell International Institute for Food, Agriculture, and Development, Cornell University.

Cousins, R.J. 1979. Regulatory aspects of zinc metabolism in liver and intestine. *Nutritional Review* 37: 97–103.

De Brouwer, K. and E. Smolders. 2006. Yield response of crops amended with sewage sludge in the field is more affected by sludge properties than by final soil metal concentration. *European Journal of Soil Science* 57: 858–867.

Dong, B., Z. Rengel, and R. D. Graham. 1995. Effects of herbicide chlorsulfuron on growth and nutrient uptake parameters of wheat genotypes differing in Zn-efficiency. *Plant and Soil* 173: 275–282.

Erdal, I., A. Yilmaz, S. Taban, S. Eker, B. Torun, and I. Cakmak. 2002. Phytic acid and phosphorus concentrations in seeds of wheat cultivars grown with and without zinc fertilization. *Journal of Plant Nutrition* 25: 113–127.

Fawzi, A. F. A., M. M. El-Fouly, and Z. M. Moubarak. 1993. The need of grain legumes for iron, manganese and zinc fertilization under Egyptian soil conditions: Effect and uptake of metalosates. *Journal of Plant Nutrition* 16: 813–823.

Frossard, E., M. Bucher, F. Mächler, A. Mozafar, and R. Hurrell. 2000. Potential for increasing the content and bioavailability of Fe, Zn and Ca in plants for human nutrition. *Journal of the Science of Food and Agriculture* 80: 861–879.

Gahoonia, T. S., O. Ali, A. Sarker, N. E. Nielsen, and M. M. Rahman. 2006. Genetic variation in root traits and nutrient acquisition of lentil genotypes. *Journal of Plant Nutrition* 29: 643–655.

Gao, X., C. Zou, X. Fan, F. Zhang, and E. Hoffland. 2006. From flooded to aerobic conditions in rice cultivation: Consequences for zinc uptake. *Plant and Soil* 280: 41–47.

Genc, Y., G. K. McDonald, and R. D. Graham. 2004. Differential expression of zinc efficiency. *Plant and Soil* 263: 273–282.

Gibson, R. S. 2006. Zinc: The missing link in combating micronutrient malnutrition in developing countries. *Proceedings of the Nutrition Society* 65: 51–60.

Gonzalez, D., A. Obrador, and J. M. Alvarez. 2007. Behavior of zinc from six organic fertilizers applied to a navy bean crop grown in a calcareous soil. *Journal of Agricultural and Food Chemistry* 55: 7084–7092.

Graham, R. D., J. S. Ascher, and C. S. Hynes. 1992. Selecting Zn-efficient cereal genotypes for soils of low zinc status. *Plant and Soil* 146: 241–250.

Graham, R. D., J. M. Humphries, and J. L. Kitchen. 2000. Nutritionally enhanced cereals: A sustainable foundation for a balanced diet. *Asia Pacific Journal of Clinical Nutrition* 9 (Suppl.): S91–S96.

Graham, R. D., R. M. Welch, and H. E. Bouis. 2001. Addressing micronutrient malnutrition through enhancing the nutritional quality of staple foods: Principles, perspectives and knowledge gaps. *Advances in Agronomy* 70: 77–142.

Grant, C. A., L. D. Bailey, J. T. Harapiak, and N. A. Flore. 2002. Effect of phosphate source, rate and cadmium content and use of *Penicillium bilaii* on phosphorus, zinc and cadmium concentration in durum wheat grain. *Journal of the Science of Food and Agriculture* 82: 301–308.

Gunes, A., A. Inal, M. S. Adak et al. 2007. Mineral nutrition of wheat, chickpea and lentil as affected by mixed cropping and soil moisture. *Nutrient Cycling in Agroecosystems* 78: 83–96.

Harland, B. F. and E. R. Morris. 1995. Phytate: A good or a bad food component? *Nutrition Research* 15: 733–754.

Hart, J.J., R.M. Welch, W.A. Norvell, and L.V. Kochian. 2002. Transport interactions between cadmium and zinc in roots of bread and durum wheat seedlings. *Physiologia Plantarum* 116: 73–78.

Honma, Y. and H. Hirata. 1978. A noticeable increase in Cd absorption by Zn deficient rice plants. *Soil Science and Plant Nutrition* 24: 295–297.

House, W. A. 1999. Trace element bioavailability as exemplified by iron and zinc. *Field Crops Research* 60: 115–141.

House, W.A. and R.M. Welch. 1989. Bioavailability of and interactions between zinc and selenium in rats fed wheat grain intrinsically labeled with Zn and Se. *Journal of Nutrition* 119: 916–921.

Hu, Y.X., C.G. Qu, and J.N. Yu. 2003. Zn and Fe fertilizers' effects on wheat's output. *Chinese Germplasm* 2: 25–28.

Hurrell, R.F. 2001. Modifying the composition of plant foods for better human health. In *Crop Science: Progress and Prospects,* J. Nösberger, H.H. Geiger, and P.C. Struik (Eds.), Bristol: CABI Publishing, pp. 53–64.

Johnson, S. E., J. G. Lauren, R. M. Welch, and J. M. Duxbury. 2005. A comparison of the effects of micronutrient seed priming and soil fertilization on the mineral nutrition of chickpea (*Cicer arietinum*), lentil (*Lens culinaris*), rice (*Oryza sativa*) and wheat (*Triticum aestivum*) in Nepal. *Experimental Agriculture* 41: 427–448.

Khan, H. R., G. K. McDonald, and Z. Rengel. 2003. Zinc fertilization improves water use efficiency, grain yield and seed Zn content in chickpea. *Plant and Soil* 249: 389–400.

Khoshgoftar, A. H., H. Shariatmadari, N. Karimian, M. Kalbasi, S.E.A.T.M. van der Zee, and D.R. Parker. 2004. Salinity and zinc application effects on phytoavailability of cadmium and zinc. *Soil Science Society of America Journal* 68: 1885–1889.

Köleli, N., S. Eker, and I. Cakmak. 2004. Effect of zinc fertilization on cadmium toxicity in durum and bread wheat grown in zinc-deficient soil. *Environmental Pollution* 131: 453–459.

Lee, D. Y., J. Schroeder III, and D. T. Gordon. 1988. Enhancement of copper bioavailability in the rat by phytic acid. *Journal of Nutrition* 118: 712–717.

Li, L., C. Tang, Z. Rengel, and F. S. Zhang. 2003. Chickpea facilitates phosphorus uptake by wheat from an organic phosphorus. *Plant and Soil* 248: 297–303.

Li, S.M., L. Li, F.S. Zhang, and C. Tang. 2004. Acid phosphatase role in chickpea/maize intercropping. *Annals of Botany* 94: 1–7.

Lindsay, W. L. 1979. *Chemical Equilibria in Soils*. New York: Wiley Interscience, pp. 449.

Liu, Z. 1994. The soil zinc distribution in China. *Chinese Agricultural Science* 27: 30–37.

Maftoun, M. and N. Karimian. 1989. Relative efficiency of two zinc sources for maize (*Zea mays* L.) in two calcareous soils from an arid area of Iran. *Agronomie* 9: 771–775.

Marschner, H. 1995. *Mineral Nutrition of Higher Plants*, 2nd ed., London: Academic Press, pp. 889.

McKenna, I. M., R. L. Chaney, and F. M. Williams. 1993. The effects of cadmium and zinc interaction on the accumulation and tissue distribution of zinc and calcium in lettuce and spinach. *Environmental Pollution* 79: 113–120.

McLaughlin, M. J., L. T. Palmer, K. G. Tiller, T. A. Beech, and M. K. Smart. 1994. Increased soil salinity causes elevated cadmium concentrations in field-grown potato tubers. *Journal of Environmental Quality* 23: 1013–1018.

Moraghan, J. T. 1994. Accumulation of zinc, phosphorus, and magnesium by navy bean seed. *Journal of Plant Nutrition* 17: 1111–1125.

Nambiar, K.M. and D. P. Motiramani. 1983. Zinc nutrition on the yield and nutrient composition of chickpea (*Cicer arietinum* L.) as related to four soil types. *Journal of Agronomy and Crop Science* 152: 165–172.

Oliver, D. P., R. Hannam, K. G. Tiller, N. S. Wilhelm, R. H. Merry, and G. D. Cozens. 1994. The effects of zinc fertilization on cadmium concentration in wheat grain. *Journal of Environmental Quality* 23: 705–711.

Ozturk, L., M. A. Yazici, C. Yucel et al. 2006. Concentration and localization of zinc during seed development and germination in wheat. *Physiologia Plantarum* 128: 144–152.

Paterson, J. E., G. F. Berndt, D. Cameron, and W. Rowbottom. 1991. Investigation into the response of barley to applied zinc. *Journal of the Science of Food and Agriculture* 54: 387–392.

Payne, G. G., D. C. Martens, C. Winarko, and N. F. Perera. 1988. Form and availability of copper and zinc following long-term copper sulfate and zinc sulfate applications. *Journal of Environmental Quality* 17: 707–711.

Peck, N., D. L. Grunes, R. M. Welch, and G. E. MacDonald. 1980. Nutritional quality of vegetable crops as affected by phosphorus and zinc fertilizers. *Agronomy Journal* 72: 528–534.

Prasad, A. S., A. Miale, Z. Farid, H. H. Sandstead, and A. R. Schulert. 1963. Zinc metabolism in patients with syndrome of iron deficiency anemia, hepatosplenomegaly, dwarfism and hypogonadism. *Journal of Laboratory and Clinical Medicine* 61: 537–549.

Prasad, B. and M. K. Sinha. 1981. The relative efficiency of zinc carriers on growth and zinc nutrition of corn. *Plant and Soil* 62: 45–52.
Raboy, V. and D. B. Dickinson. 1984. Effect of phosphorus and zinc nutrition on soybean seed phytic acid and zinc. *Plant Physiology* 75: 1094–1098.
Rashid, A., E. Rafique, N. Bughio, and M. Yasin. 1997. Micronutrient deficiencies in rainfed calcareous soils of Pakistan. 4. Zinc nutrition of sorghum. *Communications in Soil Science and Plant Analysis* 28: 455–467.
Rengel, Z., G. D. Batten, and D. E. Crowley. 1999. Agronomic approaches for improving the micronutrient density in edible portions of field crops. *Field Crops Research* 60: 27–40.
Rengel, Z. and R. D. Graham. 1995. Importance of seed Zn content for wheat growth on Zn-deficient soil. 2. Grain yield. *Plant and Soil* 173: 267–274.
Ryan, M. H., J. W. Derrick, and P. R. Dann. 2004. Grain mineral concentrations and yield of wheat grown under organic and conventional management. *Journal of the Science of Food and Agriculture* 84: 207–216.
Shaver, T. M., D. G. Westfall, and M. Ronaghi. 2007. Zinc fertilizer solubility and its effects on zinc bioavailability over time. *Journal of Plant Nutrition* 30: 123–133.
Shute, T. and S. M. Macfie. 2006. Cadmium and zinc accumulation in soybean: A threat to food safety? *Science of the Total Environment* 37: 63–73.
Srivastava, P. C. and U. C. Gupta. 1996. *Trace elements in crop production*. Lebanon, NH: Science Publishers, pp. 356.
Tarkalson, D. D., V. D. Jolley, C. W. Robbins, and R. E. Terry. 1998a. Mycorrhizal colonization and nutrition of wheat and sweet corn grown in manure-treated and untreated topsoil and subsoil. *Journal of Plant Nutrition* 21: 1985–1999.
Tarkalson, D. D., V. D. Jolley, C. W. Robbins, and R. E. Terry. 1998b. Mycorrhizal colonization and nutrient uptake of dry bean in manure and compost manure treated topsoil and untreated topsoil and subsoil. *Journal of Plant Nutrition* 21: 1867–1878.
Torun, A., I. Gültekin, M. Kalayci, A. Yilmaz, S. Eker, and I. Cakmak. 2001. Effects of zinc fertilization on grain yield and shoot concentrations of zinc, boron, and phosphorus of 25 wheat cultivars grown on a zinc-deficient and boron-toxic soil. *Journal of Plant Nutrition* 24: 1817–1829.
Vallee, B. L. and K. H. Falchuk. 1993. The biochemical basis of zinc physiology. *Physiology Review* 73: 79–118.
Verna, T. S. and R. S. Minhas. 1987. Zinc and phosphorus interaction in a wheat–maize cropping system. *Fertilizer Research* 13: 77–86.
Welch, R. M., W. A. House, and W. H. Allaway. 1974. Availability of zinc from pea seeds to rats. *Journal of Nutrition* 104: 733–740.
Welch, R. M. 2002. The impact of mineral nutrients in food crops on global human health. *Plant and Soil* 247: 83–90.
Welch, R. M. and R. D. Graham. 1999. A new paradigm for world agriculture: Meeting human needs—Productive, sustainable, nutritious. *Field Crops Research* 60: 1–10.
Welch, R. M. and R. D. Graham. 2004. Breeding for micronutrients in staple food crops from a human nutrition perspective. *Journal of Experimental Botany* 55: 353–364.
White, J. G. and R. J. Zasoski. 1999. Mapping soil micronutrients. *Fields Crops Research* 60: 11–26.
White, P. J. and M. R. Broadley. 2005. Biofortifying crops with essential mineral elements. *Trends in Plant Science* 10: 586–593.
Yang, X.-E., W.-R. Chen, and Y. Feng. 2007. Improving human micronutrient nutrition through biofortification in the soil–plant system: China as a case study. *Environmental Geochemistry and Health* 29: 413–428.

Yasrebi, J., N. Kariman, M. Maftoun, A. Abtahi, and A. M. Sameni. 1994. Distribution of zinc forms in highly calcareous soils as influenced by soil physical and chemical properties and application of zinc-sulfate. *Communications in Soil Science and Plant Analysis* 25: 2133–2145.

Yilmaz, A., H. Ekiz, I. Gültekin et al. 1998. Effect of seed zinc content on grain yield and zinc concentration of wheat grown in zinc-deficient soils. *Journal of Plant Nutrition* 21: 2257–2264.

Yilmaz, A., H. Ekiz, B. Torun et al. 1997. Effects of different zinc application methods on grain yield and zinc concentrations in wheat grown on zinc-deficient calcareous soils in Central Anatolia. *Journal of Plant Nutrition* 20: 461–471.

Zebarth, B. J., C. J. Warren, and R. W. Sheard. 1992. Influence of the rate of nitrogen fertilization on the mineral content of winter wheat in Ontario. *Journal of Agricultural and Food Chemistry* 40: 1528–1530.

Zhu, Y. L., G. R. Zhao, and Z. X. Yu. 1997. The effect of Zn-fertilizer on rice yield. *Anhui Agricultural Science Bulletin* 3: 36–37.

7 Farming Practices and Crop Quality

Implementation on Feed and Diet Quality

Tiequan Zhang and Yutao Wang

CONTENTS

7.1 INTRODUCTION

Nowadays, many farming practices, such as fertilization, crop rotation, intercropping, conservation tillage, and irrigation, are initially employed to improve crop productivity and/or sustainability of agricultural production worldwide. No doubt, fertilization and irrigation play the fundamental roles in feeding the world by increasing crop yield. Crop rotation and intercropping have been utilized as beneficial management practices for centuries, which in many cases have demonstrated the potential for contributing to increased crop productivity, especially on nitrogen-deficient soils (Peterson and Varvel 1989; Ayisi et al. 1997; Kouyaté et al. 2000; Lithourgidis et al. 2006; Li et al. 2007; Song et al. 2007; Kaye et al. 2007). On the other hand, the increasing concern about agricultural sustainability also favors the maintenance of both crop rotation and intercropping, which improves soil physical quality and nutrient availability, and reduces weeds population and heritable infection of crop diseases and insects (Miller and Holmes 2005; Song et al. 2007; Lithourgidis et al. 2007; Kaye et al. 2007). In recent years, conservation tillage as an essential component of sustainable agriculture has earned increased attention because it can improve soil quality, reduce labor and fuel consumption, contribute to global environmental conservation with the potential for carbon sequestration in soil, and thus facilitate more intensive crop rotation (Halvorson et al. 2004; Tsuji et al. 2006).

The awareness of enhancing crop-derived feed and diet quality rises due to the widespread incidence of nutrition deficiency in human beings, for example, insufficient protein and minerals (Welch and Graham 2002; Peters et al. 2003). Consequently, evaluating the effects of popular farming practices on crop-derived feed and diet quality in recent years has become a research priority, which may help us understand the possible roles that these farming practices play in enhancing crop quality. In a broad view, an understanding of the effects of farming practices on crop quality would help us develop and integrate the beneficial farming practices into a modern agricultural production system, which not only focus on increasing crop quantity and improving the agricultural environment, but also intend to produce quality crops for feed and diet production.

This chapter will review studies dealing with the effects of fertilization, crop rotation, intercropping, tillage, and irrigation on the contents of protein, oil, minerals, and other crop components related to food and diet quality. However, not every crop component mentioned will be discussed under every farming practice due to the limited availability of literature. In addition, the review will focus on long-term impacts (over 5 years), as the effects of consistent farming practices on feed and diet quality is often temporally gradual and cumulative.

7.2 INORGANIC FERTILIZATION

7.2.1 Protein

7.2.1.1 Nitrogen

Most commercial fertilizers supply nitrogen (N) in soluble forms that provide either directly ammonium cation and/or nitrate anion, such as ammonium nitrate, or indirectly through transformation after application, such as urea. As a general pattern, plants adapted to acid soils or to low soil redox potential have a preference for ammonium. In contrast, plants with preference to calcareous soils utilize preferentially nitrate (Marschner 1995). Nitrogen application not only promotes increases of crop yields, but also greatly alters plant composition much more than any other mineral nutrients (Marschner 1995).

Numerous studies demonstrated that the protein content (PC) in crops (i.e., wheat, corn, rice, oilseed rape, and pearl millet) increases with increases of N application (Ayoub et al. 1994; López-Bellido et al. 1998; Kennedy et al. 2002; Ames et al. 2003; Leesawatwong et al. 2005; Rathke et al. 2005; Farrer et al. 2006; Miao et al. 2006). However, we selected wheat as a model cereal crop and soybean as a model legume crop to discuss the effect of N application rate, time, and method on PC in detail.

Increasing N fertilization rate contributes significantly to the increase in grain PC, especially when N fertilizer rates satisfy the requirements of both yield and protein formation (Johnson and Mattern 1987). Moreover, the favorable response of PC in wheat grain to N application rate can be either linear (Guttieri et al. 2005) or quadratic (Oikeh et al. 1998; Miao et al. 2006). Farrer et al. (2006) reported that N application accounted for the majority (51.4%) of wheat PC variability in the southeastern United States with respect to seasonal temperatures, moisture, and soil types. Therefore, there was no surprise that N application has been considered the most efficient way to improve wheat grain PC. However, others have noted that after incremental additions of fertilizer N, wheat grain PC reaches its maximum and then remains constant despite the ample N supply (Ryan et al. 1997; Lloveras et al. 2001; Barneix 2007). This phenomenon is related to two main regulatory points during grain filling when plant N status is ample (Barneix 2007). First, the N uptake transporters in the roots are depressed due to the high amino acid concentration in the tissues, which results in a low N uptake. Secondly, a high amino acid concentration keeps the cytokines level high, repressing leaf protein degradation and decreasing amino acid export to the phloem, which in turn inhibits protein synthesis. As a consequence, wheat grain PC cannot be increased under high N application. However, when fertilizer N supply is below the rate producing the maximum yield, wheat gain PC can decrease with increased fertilizer N addition due to dilution effect resulted from substantial yield increases (Bly and Woodard 2003).

It has been reported that the N application time has a significant effect on optimizing PC in wheat grain. Generally, N application later in the season is more effective than earlier applications in increasing PC. For example, fertilizer N applied in the spring is more effective at increasing PC than previous fall-applied fertilizer N in winter wheat grain (Kelley 1995). Applications of fertilizer N near anthesis are more efficient at increasing grain PC than earlier applications (Ottman et al. 2000). However, for a given amount of total spring fertilizer N, applying ≥ 80% of the fertilizer at growth-stage 25 resulted in a significant elevation in grain PC in soft red winter wheat, compared to the same application rate at growth-stage 30 (Farrer et al. 2006).

Nitrogen application methods also play a role in affecting PC in wheat grain. Ayoub et al. (1994) observed that splitting the N application (60% at seeding and 40% at anthesis) improved the PC of wheat flour by 3–8%, compared with single N application at seeding. According to Borghi et al. (1997), the beneficial effect of N splitting in the Mediterranean climate may be related to reduction of N losses and to a better translocation of preanthesis assimilates to the grain. Foliar application of urea solution at flowering has been found effective in increasing PC in wheat grain (Terman et al. 1969). Typical amounts of N supplied as late season foliar urea vary between 30 and 50 kg N ha^{-1} (Gooding and Davies 1997). A foliar N application resulted in a higher-grain PC than broadcast as dry granular fertilizer (Alkier et al. 1972; Strong 1982).

Compared with cereal crops, legume crops, for example, soybean, show inconsistent responses to N application, which may be related to their capability of N_2 fixation. Ray et al. (2006) reported that N from N_2 fixation accounts for 50 to 75% of the soybean demand for N. Soybean has been classified as a poorer responder to N fertilization compared with other grain crops, although responses have been observed in low-N testing soils (Kamprath 1974). Weber (1966) reported small increases in seed PC with 56 and 168 kg N ha^{-1} applied at planting. Purcell et al. (2004) reported a general trend of decreasing PC with 112 kg N ha^{-1} applied at six unfolded trifoliolate leaves (V6), full bloom (R2), or V6 and R2 (stages described by Fehr et al. 1971). Similarly, Ray et al. (2006) showed that fertilizer N significantly reduced PC by 1.9–2.7% in soybean seeds. However, Gutiérrez-Boem et al. (2004) found that late season applications of fertilizer N (50 or 100 kg N ha^{-1}) had no effect on PC.

7.2.1.2 Other Inorganic Fertilizers

Compared with N, limited information is available on the effects of other inorganic fertilizers on PC in crops. Across 112 field trials from 1994 to 2001, Haq and Mallarino (2005) concluded that responses of soybean grain PC to P and K application were infrequent, small, and inconsistent (positive or negative), which is in agreement with the study by Ham et al. (1973). This observation may be related with the fact that soybean is a poorer responder to P and K fertilization compared with other grain crops (Kamprath 1974). However, Gobarah et al. (2006) found that increased addition of fertilizer P from 30 to 60 kg P_2O_5 ha^{-1} significantly increased PC of seed groundnut, similarly observed with a foliar application of zinc (Zn) from 0.5 to 1.0 g L^{-1}.

The effect of sulfur (S) fertilization on PC in oil crops has been studied. For example, Malhi et al. (2007) found that PC in the seeds of four oilseed cultivars increased significantly with increasing S rate up to 30 kg S ha^{-1}. Likewise, Malhi and Gill (2006) concluded that PC in seeds tended to increase in response to S fertilization up to 15 kg S ha^{-1} averaged across four canola cultivars. Earlier studies with rapeseed in Saskatchewan, Canada, have also shown PC increases with application of fertilizer S on S-deficient soils (Malhi et al. 2007). However, some studies with canola showed a slight increase to some extent of decrease of PC with application of fertilizer S even on S-deficient soils (Asare and Scarisbrick 1995; Malhi and Gill 2002; Grant et al. 2003). Others believed that the effect of S on PC may depend on the N:S ratio (Grant et al. 2003). Activities of ATP-sulphurylase and nitrate reductase, which are enzymes involved in sulphate and nitrate assimilation, respectively, increased by the co-application of S and N as compared with N alone (Grant et al. 2003). Improper or unfavorable ratios of N to S decreased the activities of both ATP-sulphurylase and nitrate reductase.

With integrated application of N, P, and K, Zhang et al. (2007) found that 47-year consistent fertilization (16.8 kg N ha^{-1}, 67.2 kg P ha^{-1}, and 33.6 kg K ha^{-1} applied at planting every year) significantly increased crude PC in corn grain, but had no effect on crude PC of oats. Poole et al. (1983) observed increases of soybean grain PC as a result of N-P-K-S foliar fertilization.

7.2.1.3 Effects of Inorganic Fertilization on Compositional Distribution of Protein

Inorganic fertilization not only affects the amounts of protein in crops, but also influences the quality of protein. In a detailed study on protein composition of 13 wheat cultivars, Wieser and Seilmeier (1998) found that increased N fertilization decreased the proportion of hydrophobic protein (ω-gliadins, low-molecular-weight subunits of glutenin) and increased the proportions of hydrophilic proteins (ω-gliadins, high-molecular-weight glutenins). Significant increases in ω-gliadins together with the reduced increases in high-molecular-weight glutenin subunits were also reported under S deficiency conditions (Wrigley et al. 1984). Leesawatwong et al. (2005) found that increased N application significantly resulted in increased glutlin, the main storage protein of rice, which is relatively rich in lysine. Nitrogen applications may thus be able to increase the nutritional quality of milled rice by increasing protein concentration. Moreover, the responses of various proteins to fertilizer addition may be related to their amino acid compositions. Tatham and Shewry (1995) classified the glutamine- and praline-rich proteins into three groups based on S content: the S-poor ω-gliadins; the high-molecular-weight glutein subunit with intermediate amounts of S; and the S-rich α-gliadins, γ-gliadins and low-molecular-weight glutenin subunit. Application of fertilizer N is reported to affect the N:S ratios of wheat, and may lead to S deficiency and impair the synthesis of S-containing amino acids (Garrido-Lestache et al. 2005). Bulman et al. (1994) observed that high N fertility caused reductions in the proportion of lysine, cysteine, and methionine in protein of spring barley grain, which reduced accordingly the protein quality.

7.2.1.4 Factors Controlling the Relationships between Fertilizer Nitrogen and Protein Content

There are other factors that have been reported to primarily affect the relationships between inorganic fertilization and crop PC. First, the effects of fertilizer N on grain PC depends on genotypes that vary in the capability of extracting soil-available N, especially during stem elongation (Deleens et al. 1994). Secondly, Campbell et al. (1993) reported a strong interaction between soil water use and crop fertilizer response in semiarid conditions. Similarly, Denys et al. (2006) investigated the effect of topography on PC in soft red winter wheat grain, and observed that grain PC was significantly different with respect to slope position, with increases from the lower to the upper position. This was likely due to proportional yield reduction arisen from lower levels of plant available water in the upper position (Manning et al. 2001). Finally, the degree of influence of N application on crop PC is related to weather conditions (i.e., annual temperature, precipitation, and solar radiation), and soil type and related properties (residual soil N, organic matter content, and hydraulic conductivity) (López-Bellido et al. 2001; Halvorson et al. 2004; Delin et al. 2005; Ray et al. 2006).

7.2.2 Oil

Inconsistent responses of oil content in crop seeds to N fertilizer application have been observed. The negative influence of N fertilization on oil content of crop seeds was reported by Mason and Brennan (1998), Cheema et al. (2001), and Jackson (2000). Fox example, a 6-year field study showed that the lowest oil content (43.8–44.1%) occurred under high N rates (240 kg ha^{-1}) in winter oilseed rape, and the highest oil yield (46.8–47.7%) were observed on rape grown on unfertilized soils (Rathke et al. 2005). However, Ray et al. (2006) found N application resulted in increases of oil concentration by 2.2–2.7% in soybean seeds. Similarly, Purcell et al. (2004) reported a general trend of increase in oil concentration of soybean seed with 112 kg ha^{-1} of N applied. At the same time, some studies showed no response of oil content in the crop seeds to N application (Mason and Brennan 1998; Brennan et al. 2000). Perhaps soil N status and amounts of N applied explain the inconsistent response of seed oil content to N application. For instance, Welch (1969) found that N fertilization significantly increased corn oil content, but there were no significant differences in oil content across the higher N rates (67, 134, 201, and 268 kg ha^{-1}).

Sulfur deficiency has been associated with reduced oil concentration in canola seeds (Jackson 2000; Ahmad et al. 2000) because of the reduction acetyl-CoA carboxylase activity (Ahmad et al. 2000). Oil concentration in oil crop seeds increased with application of fertilizer S, if soils were deficient in available forms of S (Malhi and Gill 2002; Grant et al. 2003; Malhi and Gill 2006; Malhi et al. 2007). This observation was confirmed by Jackson (2000) who found that addition of S increased oil content in canola seeds on one of five sites that was with the lowest soil available S (50 kg ha^{-1} in 0–30 cm soil depth), but had no effects on oil levels in canola seeds on the other four sites that were with sufficient soil available S (67–202 kg ha^{-1} in 0–30 cm soil depth) in the western triangle area of Montana in the United States.

With respect to soil P and K content, the responses of oil content in soybean grains were infrequent, small, or inconsistent (positive or negative) (Ham et al. 1973; Haq and Mallarino 2005), which may be due to the poor response of soybean to P and K fertilization compared with other grain crops (Kamprath 1974). However, Jones and Lutz (1971) showed decreases in oil concentration of soybean grain with increased P and K fertilization. Cheema et al. (2001) also reported that increased P application decreased oil content in canola seeds. Likewise, Poole et al. (1983) showed reduced oil concentration as a result of N-P-K-S foliar fertilization at the reproductive stages.

In regard to Zn, Khurana and Chatterjee (2001) found that the oil content of sunflower seeds grown in refined sand tended to increase at first and then decrease with increasing Zn supply levels between 0.65 and 65000 μg Zn L^{-1}, with the highest oil content at 0.65 mg Zn L^{-1}. Similarly, Mirzapour and Khoshgofar (2006) observed that the oil content of sunflower seeds was significantly increased with soil application of up to 20 kg Zn ha^{-1}, above which the oil content decreased. However, the sunflower seed soil content was unaffected by foliar spraying of 0.5 kg Zn ha^{-1} using $ZnSO_4$.

7.2.3 Minerals

In general, large increments in crop yield due to N fertilization tend to dilute the mineral content in crops. Feil et al. (2005) reported that application of N fertilizer reduced the concentrations of Ca and Zn in maize grains. Hopkins et al. (1994) concluded that applied N reduced concentrations of Ca, Mn, and S in herbage. McKenzie and Jacobs (2002) and Lambert and Toussaint (1978) showed that adding N fertilizers decreased Ca and P concentrations, respectively, in herbage, while Pietz et al. (1978) observed highly significant negative correlations between applied fertilizer N and K and Na concentrations in corn grain.

At the same time, the increased concentrations of some minerals have also been observed with N application, for example, P, K, S, and Mg content in herbage (McKenzie and Jacobs 2002) and Mn in corn grains (Pietz et al. 1978; Feil et al. 2005). According to Feil et al. (2005) and Hopkins et al. (1994), the effect of N fertilization on mineral composition of plant tissues may depend on the N forms. Generally, the bioavailability of some micronutrients, for example, Mn, Cu, and Zn, to crops depends on soil pH; a decrease can consequently stimulate their uptake (Srikumar and Öckerman 1990). The form of N supply has the most predominant influence on rhizosphere pH (Marschner 1995). Ammonium supply decreases pH in the rhizosphere, while nitrate supply increases pH in the rhizosphere. Therefore, application of ammonium is considered to have positive effect on uptake of minerals by plants (Srikumar and Öckerman 1990). Feil et al. (2005) pointed out that N fertilization leads to higher concentrations of grain Mn, given that the dilution effect caused by increased increments in yield is less than the ammonium effect. Nevertheless, no response of minerals to N application has been reported. For example, Feil et al. (2005) concluded that the concentrations of most mineral nutrients (e.g., P, K, Mg, and Cu) in maize grains did not respond to the increase of N rates using ammonium sulphate as N source. Especially, little or no clear effect of N application on the concentration of P and K content in maize grain has been observed by many researchers (Thiraporn et al. 1992; Feil et al. 1993; Ahmadi et al. 1993; Alfoldi et al. 1994; Feil et al. 2005).

The effect of K application on mineral content has been investigated in grass forages. In most cases, the increases of K concentrations of grass in response to increased K fertilizer have been shown in many studies (Whitehead 2000; Swift et al. 2007). However, the responses of other minerals to K fertilization in grass were inconsistent. Mosquera-Losada et al. (2004) and Roche et al. (2002) found that K fertilization did not affect P, Ca, Mg, Zn, and Cu content of pastures planted to white clover (*Trifolium repens* L.) and perennial ryegrass (*Lolium perenne* L.) sward. At the same time, the inconsistent effect of K supply on S content in grass forages was reported by Swift et al. (2007), Roche et al. (2002), and McLaren (1976). One of the reasons for the inconsistent effect of K application on minerals may be related to the balance between K requirement of plants and K application rates. For example, Whitehead (2000) found that application of K fertilizer in amounts greater than required for maximum yield decreased the Mg concentration in forage.

Meanwhile, S and Zn fertilization may have effects on contents of some minerals in crops. Grant et al. (2003) concluded that canola seed S concentration generally increased with increases in soil available S. However, Jackson (2000) reported that S application increased canola seed S concentrations only at sites where significant seed yield responded to S application, and not in seed harvested from sites where S application had no effects on canola seed yields. Khan et al. (2003) reported that increasing Zn levels of nutrient solutions greatly increased seed Zn concentration of chickpea under both well-watered and water-stress conditions, which is a beneficial nutritional trait for human nutrition.

Zhang et al. (2007) observed that long-term (47 years) consistent fertilization of combining N, P, and K application increased contents of P, Mg, and Mn in oats, while it decreased Fe and Cu contents in oats compared with nonfertilized treatments. Srikumar and Öckerman (1990) found that an increased application of inorganic N-P-K fertilizer increased the content of Fe in potato but decreased the content of Zn, with no effect on the concentrations of Cu and Mg. Moreover, they also found that the addition of ammonium nitrate along with an increased application of N-P-K multielement fertilizer increased the contents of Zn and Fe, whereas the Cu content remained unchanged.

7.3 CROP ROTATION

Numerous studies have demonstrated the positive eff ects of legume crops, such as soybean, on increasing grain crude PC of subsequent cereals in the long-term legume-cereal rotations (Borghi et al. 1995; López-Bellido et al. 2001; Rathke et al. 2005). For example, López-Bellido et al. (2001) evaluated the effect of crop rotation, including wheat–sunflower, wheat–chickpea, wheat–faba bean, wheat–fallow, and continuous wheat-on-wheat PC over a 6-year period. They found that wheat PC was enhanced when a legume was included in the rotation, particularly with faba bean. Zentner et al. (2001) evaluated the effects of different cropping systems (including continuous wheat and wheat–lentil) on wheat grain PC over the 1979 to 1997 period, and found that spring wheat PC was higher (average 13.4 g kg^{-1} more) in 11 of these 18 years when grown on lentil stubble compared with wheat stubble in Swift Current, SK, Canada. Evans et al. (1991) reported that wheat grown after pea in the semiarid southeastern Australia produced an average of 12% more grain N than growing continuous wheat. Moreover, the harvesting time of pea may play a significant role in enhancing PC of subsequent cereal crops. Miller et al. (2006) implied that midseason pea harvest timing resulted in 9% greater grain protein of the subsequent wheat, compared with the mature pea harvest timing.

In contrast, limited information is available on the effect of legume-excluded rotation systems on crop PC. Miller et al. (2002a) and Gan et al. (2003) observed a 5–6% PC increase in wheat after oilseed, compared with wheat after wheat. However, Gan et al. (2003) indicated that pulse crops increased greater PC in subsequent wheat than did oilseed. Zhang et al. (2007) observed that crude PC of corn grain grown in a long-term rotation system, corn–oats–alfalfa–alfalfa, increased by 10 and 11% in nonfertilized and continuously fertilized soils, respectively. However, no significant cropping sequence effects were observed on forage crude PC.

Increased PC in crop rotation systems are likely due to the following reasons: increases in symbiotically fixed N contained in legume crop residues, gradual release of mineralizable N as crop residues decomposition during the growing season, and increased availability of soil water for the subsequent crops (López-Bellido et al. 2001; Miller et al. 2002a; Gan et al. 2003; Rathke et al. 2005; Miller et al. 2006). Miller et al. (2006) observed that midseason harvest timing of pea increased soil available N by 30–39 kg NO_3-N ha^{-1}, and soil available water by 19–39 mm to wheat in the subsequent spring conditions, which led to 9% greater grain protein. According to Campbell et al. (1992), there was a cumulative enhancement of soil N supplying power after growing lentil due to the pulse residual contribution in a long-term wheat–lentil rotation. Rathke et al. (2005) found that incorporation of pea into crop rotation instead of winter barley resulted in saving of fertilizer N to winter oilseed rape by 25 kg ha^{-1}. Jensen (1994) reported that winter oilseed rape recovered 13% of the pea residue-N during the first three months of growth. However, Miller et al. (2002b) and Miller and Holmes (2005) indicated that the positive benefits of pulse crops on increasing PC of subsequent crops resulted from the increased soil N rather than the increased available water. Nevertheless, Gan et al. (2003) indicated that these two factors accounted for 12 to 24% of the PC increase. Therefore, other factors might contribute to the protein increases in subsequent crops with a crop rotation system, such as increased soil labile organic matter and improved rhizosphere microbial activity (Biederbeck et al. 1994, 2000).

Few studies have dealt with the effects of crop rotation on other crop quality parameters, for example, oil and minerals, besides PC. Rathke et al. (2005) observed that rapeseed grown after winter barley produced significantly higher oil content than grown after pea. Zhang et al. (2007) found that the 47-year crop rotation (corn–oats–alfalfa–alfalfa) increased the contents of P and K, and decreased Mg, Fe, and Zn contents in corn kernels. Khoshgoftarmanesh and Chaney (2007) found that Cd concentrations in wheat grain grown were seven times lower after sunflower than those grown after cotton. However, Zn concentrations in wheat grains were significantly greater when grown after sunflower than after cotton in salt-stressed soils of central Iran.

7.4 INTERCROPPING

7.4.1 Protein

In most cases, leguminous crops (i.e., pea, lentil) demonstrate superior potential for enhancing PC in cereal crops (i.e., oat, wheat, barley) when they are intercropped (Robinson 1960; Walton 1975; Droushiotis 1989; Chapko et al. 1991; Ayisi et al. 1997; Carr et al. 2004). For instance, Carr et al. (2004) observed that forage CP was 35 g kg^{-1} more for barley–pea or oat–pea intercrops, compared with a monocrop of either cereal species. Sharma and Gupta (2002) found that pearl millet as a monocrop accumulated significantly less protein in grains as compared to that when it was intercropped with legumes, including cluster bean, cowpea, and mungbean. Similarly, Hauggaard-Nielsen et al. (2006) observed that the content of protein and gluten in wheat intercropped with pea was higher than that in wheat grown as monoculture. Moreover, as the proportion of legumes in a cereal–legume intercropped system increased, the content of protein and gluten in cereal crop increased (Caballero et al. 1995; Hauggaard-Nielsen et al. 2006; Lithourgidis et al. 2007), which indicated that the relative cropping frequency of an intercropping system may be used as a regulator to obtain specific PC. Thus, it is expected that the highest crude protein was measured when the legume was grown as a monocrop or when it composed a high proportion as an intercrop with cereals (Caballero et al. 1995; Lithourgidis et al. 2007).

On the other hand, legume as an intercrop does not always enhance cereal forage PC. Carr et al. (1998) observed that oat–pea and barley–pea intercrops did not increase forage PC grown in soils with more than 55 kg N ha^{-1} in top soil. However, forage PC increased when grown in soils with less than 35 kg N ha^{-1} in the depth of 0–6 cm, which is similar to the results reported by Robinson (1960). Lauriault and Kirksey (2004) found that hairy vetch had no effect on PC of rye, barley, wheat and triticale forages, although it enhanced oat PC in intercropped systems. However, both hairy vetch and winter pea increased PC of the grass–legume mixture when intercropped with wheat, triticale, and oat forages. The findings indicate that soil N status and intercropping species might play significant roles in influencing cereal forage PC in a cereal–legume intercropping system.

7.4.2 Fiber

Studies have demonstrated that intercropping has inconsistent effect on fiber content of pastures. Lithourgidis et al. (2007) observed no differences when wheat or barley was grown either as a monocrop or as intercrops with vetch for neutral detergent fiber (NDF), acid detergent fiber (ADF), or hemicellulose content. Velazquez-Beltran et al. (2002) reported that there were no differences in ADF composition among monocrop of oat, common vetch–oat, and common vetch–oat–maize intercrops. Similarly, Carr et al. (2004) found that intercropping pea with barley did not affect forage ADF concentration compared with a barley mono-crop. However, Chapko et al. (1991) stated that forage intake potential, largely determined by NDF, was increased by intercropping pea with barley or oat. In contrast, Carr et al. (2004) reported that intercropping pea with oat or barley reduced the NDF concentration of forage compared with an oat or barley monocrop. Others have also noted that the increasing common vetch proportion in forage was usually associated with a decrease in NDF (Caballero et al. 1995; Haj-Ayed et al. 2000).

7.4.3 Minerals

There is limited information on how the mineral composition of crops is affected by intercropping. Gunes et al. (2007) observed that intercropping of wheat with chickpea improved the concentrations of N, P, K, and Fe in wheat grains and N, P, K, Fe, Zn, and Mn in chickpea seeds. Carr et al. (2004) found Ca was 2.78 g kg^{-1} greater in forage produced by a barley–pea intercrop than a monoculture of oat, and 3.45 g kg^{-1} greater in forage produced by an oat–pea intercrop than a monoculture of barley.

7.5 TILLAGE

In most cases, the PC in crops tends to decrease under no-tillage (NT) and reduced tillage (RT) conditions compared with conventional tillage (CT) (Johnston et al. 1995; López-Bellido et al. 1998; López-Bellido et al. 2001; Vita et al. 2007). Vita et al. (2007) noted that PC (15.5–19.6%) in durum wheat grain under CT was higher than wheat (11.4–14.7%) grown under NT in southern Italy. Some researchers attributed the reduction of PC under NT to the depression of soil available N, compared to more available N under CT (Johnston et al. 1995; López-Bellido et al. 2001). In general, soil N availability declines when tillage is first reduced (Carr et al. 2003). López-Bellido et al. (1996) revealed higher NO_3^--N content within the top 60 cm of soil for CT than for NT. The reduction in soil N availability under NT may result from the accumulation of crop residue at the soil surface and subsequently cool spring soil temperature. Both of these factors (PC and NT) may contribute to the slowdown of microbial conversion of organic N to plant-available forms and immobilization of soil N (Lamb et al. 1985; Westfall et al. 1996). In addition, other studies indicated that the protein depression under NT in wet years may be related to denitrification early in the growing season, which reduced available soil N (Carr et al. 2003). Nevertheless, greater amounts of soil N may occur in the RT system following an adjustment period, as occurred during the last 14 years of a 36-year study (Johnston et al. 1995). In fact, due to the complex interactions between climatic and edaphic factors altering the impact of tillage on soil N content, soil N may not increase when tillage is reduced in some environments (Carr et al. 2003). Some studies have also indicated that crop PC was not affected by tillage system (Campbell et al. 1998; McConkey et al. 1996; Carr et al. 2003; Grant et al. 2003; Temperly and Borges 2006), which may be partly explained by the observation that soil N was unaffected by tillage systems (Carr et al. 2003).

7.6 IRRIGATION

In general, the effect of moisture stress on PC is mainly through its influence on crop yield (Rao et al. 1993; Ottman et al. 2000). Water deficits, in most cases, have been associated with increased PC due to the dilution effect (Entz and Fowler 1989; Johnson and Mattern 1987; Guttieri et al. 2000), which was supported by analyzing the relationship between climate and the PC of wheat flour over the 10-year period (Rao et al. 1993). PC tends to decrease as the amount of water applied increases (Nielsen 2004; Guttieri et al. 2005). López-Bellido et al. (1998) observed in a 3-year study that the wheat grain PC was always inversely related with the amount of rainfall. Nielsen (2004) found that PC of knave forage tended to decrease from 279 g kg^{-1} to 163 g kg^{-1} with elevating water availability. According to Gooding and Davies (1997), a rainfall prior to grain filling has been thought to encourage dilution of early N reserves by vegetative proliferation in arid environments, whereas a later rainfall during grain filling can cause N dilution by extending leaf life and maintaining photosynthesis and carbohydrate translocation. However, excess irrigation has also been reported to decrease PC (Ramon Fernandez and Laird 1959; Robinson et al. 1979). Robinson et al. (1979) attributed the decreased PC to nitrate leaching caused by excessive irrigation. In contrast, irrigation has been reported to enhance PC in some cases (Yamada et al. 1972; Rajeswara Rao and Prasad 1987; Puri et al. 1989). According to Slukhai and Zhabits (1970), the elevation of grain protein was attributed to mobilizing surface-applied fertilizer under irrigation conditions.

Harder et al. (1982) imposed various moisture stress treatments on maize after silking, and found that even with a reduction of grain yield by up to 33%, the concentrations of grain P and K remained unchanged. Similarly, in spite of high grain yield of maize, Feil et al. (2005) did not observe a decrease in the mineral contents of P, K, Mg, Ca, Mn, Zn, and Cu in the grain across the different levels of water supplied at preanthesis. Additionally, increases in contents of NDF (229 to 478 g kg^{-1}) and ADF (168 to 314 g kg^{-1}) with increased water use were observed (Nielsen 2004).

7.7 CONCLUSIONS

Farming practices can greatly impact crop quality, which implies that nutritional quality of crop-derived diet and feed may be agronomically manageable through inorganic fertilization, rotation, intercropping, tillage, and irrigation to satisfy the health needs of animals and human beings. However, additional questions on producing high nutritional feed and diet by optimizing farming practices will need to be addressed with future studies. For example, research is needed to fine-tune the long-term effects of farming practices because their effects on crop growth can be temporarily gradual and consequently cumulative through changes in soil chemical, physical, and nutritional properties. Secondly, a series of field trails adapted to different site conditions are needed to determine the beneficial farming practices for a crop with high feed and diet quality within a particular region because many factors, for example, local weather conditions, soil type, and crop genotype, may interact with the direct effects of management practices on crop quality. Finally, future studies are also needed to explore the reasons explaining the effect of farming practices on crop feed and diet quality, which can be useful to help us understand, develop, and integrate these practices into a sustainable agricultural production system that not only focuses on increasing crop quantity and improving agricultural environment, but also produces crops with improved quality.

REFERENCES

Ahmad, A., I. Khan, and M. Z. Abdin. 2000. Effect of sulphur fertilisation on oil accumulation, acetyl-CoA concentration, and acetyl-CoA carboxylase activity in the developing seeds of rapeseed (*Brassica campestris* L.). *Australian Journal of Agricultural Research* 51: 1023–1029.

Ahmadi, M., W. J. Wiebold, and J. E. Beuerlein. 1993. Grain yield and mineral composition of corn as influenced by endosperm type and nitrogen. *Communications in Soil Science and Plant Analysis* 24: 2409–2426.

Alfoldi, Z., L. Pinter, and B. Feil. 1994. Nitrogen, phosphorus and potassium concentrations in developing maize grains. *Journal of Agronomy and Crop Science* 172: 200–206.

Alkier, A. C., G. J. Racz, and R. J. Soper. 1972. Effects of foliar- and soil-applied nitrogen and soil nitrate-nitrogen level on the protein content of Neepawa wheat. *Canadian Journal of Soil Science* 52: 301–309.

Ames, N. P., J. M. Clarke, J. E. Dexter, S. M. Woods, F. Selles, and B. Marchylo. 2003. Effects of nitrogen fertilizer on protein quantity and gluten strength parameters in durum wheat (*Triticum turgidum* L. var. durum) cultivars of variable gluten strength. *Cereal Chemistry* 80: 203–211.

Asare, E. and D.H. Scarisbrick. 1995. Rate of nitrogen and sulphur fertilizers on yield, yield components and seed quality of oilseed rape (*Brassica napus* L.). *Field Crops Research* 44: 41–46.

Ayisi, K. K., D. H. Putnam, C. P. Vance, M. P. Russelle, and D. L. Allan. 1997. Strip intercropping and nitrogen effects on seed, oil, and protein yields of canola and soybean. *Agronomy Journal* 89: 23–29.

Ayoub, M., S. Guertin, J. Fregeau-Reid, and D. L. Smith. 1994. Nitrogen fertilizer effect on breadmaking quality of hard red spring wheat in eastern Canada. *Crop Science* 34: 1346–1352.

Barneix, A. J. 2007. Physiology and biochemistry of source-regulated protein accumulation in the wheat grain. *Journal of Plant Physiology* 164: 581–590.

Biederbeck, V. O., H. H. Janzen, C. A. Campbell, and R. P. Zentner. 1994. Labile soil organic matter as influenced by cropping practices in an arid environment. *Soil Biology and Biochemistry* 26: 1647–1656.

Biederbeck, V. O., N. Z. Lupwayi, K. G. Hanson, W. A. Rice, and R. P. Zentner. 2000. Effect of long-term rotation with lentils on rhizosphere ecology and on entophytic rhizobia in wheat. 80. In *Abstracts*, North Am. Conf. on Symbiotic Nitrogen Fixation, 17th, 80. July 23–28, 2000. Laval Univ., Sainte-Foy, QC, Canada.

Bly, A. G. and H. J. Woodard. 2003. Foliar nitrogen application timing influence on grain yield and protein concentration of hard red winter and spring wheat. *Agronomy Journal* 95: 335–338.

Borghi, B., M. Corbellini, C. Minoia, M. Palumbo, N. Di Fonzo, and M. Perezin. 1997. Effects of Mediterranean climate on wheat bread-making quality of wheat (*Triticum aestivum* L.). *European Journal of Agronomy* 4: 145–154.

Borghi, B., Giordani, G., Corbellini, M., P. Vaccino, M. Guermandi, and G. Toderi. 1995. Influence of crop rotation manure and fertilizers on bread making quality of wheat (*Triticum aestivum* L.). *European Journal of Agronomy* 4: 37–45.

Brennan, R. F., M. G. Mason, and G. H. Walton. 2000. Effect of nitrogen fertilizer on the concentrations of oil and protein in canola (*Brassica napus*) seed. *Journal of Plant Nutrition* 23: 339–348.

Bulman, P., C. G. Zarkadas, and D. L. Smith. 1994. Nitrogen fertilizer affects amino acid composition and quality of spring barley grain. *Crop Science* 34: 1341–1346.

Caballero R., E. L. Goicoechea, and P. J. Hernaiz. 1995. Forage yields and quality of common vetch and oat sown at varying seeding ratios and seeding rates of common vetch. *Field Crops Research* 41: 135–140.

Campbell, C. A., R. P. Zentner, F. Selles, B. G. McConkey, and F. B. Dyck. 1993. Nitrogen management for spring wheat grown annually on zero-tillage: yield and nitrogen use efficiency. *Agronomy Journal* 85: 107–114.

Campbell, C. A., A. G. Thomas, V. O. Biederbeck et al. 1998. Converting from no-tillage to pre-seeding tillage: influence on weeds, spring wheat grain yields and N, and soil quality. *Soil and Tillage Research* 46: 175–185.

Campbell, C.A., R. P. Zentner, F. Selles, V. O. Biederbeck, and A. J. Leyshon. 1992. Comparative effects of grain lentil-wheat and monoculture wheat on crop production, N economy and N fertility in a brown chernozem. *Canadian Journal of Plant Science* 72: 1091–1107.

Carr, P. M., G. B. Martin, J. S. Caton, and W. W. Poland. 1998. Forage and nitrogen yield of barley–pea and oat–pea intercrops. *Agronomy Journal* 90: 79–84.

Carr, P. M., R. D. Horsley, and W. W. Poland. 2003. Tillage and seeding rate effects on wheat cultivars: I. Grain production. *Crop Science* 43: 202–209.

Carr, P. M., R. D. Horsley, and W. W. Poland. 2004. Barley oat and cereal–pea mixture as dryland forages in the northern Great Plains. *Agronomy Journal* 96: 677–684.

Chapko, L. B., M. A. Brinkman, and K. A. Albrecht. 1991. Oat, oat–pea, barley, and barley–pea for forage yield, forage quality, and alfalfa establishment. *Journal of Production Agriculture* 4: 486–491.

Cheema, M. A., M. A. Malik, A. Hussain, S. H. Shah, and S. M. A. Basra. 2001. Effects of time and rate of nitrogen and phosphorus application on the growth and the seed and oil yields of canola (*Brassica napus* L.). *Journal of Agronomy and Crop Science* 186: 103–110.

Deleens, E., L. Barthes, L. Hannachi, and A. Bousser. 1994. Un nouveau outil: l'analyze de la composition isotopique. Perspect. *Agricoles* 190: 108–122.

Delin, S., B. Lindén, and K. Berglund. 2005. Yield and protein response to fertilizer nitrogen in different parts of a cereal field: Potential of site-specific fertilization. *European Journal of Agronomy* 22: 325–336.

Denys, C. J., I. P. O'Halloran, and J. D. Lauzon. 2006. Effect of nitrogen fertilization and topography on soft red winter wheat yield and protein content in two Ontario landscapes. *Canadian Journal of Soil Science* 86: 729–739.

Droushiotis, D. N. 1989. Mixtures of annual legumes and small-grain cereals for forage production under low rainfall. *Journal of Agricultural Science* 113: 249–253.

Entz, M. H. and D. B. Fowler. 1989. Response of winter wheat to N and water: Growth, water use, yield and grain protein. *Canadian Journal of Plant Science* 69: 1135–1147.

Evans, J., N. A. Fettell, D. R. Coventry et al. 1991. Wheat response after temperate crop legumes in southeastern Australia. *Australian Journal of Agricultural Research* 42: 31–43.

Farrer, D. C., R. Weisz, R. Heiniger, J. P. Murphy, and J. G. White. 2006. Minimizing protein variability in soft red winter wheat: Impact of nitrogen application timing and rate. *Agronomy Journal* 98: 1137–1145.

Fehr, W. R., C. E. Caviness, D. T. Burmood, and J. S. Pennington. 1971. Stage of development descriptions for soybeans, *Glycine max* (L.) Merrill. *Crop Science* 11: 929–931.

Feil, B., R. Thiraporn, and H. R. Lafitte. 1993. Accumulation of nitrogen and phosphorus in the grain of tropical maize cultivars. *Maydica* 38: 291–300.

Feil, B., S. B. Moser, S. Jampatong, and P. Stamp. 2005. Mineral composition of the grains of tropical maize varieties as affected by pre-anthesis drought and rate of nitrogen fertilization. *Crop Science* 45: 516–523.

Gan, Y. T., P. R. Miller, B. G. McConkey, R. P. Zentner, F. C. Stevenson, and C. L. McDonald. 2003. Influence of diverse cropping sequences on durum wheat yield and protein in the semiarid northern Great Plains. *Agronomy Journal* 95: 245–252.

Garrido-Lestache, E., R. J. López-Bellido, and L. López-Bellido. 2005. Durum wheat quality under Mediterranean conditions as affected by N rate, timing and splitting, N form and S fertilization. *European Journal of Agronomy* 23: 265–278.

Gobarah, M. E., M. H. Mohamed, and M.M. Tawfik. 2006. Effect of phosphorus fertilizer and foliar spraying with zinc on growth, yield and quality of groundnut under reclaimed sandy soils. *Journal of Applied Science Research* 2: 491–496.

Gooding, M. J. and W. P. Davies. 1997. *Wheat Production and Utilization.* CAB International, Wallingford, UK.

Grant, C. A., G. W. Clayton, and A. M. Johnston. 2003. Sulphur fertilizer and tillage effects on canola seed quality in the Black soil zone of western Canada. *Canadian Journal of Plant Science* 83: 745–758.

Gunes, A., A. Inal, M. S. Adak et al. 2007. Mineral nutrition of wheat, chickpea and lentil as affected by mixed cropping and soil moisture. *Nutrient Cycling in Agroecosystems* 78: 83–96.

Gutiérrez-Boem, F. H., J. D. Scheiner, H. Rimski-Korsakov, and R. S. Lavado. 2004. Late season nitrogen fertilization of soybeans: Effects on leaf senescence, yield and environment. *Nutrient Cycling in Agroecosystems* 68: 109–115.

Guttieri, M. J., R. Ahmad, J. C. Stark, and E. Souza. 2000. End-use quality of six hard red spring wheat cultivars at different irrigation levels. *Crop Science* 40: 631–635.

Guttieri, M. J., R. McLean, J. C. Stark, and E. Souza. 2005. Managing irrigation and nitrogen fertility of hard spring wheat for optimum bread and noodle quality. *Crop Science* 45: 2049–2059.

Haj-Ayed, M., J. Gonzalez, R. Caballero, and M. R. Alvir. 2000. Nutritive value of on-farm common vetch-oat hays. II. Ruminal degradation of dry matter and crude protein. *Annales De Zootechnie* 49: 391–398.

Halvorson, A. D., D. C. Nielsen, and C. A. Reule. 2004. Nitrogen fertilization and rotation effects on no-till dryland wheat production. *Agronomy Journal* 96: 1196–1201.

Ham, G. E., W. W. Nelson, S. D. Evans, and R. D. Frazier. 1973. Influence of fertilizer placement on yield response of soybeans. *Agronomy Journal* 65: 81–84.

Haq, M. U. and A. P. Mallarino. 2005. Response of soybean grain oil and protein concentrations to foliar and soil fertilization. *Agronomy Journal* 97: 910–918.

Harder, H. J., R. E. Carlson, and R. H. Shaw. 1982. Yield, yield components, and nutrient content of corn grain as influenced by post-silking moisture stress. *Agronomy Journal* 74: 275–278.

Hauggaard-Nielsen, H., M. T. Knudsen; J. R. Jørgensen, and E. S. Jensen. 2006. Intercropping wheat with pea for improved wheat baking quality. Paper presented at Joint Organic Congress, Odense, Denmark, May 30–31, 2006.

Hopkins, A., A. H. Adamson, and P. J. Bowling. 1994. Response of permanent and reseeded grassland to fertilizer nitrogen. 2. Effects on concentrations of Ca, Mg, K, Na, S, P, Mn, Zn, Cu, Co and Mo in herbage at a range of sites. *Grass and Forage Science* 49: 9–20.

Jackson, G. D. 2000. Effect of nitrogen and sulphur on canola yield and nutrient uptake. *Agronomy Journal* 92: 644–649.

Jensen, E. S. 1994. Availability of nitrogen in N-15-labeled mature pea residues to subsequent crops in the field. *Soil Biology and Biochemistry* 26: 465–472.

Johnson, V. A. and P. J. Mattern. 1987. Wheat, rye, and triticale. In *Nutritional quality of cereal grains: genetic and agronomic improvement*, R. A. Olson and K. J. Frey (Eds.), Crop Science Society of America and Soil Science Society of America. Agronomy Monograph No. 28, Madison, WI. pp. 133–182.

Johnston, A. M., F. J. Larney, and C.W. Lindwall. 1995. Spring wheat and barley response to long-term fallow management. *Journal of Production Agriculture* 8: 264–268.

Jones, G. D. and J. A. Lutz, Jr. 1971. Yield of wheat and soybeans and oil and protein content of soybean as affected by fertility treatments and deep placement of limestone. *Agronomy Journal* 63: 931–934.

Kamprath, E. J. 1974. Nutrition relationship to soybean fertilization, soybean production, marketing and use. In *Tennessee Valley Authority Bulletin* Y-69, 28–32. Tennessee Valley Authority, Knoxville, KY.

Kaye, N. M., S. C. Mason, D. S. Jackson, and T. D. Galusha. 2007. Crop rotation and soil amendment alters sorghum grain quality. *Crop Science* 47: 722–729.

Kelley, K. W. 1995. Rate and time of N application for wheat following different crops. *Journal of Production Agriculture* 8: 339–345.

Kennedy, C., P. Bell, D. Caldwell, B. Habetz, J. Rabb, and M. A. Alison. 2002. Nitrogen application and critical shoot nitrogen concentration for optimum grain and seed protein yield of pearl millet. *Crop Science* 42: 1966–1973.

Khan, H. R., G. K. McDonald, and Z. Rengel. 2003. Zn fertilization improves water use efficiency, grain yield and seed Zn content in chickpea. *Plant and Soil* 249: 389–400.

Khoshgoftarmanesh, A. H. and R. L. Chaney. 2007. Preceding crop affects grain cadmium and zinc of wheat grown in saline soils of central Iran. *Journal of Environmental Quality* 36: 1132–1136.

Khurana, N. and C. Chatterjee. 2001. Influence of variable zinc on yield, oil content, and physiology of sunflower. *Communications in Soil Science and Plant Analysis* 32: 3023–3030.

Kouyaté, Z., K. Franzluebbers, A. S. R. Juo, and L. R. Hossner. 2000. Tillage, crop residue, legume rotation, and green manure effects on sorghum and millet yields in the semiarid tropics of Mali. *Plant and Soil* 225: 141–151.

Lamb, J. A., G. A. Peterson, and C.R. Ferster. 1985. Wheat fallow tillage systems' effect on a newly cultivated grassland soils' nitrogen budget. *Soil Science Society of America Journal* 49: 352–356.

Lambert, J. and B. Toussaint. 1978. An investigation of the factors influencing the phosphorus content of herbage. *Phosphorus in Agriculture* 73: 1–12.

Lauriault, L.M. and R.E. Kirksey. 2004. Yield and nutritive value of irrigated winter cereal forage grass-legume intercrops in the southern high plains, USA. *Agronomy Journal* 96: 352–358.

Leesawatwong, M., S. Jamjod, J. Kuo, B. Dell, and B. Rerkasem. 2005. Nitrogen fertilizer increases seed protein and milling quality of rice. *Cereal Chemistry* 82: 588–593.

Li, L., S. M. Li, J. H. Sun et al. 2007. Diversity enhances agricultural productivity via rhizosphere phosphorus facilitation on phosphorus-deficient soils. *Proceedings of the National Academy of Sciences of the United States of America* 104: 1192–1196.

Lithourgidis A. S., I. B. Vasilakoglou, K. V. Dhima, C. A. Dordas, and M. D. Yiakoulaki. 2006. Forage yield and quality of common vetch mixtures with oat and triticale in two seeding ratios. *Field Crops Research* 99: 106–113.

Lithourgidis, A. S., K. V. Dhima, I. B. Vasilakoglou, C. A. Dordas, and M. D. Yiakoulaki. 2007. Sustainable production of barley and wheat by intercropping common vetch. *Agronomy for Sustainable Development* 27: 95–99.

Lloveras, J., A. López, J. Ferrán, S. Espachs, and J. Solsona. 2001. Bread-making wheat and soil nitrate as affected by nitrogen fertilization in irrigated Mediterranean conditions. *Agronomy Journal* 93: 1190–1193.

López-Bellido L., M. Fuentes, J. E. Castillo, F. J. López-Garrido, and E. J. Fernández. 1996. Long-term tillage, crop rotation, and nitrogen fertilizer effects on wheat yield under Mediterranean conditions. *Agronomy Journal* 88: 783–791.

López-Bellido, L., M. Fuentes, J. E. Castillo, and F. J. López-Garrido. 1998. Effects of tillage, crop rotation and nitrogen fertilization on wheat grain quality grown under rain fed Mediterranean conditions. *Field Crops Research* 57: 265–276.

López-Bellido, L., R. J. López-Bellido, J. E. Castillo, and F. J. López-Bellido. 2001. Effects of long-term tillage, crop rotation and nitrogen fertilization on bread-making quality of hard red spring wheat. *Field Crops Research* 72: 197–210.

Malhi, S. S. and K. S. Gill. 2002. Effectiveness of sulphate-S fertilization at different growth stages for yield, seed quality and S uptake of canola. *Canadian Journal of Plant Science* 82: 665–674.

Malhi, S. S. and K. S. Gill. 2006. Cultivar and fertilizer S rate interaction effects on canola yield, seed quality and S uptake. *Canadian Journal of Plant Science* 86: 91–98.

Malhi, S. S., Y. Gan, and J. P. Raney. 2007. Yield, seed quality, and sulphur uptake of Brassica Oilseed crops in response to sulphur fertilization. *Agronomy Journal* 99: 570–577.

Manning, G., L. G. Fuller, D. N. Flaten, and R. G. Eilers. 2001. Wheat yield and grain protein variation within an undulating soil landscape. *Canadian Journal of Soil Science* 81: 459–467.

Marschner, H. 1995. *Mineral Nutrition of Higher Plants*, 2nd ed. London: Academic Press.

Mason, M. G. and R. F. Brennan. 1998. Comparison of growth response and nitrogen uptake by canola and wheat following application of nitrogen fertilizer. *Journal of Plant Nutrition* 21: 1483–1499.

McConkey, B. G., C. A. Campbell, R. P. Zentner, F. B. Dyck, and F. Selles. 1996. Long-term tillage effects on spring wheat production on three soil textures in the Brown soil zone. *Canadian Journal of Plant Science* 76: 747–756.

McKenzie, F. R. and J. L. Jacobs. 2002. Effects of application of nitrogen fertilizer on concentrations of P, K, S, Ca, Mg, Na, Cl, Mn, Fe, Cu and Zn in perennial ryegrass/white clover pastures in south-western Victoria, Australia. *Grass and Forage Science* 57: 48–53.

McLaren, R. G. 1976. Effects of fertilizers on the sulphur content of herbage. *Journal of the British Grassland Society* 31: 99–103.

Miao, Y., D. J. Mulla, P. C. Robert, and J. A. Hernandez. 2006. Within-field variation in corn yield and grain quality responses to nitrogen fertilization and hybrid selection. *Agronomy Journal* 98: 129–140.

Miller, P. R. and J. A. Holmes. 2005. Cropping sequence effects of four broadleaf crops on four cereal crops in the northern Great Plains. *Agronomy Journal* 97: 189–200.

Miller, P. R., B. G. McConkey, G. W. Clayton et al. 2002a. Pulse crop adaptation in the northern great plains. *Agronomy Journal* 94: 261–272.

Miller, P. R., J. Waddington, C. L. McDonald, and D. A. Derksen. 2002b. Cropping sequence affects wheat productivity on the semiarid northern Plains. *Canadian Journal of Plant Science* 82: 307–318.

Miller, P. R., R. E. Engel, and J. A. Holmes. 2006. Cropping sequence effect of pea and pea management on spring wheat in the northern great plains. *Agronomy Journal* 98: 1610–1619.

Mirzapour, M. H. and A. H. Khoshgoftar. 2006. Zinc application effects on yield and seed oil content of sunflower grown on a saline calcareous soil. *Journal of Plant Nutrition* 29: 1719–1727.

Mosquera-Losada, M. R., A. González-Rodríguez, and A. R. Rodríguez. 2004. Fertilization with nitrogen and potassium on pastures in temperate areas. *Journal of Range Management* 57: 280–290.

Nielsen, D. C. 2004. Kenaf forage yield and quality under varying water availability. *Agronomy Journal* 96: 204–213.

Oikeh, S. O., J. G. Kling, and A. E. Okoruwa. 1998. Nitrogen fertilizer management effects on maize grain quality in the West African moist Savanna. *Crop Science* 38: 1056–1061.

Ottman, M. J., H. A. Doerge, and E. C. Martin. 2000. Gurum grain quality as affected by nitrogen fertilization near anthesis and irrigation during grain fill. *Agronomy Journal* 92: 1035–1041.

Peters, C. J., G. W. Fick, and J. L. Wilkins. 2003. Cultivating better nutrition: Can the food pyramid help translate dietary recommendations into agricultural goals? *Agronomy Journal* 95: 1424–1431.

Peterson, T. A. and G. E. Varvel. 1989. Crop yield as affected by crop rotation and N rate. II. Sorghum. *Agronomy Journal* 81: 727–731.

Pietz, R. I., J. R. Peterson, C. Lue-Hing, and L. F. Welch. 1978. Variability in the concentration of twelve elements in corn grain. *Journal of Environmental Quality* 7: 106–110.

Poole, W. D., G. W. Randall, and G. E. Ham. 1983. Foliar fertilization of soybean: I. Effect of fertilizer sources, rates, and frequency of application. *Agronomy Journal* 75: 195–200.

Purcell, L. C., R. Serraj, T. R. Sinclair, and A. De. 2004. Soybean N_2 fixation estimates, ureide concentration, and yield responses to drought. *Crop Science* 44: 484–492.

Puri, Y. P., M. F. Miller, R. N. Sah, K. G. Baghott, E. Fereres-Castel, and R. D. Meyer. 1989. Response surface analysis of the effects of seeding rates, N-rates and irrigation on durum wheat. II. Protein yield and grain quality. *Phyton (Buenos Aires)* 49: 41–59.

Rajeswara, Rao B. R. and R. Prasad. 1987. Influence of competition, irrigation levels and N fertilization on protein content and protein yield of three spring wheat (*Triticum aestivum* L.) cultivars. *Plant Foods for Human Nutrition* 37: 127–131.

Ramon Fernandez, G. and R. J. Laird. 1959. Yield and protein content of wheat in central Mexico as affected by available soil moisture and N fertilization. *Agronomy Journal* 51: 33–36.

Rao, A. C. S., J. L. Smith, V. K. Jandhyala, R. I. Papendick, and J. F. Parr. 1993. Cultivar and climatic effects on the protein content of soft white winter wheat. *Agronomy Journal* 85: 1023–1028.

Rathke, W., O. Christen, and W. Diepenbrock. 2005. Effects of nitrogen source and rate on productivity and quality of winter oilseed rape (*Brassica napus* L.) grown in different crop rotations. *Field Crops Research* 94: 103–113.

Ray, J. D., F. B. Fritschi, and L. G. Heatherly. 2006. Large applications of fertilizer N at planting affects seed protein and oil concentration and yield in the early soybean production system. *Field Crops Research* 99: 67–74.

Robinson, F. E., D. W. Cudney, W. F. Lehman. 1979. Nitrogen fertilizer timing, irrigation, protein, and yellow berry in durum wheat. *Agronomy Journal* 71: 304–308.

Robinson, R. G. 1960. Oat-pea or oat-vetch mixtures for forage or seed. *Agronomy Journal* 52: 546–549.

Roche, J. R., J. Morton, and E. S. Kolver. 2002. Sulphur and chlorine play a non-acid base role in periparturient calcium homeostasis. *Journal of Dairy Science* 85: 3444–3453.

Ryan, J., N. Nsarellah, and M. Mergoum, 1997. Nitrogen fertilization of durum wheat cultivars in the rainfed area of Morocoo. Biomass, yield and quality considerations. *Cereal Research Communications* 25: 85–90.

Sharma, O. P. and A. K. Gupta. 2002. Nitrogen-phosphorus nutrition of pearmillet as influenced by intercrop legumes and fertilizer levels. *Journal of Plant Nutrition* 25: 833–842.

Slukhai, H. I. and P. F. Zhabits. 1970. Influence of N nutrition and irrigation on the indices of the water system, yield, and quality of the grain. *Physiology and Biochemistry of Cultivated Plants* (Engl. Transl.) 2: 493–498.

Song, Y. N., F. S. Zhang, P. Marschner et al. 2007. Effect of intercropping on crop yield and chemical and microbiological properties in rhizosphere of wheat (*Triticum aestivum* L.), maize (*Zea mays* L.), and faba bean (*Vicia faba* L.). *Biology and Fertility of Soils* 43: 565–574.

Srikumar, T. S. and P. A. Öckerman. 1990. The effects of fertilization and manuring on the content of some nutrients in potato (var. Provita). *Food Chemistry* 37: 47–60.

Strong, W. M. 1982. Effect of late application of nitrogen on the yield and protein content of wheat. *Australian Journal of Experimental Agriculture and Animal Husbandry* 22: 54–61.

Swift, M. L., S. Bittman, D. E. Hunt, and C. G. Kowalenko. 2007. The effect of formulation and amount of potassium fertilizer on macromineral concentration and cation-anion difference in tall fescue. *Journal of Dairy Science* 90: 1063–1072.

Tatham, A. S. and P. R. Shewry. 1995. The S-poor prolamins of wheat, barley and rye. *Journal of Cereal Science* 22: 1–16.

Temperly, R. J. and R. Borges. 2006. Tillage and crop rotation impact on soybean grain yield and composition. *Agronomy Journal* 98: 999–1004.

Terman, G. L., R. E. Raming, A. F. Dreier, and R. A. Olson. 1969. Yield-protein relationships in wheat grain, as affected by N and water. *Agronomy Journal* 61: 755–759.

Thiraporn, R., B. Feil, and P. Stamp. 1992. Effect of nitrogen fertilization on grain yield and accumulation of nitrogen, phosphorus and potassium in the grains of tropical maize. *Journal of Agronomy and Crop Science* 169: 9–16.

Tsuji, H., H. Yamamoto, K. Matsuo, and K. Usuki. 2006. The effect of long-term conservation tillage, crop residents and P fertilizer on soil conditions and response of summer and winter crops on an Andosol in Japan. *Soil and Tillage Research* 89: 167–176.

Velazquez-Beltran, L.G., Y. E. Felipe-Perez, and C.M. Arriaga-Jordan. 2002. Common vetch (*Vicia sativa*) for improving the nutrition of working equids in campesino systems on hill slopes in central Mexico. *Tropical Animal Health and Production* 34: 169–179.

Vita, P. D., E. Di. Paolo, G. Fecondo, N. D. Fonzo, and M. Pisante. 2007. No-tillage and conventional tillage effects on durum wheat yield, grain quality and soil moisture content in southern Italy. *Soil and Tillage Research* 92: 69–78.

Walton, P. D. 1975. Annual forages seeding rates and mixtures for central Alberta. *Canadian Journal of Plant Science* 55: 987–993.

Weber, C. R. 1966. Nodulating and non-nodulating soybean isolines: I. Agronomic and chemical attributes. *Agronomy Journal* 58: 43–46.

Welch, L. F. 1969. Effect of N, P, and K on percent and yield of oil in corn. *Agronomy Journal* 61: 890–891.

Welch, R. M. and R. D. Graham. 2002. Breeding crops for enhanced micronutrient content. *Plant and Soil* 245: 205–214.

Westfall, D. G., J. L. Havlin, G. W. Hergert, and W. R. Raun. 1996. Nitrogen management in dryland cropping systems. *Journal of Production Agriculture* 9: 192–199.

Whitehead, D. C. 2000. *Nutrient Elements in Grassland: Soil-Plant-Animal Relationships.* CABI Publishing, New York.

Wieser, H. and W. Seilmeier. 1998. The influence of nitrogen fertilization on quantities and proportions of different protein types in wheat flour. *Journal of the Science of Food and Agriculture* 76: 49–55.

Wrigley, C. W., D. L. Du Cros, J. G. Fullington, and D. D. Kasarda. 1984. Changes in polypeptide composition and grain quality due to sulphur deficiency in wheat. *Journal of Cereal Science* 2: 15–24.

Yamada, H., J. St, Andre, and R. M. Hoover. 1972. Effects of irrigation and fertilizer on INIA 66 wheat. *California Agriculture* 26: 9–10.

Zentner, R. P., C. A. Campbell, V.O. Biederbeck, P. R. Miller, F. Selles, and M. R. Fernandez. 2001. In search of a sustainable cropping system for the semiarid Canadian prairies. *Journal of Sustainable Agriculture* 18: 117–136.

Zhang, T. Q., C. F. Drury, C. S. Tan, and W. D. Reynolds. 2007. Effects of long-term (47 years) consistent fertilization and crop rotation on feed and diet productivity and quality. In *Biogeochemistry of Trace Elements: Environmental Protection, Remediation and Human Health*, Y. G. Zhu, N. Lepp, and R. Naidu, (Eds.), Beijing: Tsinghua University Press, pp. 297–299.

8 Effects of Long-Term Cattle Manure Applications on Soil, Water, and Crops

Implications for Animal and Human Health

Mônica Batista Benke, Xiying Hao, and Chi Chang

CONTENTS

8.1 INTRODUCTION

In the last 30 to 40 years, livestock production has rapidly increased, first in developed and lately in developing countries (Delgado et al. 1999; Gerber et al. 2005). The total number of cattle reached 97 million in the United States in 2007 (USDA 2007) and 15.9 million in Canada (Statistics Canada 2007). Among the developing countries, those in Asia have the fastest growing livestock sector with China in the lead (Gerber et al. 2005). Associated with growth in the livestock industry is an increased intensity of livestock operations. Some feedlots now hold 40,000 head of cattle in a relatively small area, which is increasing the amount of manure applied to nearby cropped land.

Livestock manure is a source of plant nutrients that can make a valuable contribution to soil organic matter and improve physical fertility (Blair et al. 2006). However, high rates of livestock manure land application are often followed by nutrient imbalance and contamination of soil, surface water, and groundwater (Edwards and Daniel 1992; Nicholson et al. 2003; Dach and Starmans 2005). Nicholson et al. (2003) estimated that 37 to 40% of total copper (Cu) and zinc (Zn) input to cropped land in England is from livestock manure applications. Increases in nitrate (NO_3^-) leaching, phosphorus (P) accumulation in topsoil, P runoff, and metal content (e.g., Cu and Zn) are common problems associated with high rates of livestock manure application to agricultural land. One of the major sources of lake and river eutrophication in the U.S. is P runoff from agricultural land (Daniel et al. 1998).

Changes in food quality have been reported for crops grown on land receiving high rates of livestock manure (Mathers and Stewart 1974; Liebhardt 1976; Gilbertson et al. 1979). High levels of NO_3^- in feed can cause acute poisoning in cattle, sheep, and other livestock, which may result in death (Wright and Davison 1964). Grass tetany [caused by low magnesium (Mg) levels in blood] has been reported in cattle grazed on pasture that received high rates of poultry manure, which has a high potassium (K) content (Gilbertson et al. 1979).

These elevated levels of nutrients and trace metals in manured soils could also impact human health. Consumption of water contaminated with NO_3^- is linked to the development of several types of human cancers (Ward et al. 2005). Cancer has also been sometimes linked to high levels of Cu and Zn in soils (Marjanen 1980; Senesil et al. 1999), whereas multiple sclerosis cases are associated with high levels of Zn in soil and water (Schiffer et al. 2001). This chapter summarizes recent research on the impact of long-term cattle manure applications on soil, water, and crops quality, and the resulting implications for animal and human health.

8.2 IMPACT OF LIVESTOCK PRODUCTION AND MANURE LAND APPLICATION ON SOIL AND WATER RESOURCES

8.2.1 NITROGEN

Livestock manure is a valuable source of nitrogen (N) for plants. For example, in the United States alone, manure produced by the beef cattle industry generates about 529,900 metric tons N each year (USDA 1997). However, because a significant portion of total N in livestock manure is in ammoniacal form [NH_4^+ + dissolved ammonia (NH_3)], it can be rapidly lost through NH_3 volatilization during manure handling, storage, and application to soil (Sims and Wolf 1994; Williams 1995). Volatilization of NH_3 contributes to acid rain (Williams 1995) and atmospheric NH_3 deposition to soil and plants. This affects N status and cycling in ecosystems, leading to acidification of soils (Ferm 1998), eutrophication of coastal marine waters (Burkart and James 1999), and a reduction in vegetation species diversity (Woodin and Farmer 1993).

Ammonia volatilization from livestock manure could be reduced by acidifying the slurry to a pH level of about 5 prior to soil application (Williams 1995) or by including low-pH natural feed additives in animal diets (Hao et al. 2007). The incorporation of manure to soil using conventional tillage decreases NH_3 volatilization as well. Schilke-Gartley and Sims (1993) reported that 20% of total N was lost through NH_3 volatilization when poultry litter manure was applied to nontilled soils. Immediate incorporation of the manure into soils decreased NH_3 volatilization loss to 3% of the total N.

Several studies have showed that manure amendments can cause high concentrations of NO_3^- in soil and increase the risk of NO_3^- leaching and/or runoff loss (Liebhardt et al. 1979; Hao et al. 2003; Basso and Ritchie 2005; Jones et al. 2007). Liebhardt et al. (1979) found NO_3^--N concentrations as high as 173 mg L^{-1} at 3 m depth in a loamy sand soil that received 179 Mg ha^{-1} yr^{-1} of poultry manure. Hao et al. (2003) reported significant increases in NO_3^- content throughout the 0–150 cm depth in a cultivated clay loam soil after 25 annual cattle manure applications. Basso and Ritchie (2005) studied the influence of manure and inorganic-N fertilizer on NO_3^- leaching in cropped plots for 6 years using undisturbed drainage lysimeters and observed that NO_3^- leaching was consistently higher in manure treatments than nonmanured controls.

After monitoring small agricultural catchments for several years in Sweden, Kyllmar et al. (2006) reported that high animal density associated with intensive cropping systems was a contributing factor to large N loads in stream outlets. A study involving 631 water wells in over 600 swine farms distributed across 18 states in the United States revealed that 53.6% of the wells contained detectable levels of NO_3^- (Bruning-Fann et al. 1994). In 11.7% of the wells studied, NO_3^--N concentration exceeded the limit of 10 mg L^{-1} (equivalent to 45 mg L^{-1} as NO_3^- for human consumption, and 4.3% exceeded the limit of 22.6 mg L^{-1} (equivalent to 100 mg L^{-1} as NO_3^-) for animal consumption set by the United States Environmental Protection Agency (USEPA 1973). Natural or background levels of NO_3^--N in groundwater are normally very low (< 3 mg L^{-1}) (Bouchard et al. 1992). Higher NO_3^--N levels above the background are usually attributed to anthropogenic activities.

8.2.2 Phosphorus

The amount of manure applied to soils is often based on crop N requirements. However, application of manure based on N requirements usually supplies more of other nutrients, especially P, than is required by crops (Liebhardt 1976; Edwards and Daniel 1992; Sims and Wolf 1994). For example, the N-based cattle manure application rates of 30 and 60 Mg^{-1} ha^{-1} yr^{-1} (wet weight) under rainfed and irrigated conditions, respectively, used for barley crops in southern Alberta, Canada, provide six times more P than is recommended (Whalen and Chang 2001). Plants have a higher demand for N compared to P. The N:P grain uptake ratio varies between 4.5:1 and 8.8:1 for most cultivated plants (Gilbertson et al. 1979), whereas manure usually has a lower N:P ratio. The average N:P ratio in cattle manure is about 2.6:1 (Whalen and Chang 2001).

Nitrogen-based manure applications not only seem to increase soil total P content but have been shown to change the soil P pool by significantly increasing the proportion of soluble forms. Hao et al. (2008) reported significant increases in the soil test P (STP) fraction after long-term cattle manure applications to Canadian prairie soils. Sharpley et al. (1984) reported that the application of 67 Mg^{-1} ha^{-1} yr^{-1} of beef feedlot manure over 8 years to a clay loam soil increased total P from 0.35 (control) to 1.0 g kg^{-1}, and available P from 15 (control) to 230 mg kg^{-1} in the 0–30 cm depth of soil. In the Sanborn Field experimental plots in Missouri, 110-year manure (horse or dairy cow) applications significantly increased total P level, with most of the increases occurring in the soluble inorganic P fraction (Motavalli and Miles 2002).

Excessive P accumulation in soils (especially in the soluble form) increases the risk of P runoff into surface water through manure applications (Daniel et al. 1998). Eghball et al. (1996) studied P movement in soil after 40 years of manure and chemical fertilizer applications, and reported greater amounts of available P with depth (up to 1.8 m) in manured plots than those receiving fertilizer. Whalen and Chang (2001) observed substantial P loss in a study of P balance in irrigated soils after long-term cattle manure applications, and attributed that loss to leaching and wind erosion. Phosphorus concentration in surface water is often the key element in limiting aquatic biological productivity (Daniel et al. 1998).

8.2.3 Soluble Salts

Applying N-based rates of animal manure often supplies high amounts of soluble salts as well. Significant increases in soluble Na^+, K^+, Mg^{2+}, Cl^-, HCO_3^-, and SO_4^{2-} were observed in a clay loam Chernozemic soil following long-term additions (Hao and Chang 2003). Whereas soluble K^+, Mg^{2+}, HCO_3^-, and SO_4^{2-} increased only near the soil surface, Na^+ and Cl^- increased throughout the 0–150 cm depth. Accumulation of soluble salts in the soil elevates the risk of soil salinity, whereas leaching of salts to groundwater could result in contamination. Rodvang et al. (2004) reported significant increases in Cl^- in shallow groundwater near irrigated agricultural areas receiving manure in southern Alberta.

Generally, K^+ is the dominant soluble cation in animal manure. Long-term manure applications increase the K/(Ca + Mg) ratio in solution and at the exchange sites in soils (Liebhardt and Shortall 1974; Hao and Chang 2002, 2003). For example, Hao and Chang (2003) reported that the K/(Ca + Mg) ratio in the surface soil increased from 0.3 (control) to 1.3, 1.6, and 2.1 when cattle manure was applied to soil at 30, 60, and 90 metric tons (Mg) ha^{-1} yr^{-1} for 25 years, respectively. Decreases in extractable Ca^{2+} levels and increases in extractable K were also observed in soils that received poultry manure (Liebhardt and Shortall 1974; Walker and Bernal 2008). A higher level of K^+ in soil has been shown to interfere with crop uptake of other nutrients, such as Mg and Ca (Gallaher et al. 1975; Stout and Baker 1981; Daliparthy et al. 1994; Callan and Westcott 1996).

8.2.4 Trace Metals

The content of trace metals in farm-animal manure is lower than in industrial, municipal, and sewage wastes, and in most cases similar to or lower than the maximum permissible metal additions to uncontaminated soils in effect in Canada and Europe (Ihnat and Fernandes 1996; Nicholson et al. 1999; Kabata-Pendias and Pendias 2001). Nevertheless, farm animal manure applications to soil may result in trace metal accumulation due to long-term high application rates (Nicholson et al. 2003; Dach and Starmans 2005).

In Austria, Cu and Zn input to agricultural soils by farm-animal manure exceeded the threshold limits for soil contamination in certain areas (Sager 2007). A study conducted to investigate the source of trace metal input to agricultural soils in England indicated that the majority of Cu and Zn came from livestock manure (37 to 40%) rather than from sewage sludge (8 to 17%) (Nicholson et al. 2003). In the Netherlands, the trace metal input to agricultural soil by animal manure is much higher than from mineral fertilizers. On average, manure fertilization adds 12 times more Cu and 25 times more Zn than mineral fertilizers (N, P, K, and Ca) to Dutch cropland (Dach and Starmans 2005). Continuous manure applications for 25 years increased both Cu and Zn content in the surface soil under irrigated conditions in southern Alberta (Benke et al. 2008). Total Cu varied from 26.3 to 34.0 mg kg^{-1} in the manured soil, which is lower than the maximum allowed concentration set by both the United States and the European Community (EC), and total Zn content ranged from 115.5 to 187.5 mg kg^{-1}, also within the recommended EC guideline (Kabata-Pendias and Pendias 2001). In this study, significant increases in available Cd and Zn contents (EDTA-extractable) were also observed. This observation is consistent with results obtained from a grassland soil receiving long-term cattle and pig slurry in Northern Ireland (Christie and Beattie 1989). The increases in EDTA-extractable metals are a concern as they are more bioavailable and soluble.

8.3 IMPACT OF MANURE LAND APPLICATIONS ON VEGETATION AND WATER—IMPLICATIONS FOR ANIMAL FEED QUALITY

8.3.1 Yield

Long-term manure applications to a Chernozemic soil in the Canadian prairies did not significantly affect triticale (*Triticosecale* L.), barley (*Hordeum vulgare* L.), and corn (*Zea Mays* L.) forage yields when manure was applied at recommended agronomic rates of 30 Mg ha^{-1} yr^{-1} under rainfed, and 60 Mg ha^{-1} yr^{-1} under irrigated conditions (Hao and Chang unpublished data). However, a significant reduction in triticale forage yield was observed when higher manure rates were used under rainfed conditions. This yield reduction was attributed to the increased salt accumulation in soil when manure rates are high (Hao and Chang 2007). Richards et al. (1987) also observed declines in biomass yield of several varieties of barley, triticale, and wheat when cultivated in soils with a Na-dominated salinity of > 3 dS m^{-1}.

8.3.2 Protein

Cattle fed a nutritionally deficient diet, particularly lacking in protein, can voluntarily reduce feed intake and consequently consume neither adequate energy nor protein (Dunn and Moss 1992). Cattle respond to protein supplementation when forage crude protein content is less than 6 to 8% (NRC 2000). For dairy cows producing > 30 kg d^{-1} of milk, the recommended crude protein diet content is 16% (Jordan and Swanson 1978). However, excess protein in feed could reduce dairy cattle reproduction. Cattle fed diets containing 19% or more crude protein have lower conception rates (Jordan and Swanson 1978; Canfield et al. 1990).

Manure applications at the recommended agronomic rates of 30 Mg $ha^{-1}yr^{-1}$ (rainfed) or 60 Mg $ha^{-1}yr^{-1}$ (irrigated) resulted in a significant increase in crude protein content in barley (Benke et al., unpublished data) and triticale forage (Hao and Chang 2007). The average protein content in barley (16.0%) and triticale (14.6%), under these conditions, were within the recommended levels for dairy cows. However, increasing the manure applications above the recommended agronomic rates resulted in crude protein content above 19% in barley forage under irrigated conditions. Elevated crude protein levels should be taken into consideration when formulating cattle diets. The implication is that less protein supplement would be needed if using feedstuff from highly manured soil.

8.3.3 Nitrate

Elevated levels of NO_3^- in water and forage can cause toxicity in ruminant animals through NO_3^- conversion to more toxic nitrites in the rumen. Nitrites react with hemoglobin, forming methemoglobin. Oxygen does not bind to methemoglobin, interfering with the transport of oxygen in blood. This could be lethal to the animals (Wright and Davison 1964; Reid 1994; Ward et al. 2005).

Generally, NO_3^- levels in plant tissue are very low because NO_3^- is rapidly metabolized to other compounds. However, under certain conditions, such as heavy N fertilization, herbicide application, environmental stress (e.g. drought, hail, shade, and disease), or any other factors affecting photosynthesis, NO_3^- can accumulate in plant tissue (Wright and Davison 1964; Maynard et al. 1976). Extremely high amounts of $NO_3^-N^-$ (e.g., 26 to 40 g kg^{-1}) in feed will poison cattle, sheep, and other livestock, and may lead to their death (Wright and Davison 1964). These studies also indicated that abortions were observed in animals fed excessive NO_3^- . Concentrations of 2–4 g kg^{-1} NO_3^- -N in feed are not recommended for pregnant animals. Storing forage with high levels of NO_3^- in silos has also been known to release enough poisonous gases (e.g., NO, NO_2, and N_2O_4) to injure or even kill livestock (Wright and Davison 1964). Chronic poisoning due to NO_3^- in feed has also been reported in the Netherlands. Reduced growth, reproduction, and milk production were observed in dairy cattle after consumption of NO_3^--rich diets (Reid 1994). This was related to the fact that high NO_3^- concentrations result in low cattlefeed intake. Nitrate toxicity seems also to be influenced by the rate of NO_3^- intake and nature of the feed (Geurink et al. 1979).

In the Netherlands, the maximum NO_3^--N concentrations permitted are 1.67 g kg^{-1} for prewilted silage and hay, or 3.39 g kg^{-1} for grazing grass fed twice daily (Prins 1984). Mayland et al. (2007) suggested that forage should be limited to 50% of total dry matter feed when NO_3^--N concentrations are in the range of 1.5 to 2.0 g N kg^{-1}. Animals fed tall fescue grass containing 3.3 g kg^{-1} of NO_3^--N have developed grass tetany and fat necrosis (Stuedemann et al. 1975). The fescue grass was grown on soil receiving annual applications of 18 Mg ha^{-1} of poultry litter for 6 years.

Significant increases in the level of NO_3^--N in corn and triticale (Hao and Chang 2007) and barley (Benke et al., unpublished data) were observed following long-term cattle manure applications. In this study, the NO_3^--N content in triticale was 0.45 g kg^{-1} (control), but increased to 1.23, 1.76, and 2.24 g kg^{-1} when cattle manure was applied for 25 years at 30, 60, and 90 Mg $ha^{-1}y^{-1}$, respectively, under rainfed conditions. The NO_3^--N content was higher than the maximum concentration permitted in the Netherlands of—1.67 g kg^{-1} for the two highest manure rates. The level of NO_3^--N in irrigated corn was much lower than this maximum level, perhaps due to better NO_3^- metabolism and lower soil NO_3^- content under irrigation (Hao and Chang 2003). Another factor that may have contributed to the higher NO_3 levels in manure-treated crops is the abundant supply of K in animal manure (Edwards and Daniel 1992; Hao and Chang 2003). High K content in soils stimulates NO_3^- absorption (Maynard et al. 1976), which allows for luxury consumption by plants to occur. The mutual absorption preserves electrical neutrality in the plant cells (Wright and Davison 1964).

8.3.4 Minerals

Mineral content in forage is an important factor in preventing livestock diseases. Ruminal microbial activity is also influenced by the forage mineral content (Buxton and Fales 1994). Five selected macrominerals (Ca, Mg, P, K, and Na) and three microminerals (Cu, Mn, and Zn) are discussed in the following sections.

8.3.4.1 Macrominerals

According to NRC (2000), the Ca requirement in diets for growing and finishing cattle varies from 1.9 to 7.3 g kg^{-1}, depending on the total digestible nutrient levels in feed and the current body weight. Long-term application of cattle manure above the agronomic rate significantly decreased Ca and Mg content in corn forage and triticale forage (Hao and Chang 2007). Thus, formulating beef and dairy diets using forage produced from heavily manured soil may require greater Ca and Mg.

The minimum Mg requirement is 1.0 g kg^{-1} for growing and finishing cattle, and 2.0 g kg^{-1} for early lactation cows (NRC 2000). Magnesium deficiency can cause grass tetany (or hypomagnesium tetany), which especially affects lactating beef cows. This nutrient disorder is potentially fatal if Mg absorption in the required amounts is not restored soon after the animals show the initial symptoms (Grunes et al. 1970). Grass tetany is responsible for considerable economic losses in the beef industry (Mayland et al. 1990). The addition of poultry manure to a sandy coastal soil in Delaware during a 3-year period significantly decreased Mg concentration in corn (Liebhardt 1976).

The minimum P requirement for beef cattle feed varies from 1.3 to 3.4 g kg^{-1}, depending on the total digestible nutrient levels in feed and body weight (NRC 2000). Hao and Chang (2007) reported that the P content in both corn and triticale forage increased significantly with the cattle manure applications at recommended agronomic rates, but values were still within the NRC (2000) requirements. Another aspect that should be taken into consideration is the Ca:P ratio in animal diets. According to Smart and Cymbaluk (1997), high total P associated with low Ca intake could cause hyperparathyroidism in cattle. Research has shown that the effects of the Ca:P ratio on the absorption of Ca is critical when the ratio is < 1 (NRC 2001). These authors recommended that, for maximum milk yield and fertility, Ca:P in forage should be around 1.4:1. In a manure study by Hao and Chang (2007), the forage Ca:P ratio (0.8:1 in corn and 0.7:1 in triticale) was lower than the minimum level for beef cattle. Even lower ratios were observed when manure applications were above the recommended agronomic rates.

The minimum requirement for K is around 6.0 g kg^{-1} for feedlot cattle, and 3 to 4 g kg^{-1} for range cattle, whereas the maximum tolerable concentration has been set at 30 g kg^{-1} (NRC 2000). Manure applications resulted in a significant increase in K content in the forage (Hao and Chang 2007). However, the K levels (ranging from 28 to 33 g kg^{-1}) were at or above the maximum tolerable amount set by NRC (2000) in barley forage. Cattle fed grass with high K content could develop grass tetany even when the Mg concentration is above the minimum required levels. Excess K in the rumen inhibits Mg absorption (Grunes et al. 1970; Mayland et al. 1990; Cherney et al. 2002). Development of grass tetany has been observed in ruminants grazing pastures with a K/(Ca + Mg) ratio equal to or greater than 2.2 in the feed (Grunes et al. 1970). The K/(Ca + Mg) ratio in corn and barley cultivated in Canadian Chernozemic soils that received one, two, and three times the agronomic cattle manure rate surpassed 2.2 (Hao and Chang unpublished data). This observation is related to the fact that long-term manure applications led to much higher soil K content than Ca and Mg (Hao and Chang 2003), as excess K levels in soil inhibits plant absorption of Ca and Mg (Mayland et al. 1990).

The Na requirement in beef cattle diets varies from 0.6 to 1.0 g kg^{-1} (NRC 2000). Even though manure applications resulted in a significant increase in soil Na content under both rainfed and irrigated conditions, the Na content in triticale (0.29 to 0.88 g kg^{-1}) and corn (0.07 to 0.29 g kg^{-1}) were below the required minimum values (Hao and Chang 2003). Thus, Na supplementation is needed for dairy cows that have a higher minimum Na requirement (1.0 g kg^{-1}) even for feed grown from soil receiving a heavy rate of manure applications.

8.3.4.2 Microminerals

The minimum Cu requirement in beef cattle diets has been set at 10 mg kg^{-1} with a maximum tolerable level of 100 mg kg^{-1} (NRC 2000). Forage diets with < 10 mg Cu kg^{-1} may not meet an animal's needs because Cu in forage is less available than inorganic sources (NRC 2000). Long-term feedlot cattle manure applications at an agronomic rate of 60 Mg $ha^{-1}yr^{-1}$ to irrigated crops in Southern Alberta significantly increased Cu from 15.3 in the control to 20.1 g kg^{-1} in irrigated corn (Hao and Chang 2007). However, manure applications did not affect Cu content in the triticale and barley forage grown in these experimental plots. Christie and Beattie (1989) also reported an increased Cu content in perennial ryegrass (*Lolium perenne* L.) after long-term pig slurry applications to an Irish soil.

Depending on the soil, forage Fe content could be as low as 70 or as high as 500 mg kg^{-1} (NRC 2000). The Fe content in the forage study by Hao and Chang (2007) ranged from 50 to 200 mg kg^{-1} and was not significantly affected by the long-term cattle manure application. According to the NRC (2000) guidelines, the minimum Fe requirement in beef cattle diets is 50 mg kg^{-1}, and feedstuff produced from non-manured or long-term manured soil could meet the beef and dairy cattle minimum requirement with no supplementation needed.

The minimum Mn requirement is about 20 mg kg^{-1} for beef cattle and about 40 mg kg^{-1} for reproductive beef cattle. Diets containing less than 15.8 mg kg^{-1} of Mn could result in lower reproduction rates (NRC 2000). Manure applications had no effect on Mn content in rainfed triticale forage (20.8–22.1 mg kg^{-1}), but applying 180 Mg ha^{-1} yr^{-1} (three times the recommended level under irrigation) for 25 years significantly decreased Mn content from around 50 mg kg^{-1} in the nonmanured control to about 30 mg kg^{-1} in irrigated corn in Canadian prairie soils (Hao and Chang 2007). Based on these results, Mn supplements are still needed as Mn values in corn and triticale were below the 40 mg kg^{-1} minimum Mn levels for gestating and lactating cows. No supplement is needed in feed for beef cattle, as Mn values in triticale and corn are all above their 20 mg kg^{-1} minimum requirement.

Long-term cattle manure applications significantly increased Zn content in triticale and corn (Hao and Chang 2007) and in perennial ryegrass (Christie and Beattie 1989). Applying 60 Mg ha^{-1} yr^{-1} of manure over 25 years increased Zn content in triticale from 16.7 to 31.1 mg kg^{-1}. Similarly, applying 120 Mg ha^{-1} yr^{-1} of manure over 25 years almost tripled Zn content in corn to 48.8 mg kg^{-1} from 16.7 mg kg^{-1} (control). The minimum requirement for Zn in beef cattle diets is 30 mg kg^{-1} (NRC 2000). Therefore, manure applications had a positive effect in increasing plant Zn content to meet animal Zn requirements.

8.4 IMPACT OF MANURE LAND APPLICATIONS ON VEGETATION AND WATER—IMPLICATIONS FOR HUMAN HEALTH

8.4.1 NITRATE

Due to manure and/or inorganic fertilizer application, the levels of NO_3^--N in groundwater could be higher than the maximum recommended concentration of 10 mg L^{-1} (Bruning-Fann et al. 1994). Higher levels of NO_3^- may pose risks to human health. Nitrate-N levels in water above this critical concentration, in the presence of nitrate-reducing organisms, have been reported to cause diarrhea and formation of methemoglobin (a condition known as methemoglobinemia or "blue baby") in infants (Wright and Davison 1964). Data from 1973 indicated that there have been about 2000 cases of "blue baby" (many fatal) due to water consumption with high NO_3^- concentration in the United States and Europe (Maynard et al. 1976). Cases of infant methemoglobinemia have also been reported after consumption of baby food prepared with fresh spinach and carrots with high levels of NO_3^- and stored at room temperature (Simon 1966; Maynard et al. 1976). Simon (1966) suggested NO_3^--N in spinach given to children should not exceed 0.7 g kg^{-1}, whereas other studies estimated that a maximum should be set at 1.0 g kg^{-1} (Maynard et al. 1976). Suspicion has also been raised about the possibility that milk and meat from cattle consuming high levels of dietary NO_3^- may itself contain sufficien NO_3^- to cause health problems. However, this has not been proven (Wright and Davison 1964; Van Broekhoven et al. 1989). Even though there have been no reports of methemoglobinemia in adults caused by water contaminated with NO_3^--N (Wright and Davison 1964; Maynard et al. 1976), there have been other concerns about the relationship between dietary NO_3^--N levels and the formation of cancer-causing substances (Bruning-Fann and Kaneene 1993; Ward et al. 2005). About 20% of the NO_3^- ingested in food or water is transformed to nitrite by the mouth bacterial flora (Zeegers et al. 2006). In the stomach, nitrite may react with secondary amines and amides to form *N*-nitroso compounds, a process called *nitrosation*. Studies conducted with animals indicate that *N*-nitroso compounds are potent carcinogens (Ward et al. 2005). These compounds have resulted in cancer development in every animal species tested, and it is unrealistic to believe that humans are an exception (Lijinsky 1986).

Human ingestion of either water with NO_3^--N levels above 10 mg L^{-1} or vegetables high in NO_3^- and low in ascorbic acid increases the content of *N*-nitroso compounds in the urine (Ward et al. 2005). Positive correlations between exposure to drinking water with NO_3^- levels above the critical limit established by USEPA and incidence and mortality rates of certain types of cancer in humans have been observed in several countries (Ward et al. 2005). Chiu et al. (2007) found a significant positive relationship between bladder cancer mortality and NO_3^- exposure from drinking water in Taiwan. In a German community, a group of people supplied with drinking water with 60 mg L^{-1} NO_3^--N showed a significantly higher incidence of urological cancers compared to those consuming water with 10 mg L^{-1} NO_3^--N (Volkmer et al. 2005). In the United States and Slovakia, significant positive correlations between the levels of NO_3^- in water and non-Hodgkin's lymphoma were found (Ward et al. 1996; Gulis et al. 2002). In both the United States and Europe, higher levels of NO_3^- were found in wells located in agricultural areas (Ward et al. 2005). Manure application and fertilizer use are the main contributing factors for the higher NO_3^- levels in groundwater in agricultural areas.

Vegetables are generally the main source of NO_3^- intake by humans (Zeegers et al. 2006). Nitrate-N concentration in vegetables could be more than 50 times higher than the critical limit of 10 mg L^{-1} for drinking water (L'hirondel et al. 2006). However, research conducted in several locations indicates no relationship or a negative relationship between vegetable intake of NO_3^- and human cancers (Ward et al. 1996; Zeegers et al. 2006). This has been attributed to the presence of nitrosation inhibitors in vegetables such as antioxidants, e.g., vitamin C, (Ward et al. 2005).

8.4.2 Trace Metals

Human intake of trace metals occurs mainly by inhalation of dust, direct ingestion of soil, and consumption of food plants grown in metal-contaminated soils. Although rare, there have been cases of illness and death associated with high levels of certain trace metals in soils (Senesil et al. 1999). A well-known example is Itai Itai disease, characterized by severe bone pain, which is common in residents of the Jintsu river basin in Japan and is related to the accumulation of Cd in the body (Tackeuchi et al. 1990). Soils and water in this region are heavily contaminated with Cd through mining activity.

Cases of cancer linked to high levels of Cu and Zn in soils have been observed in England (Senesil et al. 1999) and Finland (Marjanen 1980). Marjanen (1980) suggested that an abundance of Zn and a deficiency of Mn in food prevents Se absorption. Selenium seems to play a role in preventing cancer, especially ovarian and breast (Senesil et al. 1999). High levels of Cu and Zn, together with Mn, Fe, Cd, and Pb in maternal plasma, oviduct and/or uterine fluids, also seem to be a contributing factor in inhibiting human conception (Hanna et al. 1997). Unusually high rates of multiple sclerosis in Henribourg, Saskatchewan, Canada, led to a comparison of soil and water samples between this location and others nearby (Irvine and Schiefer 1988, 1989). The soil and water samples from Henribourg had higher Zn but lower Se (water only). A cluster of multiple sclerosis cases in a small community in Illinois was also associated with high levels of Zn in soil and water (Schiffer et al. 2001).

Long-term application of cattle manure did not seem to influence the levels of Cu, Fe, and Mn in barley grain under both rainfed and irrigated conditions in Southern Alberta (Benke et al., unpublished data). However, Zn content in barley grain significantly increased from 36.4 (nonmanured or control) to 49.8, 50.4, and 60.0 mg kg^{-1} when cattle manure was applied at 60, 120, and 180 Mg $ha^{-1}yr^{-1}$ under irrigated conditions (Benke et al., unpublished data). These Zn contents are above the maximum Zn levels of 20 mg kg^{-1} in food (based on fresh weight) for human health established by the Chinese Department of Preventive Medicine (Islam et al. 2007). Southern Alberta is subject to strong Chinook winds (warm westerly winds that may gust to 120 km h^{-1}). Therefore, agricultural soils in the area are very vulnerable to wind erosion (Larney and Bullock 1994). As a consequence, relatively high levels of Zn (187.5 mg kg^{-1}) in the surface soil following high rates of long-term cattle manure applications (Benke et al. 2007) could be a hazard for human health due to dust inhalation.

8.5 CONCLUSIONS

Long-term high rates of manure applications generally affect soil properties. As a consequence, manure applications can also influence plant nutrient concentrations and surface water and groundwater quality with implications for animal and human health. Long-term continuous cattle manure N-based applications increase NO_3^-, P, K, and Zn in the soil (Hao and Chang 2002; Hao et al. 2003; Benke et al. 2007). Even though these increases were more significant at higher manure rates, accumulation of P, K, and Zn in the surface soil was also observed at recommended agronomic rates.

Elevated concentrations of these elements were also found in triticale, corn, and barley grown in manured plots (Hao and Chang unpublished data).

High levels of NO_3^- in animal forage could cause toxicity. High P content in the forage can result in a lower-than-recommended Ca:P ratio for maximum milk yield in dairy cattle. The K/(Ca + Mg) ratio in all crops from manured soil surpassed 2.2, which creates a risk of grass tetany development in cattle. However, the levels of Cu and Zn in the forage were within the required amounts for livestock. These results clearly indicate that there were changes in nutrient levels in feed produced from soil receiving long-term manure applications, and these changes should be taken into consideration when formulating beef and dairy cattle diets instead of relying only on the standard nutrient contents provided by NRC (2000, 2001).

The major adverse human health risk from high rates of long-term livestock manure applications is NO_3^- accumulating in soil and then migrating to surface water and groundwater. High levels of NO_3^- in drinking water have been positively related to several types of cancer, whereas high levels of Zn in soil and water have been linked to cancer and multiple sclerosis. Under irrigation, long-term manure applications resulted in Zn concentrations in barley grain above the critical threshold for Zn in food established by the Chinese Department of Preventive Medicine. Published data suggest that long-term high rates of manure application to agricultural land would not be sustainable. As a result, many countries, including Canada, have adopted regulations to decrease the amount of N and P added to cropped land through manure applications. Recent research has shown that long-term manure applications to soils are becoming one of the major sources of Cu and Zn in surface soils, indicating the need to include trace metal content in manure application guidelines (Nicholson et al. 2003; Bolan et al. 2004).

ACKNOWLEDGMENTS

Funding for this research program was provided by Agriculture and Agri-Food Canada. The contributions from everyone who participated in this research program since 1973 are greatly appreciated.

REFERENCES

Basso, B. and J. T. Ritchie. 2005. Impact of compost, manure and inorganic fertilizer on nitrate leaching and yield for a 6-year maize-alfalfa rotation in Michigan. *Agriculture, Ecosystems and Environment* 108: 329–341.

Benke, M. B., S. P. Indraratne, X. Hao, C. Chang, and T. B. Goh. 2008. Trace element changes in soil after long-term cattle manure applications. *Journal of Environmental Quality* 37: 798–807.

Blair, N., R. D. Faulkner, A. R. Till, M. Korschens, and E. Schulz. 2006. Long-term management impacts on soil C, N and physical fertility: Part II: Bad Lauchstadt static and extreme FYM experiments. *Soil and Tillage Research* 91: 39–47.

Bolan, N. S., D. C. Adriano, and S. Mahimairaja. 2004. Distribution and bioavailability of trace elements in livestock and poultry manure by-products. *Critical Review Environmental Science Technology* 34: 291–338.

Bouchard, D. C., M. K. Williams, and R. Y. Surampalli. 1992. Nitrate contamination of groundwater: Sources and potential health effects. *Journal of American Water Works Association* 85: 85–90.

Bruning-Fann, C. S. and J. B. Kaneene. 1993. The effects of nitrate, nitrite and *N*-nitroso compounds on human health: A review. *Veterinary and Human Toxicology* 35: 521–538.

Bruning-Fann, C., J. B. Kaneene, R. A. Miller, I. Gardner, R. Johnson, and F. Ross. 1994. The use of epidemiological concepts and techniques to discern factors associated with the nitrate concentration of well water on swine farms in the USA. *Science of the Total Environment* 153: 85–96.

Burkart, M. R. and D. E. James. 1999. Agricultural-nitrogen contributions to hypoxia in the Gulf of Mexico. *Journal of Environmental Quality* 28: 850–859.

Buxton, D. R. and S. L. Fales. 1994. Plant environment and quality. In *Forage Quality, Evaluation, and Utilization*, G. C. Fahey Jr., M. Collins, D. R. Mertens, and L. E. Moser (Eds.), Madison, WI: American Society of Agronomy, Inc., pp. 155–199.

Callan, N. W. and M. P. Westcott. 1996. Drip irrigation for application of potassium to tart cherry. *Journal of Plant Nutrition* 19: 163–172.

Canfield, R. W., C. J. Sniffen, and W. R. Butler. 1990. Effects of excess degradable protein on postpartum reproduction and energy balance in dairy cattle. *Journal of Dairy Science* 73: 2342–2349.

Cherney, J. H., E. A. Mikhailova, and D. J. R. Cherney. 2002. Tetany potential of orchardgrass and tall fescue as influenced by fertilization with dairy manure or commercial fertilizer. *Journal of Plant Nutrition* 25: 1501–1525.

Chiu, H. F., S. S. Tsai, and C. Y. Yang. 2007. Nitrate in drinking water and risk of death from bladder cancer: An ecological case-control study in Taiwan. *Journal of Toxicology and Environmental Health* 70: 1000–1004.

Christie, P. and J. A. M. Beattie. 1989. Grassland soil microbial biomass and accumulation of potentially toxic metals from long-term slurry applications. *Journal of Applied Economics* 26: 597–612.

Dach, J. and D. Starmans. 2005. Heavy metals balance in Polish and Dutch agronomy: Actual state and previsions for the future. *Agriculture, Ecosystems and Environment* 107: 309–316.

Daliparthy, J., A. V. Barker, and S. S. Mondal. 1994. Potassium fractions with other nutrients in crops: A review focusing on the tropics. *Journal of Plant Nutrition* 17: 1859–1886.

Daniel, T. C., A. N. Sharpley, and J. L. Lemunnyon. 1998. Agricultural phosphorus and eutrophication: A symposium overview. *Journal of Environmental Quality* 27: 251–257.

Delgado, C., M. Rosegrant, H. Steinfeld, S. Ehui, and C. Courbois. 1999. *Livestock to 2020: The Next Food Revolution.* Washington International Food Policy Research Institute, Food and Agriculture Organization of the United Nations, International Livestock Research Institute.

Dunn, T. G. and G. E. Moss. 1992. Effects of nutrient deficiencies and excesses on reproductive efficiency of livestock. *Journal of Animal Science* 70: 1580–1593.

Edwards, D. R. and T. C. Daniel. 1992. Environmental impacts of on-farm poultry waste disposal—A review. *Bioresource Technology* 41: 9–33.

Eghball, E., G. D. Binford, and D. D. Baltensperger. 1996. Phosphorus movement and adsorption in a soil receiving long-term manure and fertilizer application. *Journal of Environmental Quality* 25: 1339–1343.

Ferm, M. 1998. Atmospheric ammonia and ammonium transport in Europe and critical loads: A review. *Nutrient Cycling In Agroecosystems* 51: 5–17.

Gallaher, R. N., W. L. Parks, and L. M. Josephson. 1975. Some factors influencing yield and cation sum and ratios in corn. *Communications in Soil Science and Plant Analysis* 6: 51–61.

Gerber, P., P. Chilonda, G. Franceschini, and H. Menzi. 2005. Geographical determinations and environmental implications of livestock production intensification in Asia. *Bioresource Technology* 96: 263–276.

Geurink, J. H., A. Malestein, A. Kemp, and A. Th. van't Klooster. 1979. Nitrate poisoning in cattle. 3. The relationship between nitrate intake with hay or fresh roughage and the speed of intake on the formation of methemoglobin. *Netherlands Journal of Agriculture Science* 27: 268–276.

Gilbertson, C. B., F. A. Norstadt, A. C. Mathers et al. 1979. *Animal Waste Utilization on Cropland and Pastureland: A Manual for Evaluating Agronomic and Environmental Effects*. Washington, DC: USDA Utilization Res. Rep. 6. EPA-600 / 2-79-059.

Grunes, D. L., P. R. Stout, and J. R. Brownell. 1970. Grass tetany of ruminants. *Advances in Agronomy* 22: 331–374.

Gulis, G., M. Czompolyova, and J. R. Cerhan. 2002. An ecological slurry of nitrate in municipal drinking water and cancer incidence in Trnava District, Slovakia. *Journal of Environmental Research and Public Health* 88: 182–187.

Hanna, L. A., J. M. Peters, L. M. Wiley, M. S. Clegg, and C. L. Keen. 1997. Comparative effects of essential and nonessential metals on preimplantation mouse embryo development in vitro. *Toxicology* 116: 123–131.

Hao, X. and C. Chang. 2002. Effect of 25 annual cattle manure applications on soluble and exchangeable cations in soil. *Soil Science* 167: 126–134.

Hao, X. and C. Chang. 2003. Does long-term heavy cattle manure application increase salinity of a clay loam soil in semi-arid southern Alberta? *Agriculture, Ecosystems and Environment* 94: 98–104.

Hao, X. and C. Chang. 2007. Effect of long-term cattle manure application on triticale and corn feed quality. In *Biogeochemistry of Trace Elements: Environmental Protection, Remediation and Human Health*, Y. Zhu, N. Lepp, and R. Naidu (Eds.), 270–272. Beijing: Tsinghua University Press, pp. 270–272.

Hao, X., C. Chang, G. R. Travis, and F. Zhang. 2003. Soil carbon and nitrogen response to 25 annual cattle manure applications. *Journal of Plant Nutrition and Soil Science* 166: 239–245.

Hao, X., T. McAllister, and Y. Wang. 2007. Effect of a lignite-coal extract on nutrient composition and gas emissions from cattle feedlot manure. *Canadian Journal of Soil Science* 87: 281–290.

Hao, X., F. Godlinski, and C. Chang. 2008. Distribution of phosphorus forms in soil following long-term continuous and discontinuous cattle manure applications. *Soil Science Society of America Journal* 72: 90–97.

Ihnat, M. and L. Fernandes. 1996. Trace elemental characterization of composted poultry manure. *Bioresource Technology* 57: 143–156.

Irvine, D. G. and H. B. Schiefer. 1988. Geotoxicology of multiple sclerosis: The Henribourg, Saskatchewan, cluster focus. II. The soil. *Science of the Total Environment* 77: 175–188.

Irvine, D. G. and H. B. Schiefer. 1989. Geotoxicology of multiple sclerosis: The Henribourg, Saskatchewan, cluster focus. I. The water. *Science of the Total Environment* 84: 45–59.

Islam, E., X. Yang, Z. He, and Q. Mahmood. 2007. Assessing potential dietary toxicity of heavy metals in selected vegetables and food crops. *Journal of Zhejiang University Science. B* 8: 1–13.

Jones, S. K., R. M. Rees, U. M. Skiba, and B. C. Ball. 2007. Influence of organic and mineral N fertilizer on NO_2 fluxes from a temperate grassland. *Agriculture, Ecosystems and Environment* 121: 74–83.

Jordan, E. R. and L. V. Swanson. 1978. Effect of crude protein on reproductive efficiency, serum total protein, and albumin in the high-producing dairy cow. *Journal of Dairy Science* 62: 58–63.

Kabata-Pendias, A. and H. Pendias. 2001. *Trace Metals in Soils and Plants*, 3rd ed. Boca Raton: CRC Press.

Kyllmar, K., C. Carlsson, A. Gustafson, B. Ulen, and H. Johnsson. 2006. Nutrient discharge from small agricultural catchments in Sweden—Characterization and trends. *Agriculture, Ecosystems and Environment* 115: 15–26.

Larney, F. J. and M. S. Bullock. 1994. Influence of soil wetness at time of tillage and tillage implement on soil properties affecting wind erosion. *Soil and Tillage Research* 29: 83–95.

Liebhardt, W. C. 1976. Soil characteristics and corn yield as affected by previous applications of poultry manure. *Journal of Environmental Quality* 5: 459–462.

Liebhardt, W. C. and J. G. Shortall. 1974. Potassium is responsible for salinity in soils amended with poultry manure. *Communications in Soil Science and Plant Analysis* 5: 385–398.

Liebhardt, W. C., C. Golt, and J. Tupin. 1979. Nitrate and ammonium concentrations of ground water resulting from poultry manure applications. *Journal of Environmental Quality* 8: 211–215.

L'hirondel, J., Avery, A. A., and T. Addiscott. 2006. Dietary nitrate: where is the risk? *Environmental Health Perspective* 114: A458–A459.

Lijinsky, W. 1986. The significance of *N*-nitroso compounds as environmental carcinogens. *Journal of Environmental Science and Health* C4: 1–45.

Marjanen, H. 1980. On the relationship between the contents of trace elements in soils and plants and the cancer incidence in Finland. In *Geochemical Aspects in Present and Future Research*, J. Lag (Ed.), Oslo: Universitatesforlaget, pp. 149–166.

Mathers, A. C. and B. A. Stewart. 1974. Corn silage yield and soil chemical properties as affected by cattle feedlot manure. *Journal of Environmental Quality* 3: 143–147.

Mayland, H. F., P. R. Cheeke, W. Majak, and J. P. Goff. 2007. Forage-induced animal disorders. In *Forages—The Science of Grassland Agriculture*, 6th ed. R. F. Barns, C. J. Nelson, K. J. Moore, and M. Collins (Eds.), Oxford: Blackwell, pp. 687–707.

Mayland, H. F., L. W. Green, D. L. Robinson, and S. R. Wilkinson. 1990. Grass tetany: A review of Mg in the soil-plant-animal continuum. In *Proceedings of Pacific Northwest Animal Nutrition Conference*, 29–41, November 6–8, Vancouver, BC, Canada.

Maynard, D. N., A. V. Barker, P. L. Minotti, and N. H. Peck. 1976. Nitrate accumulation in vegetables. *Advances in Agronomy* 28: 71–118.

Motavalli, P. P. and R. J. Miles. 2002. Soil phosphorus fractions after 11 years of animal manure and fertilizer applications. *Biology and Fertility Soils* 36: 35–42.

Nicholson, F. A., B. J. Chambers, J. R. Williams, and R. J. Unwin. 1999. Heavy metal content of livestock feeds and animal manures in England and Wales. *Bioresource Technology* 70: 23–31.

Nicholson, F. A., S. R. Smith, B. J. Alloway, C. Carlton-Smith, and B. T. Chambers. 2003. An inventory of heavy metals inputs to agricultural soils in England and Wales. *Science of the Total Environment* 311: 205–219.

NRC (National Research Council). 2000. *Nutrient Requirements of Beef Cattle*. 7th revised ed. Washington, D.C.: National Academy Press.

NRC (National Research Council). 2001. *Nutrient Requirements of Dairy Cattle*. 7th revised ed. Washington, D.C.: National Academy Press.

Prins, W. H. 1984. Limits of nitrogen fertilizer on grassland. *Netherlands Journal of Agricultural Sciences* 32: 319–321.

Reid, R. L. 1994. Nitrogen components of forage and feedstuffs. In *Principles of Protein Nutrition of Ruminants*, J. M. Asplund (Ed.), Boca Raton: CRC Press, pp. 43–70.

Richards, R. A., C. W. Dennett, C. O. Quaslset, E. Epstein, J. D. Norlyn, and M. D. Winslow. 1987. Variation in yield of grain and biomass in wheat, barley, and triticale in a salt-affected field. *Field Crops Research* 15: 277–287.

Rodvang, S. J., D. M. Mikalson, and M. C. Ryan. 2004. Changes in ground water quality in an irrigated area of southern Alberta. *Journal of Environmental Quality* 33: 476–487.

Sager, M. 2007. Trace and nutrient elements in manure, dung and compost samples in Austria. *Soil Biology Biochemistry* 39: 1383–1390.

Schiffer, R. B., M. P. McDermott, and C. Copley. 2001. A multiple sclerosis cluster associated with a small, north-central Illinois community. *Archives of Environmental Health* 56: 389–395.

Schilke-Gartley, K. L. and J. T. Sims. 1993. Ammonia volatilization from poultry manure-amended soils. *Biology and Fertility Soils* 16: 5–10.

Senesil, G. S., G. Baldassarre, N. Senesi, and B. Radina. 1999. Trace element inputs into soils by anthropogenic activities and implications for human health. *Chemosphere* 39: 343–377.

Sharpley, A. N., S. J. Smith, B. A. Stewart, and A. C. Mathers. 1984. Forms of phosphorus in soil receiving cattle feedlot waste. *Journal of Environmental Quality* 13: 211–215.

Simon, C. 1966. Nitrite poisoning from spinach. *Lancet* 287: 872.

Sims, J. T. and D. C. Wolf. 1994. Poultry waste management: agricultural and environmental issues. *Advances in Agronomy* 52: 1–83.

Smart, M. and N. F. Cymbaluk. 1997. Role of nutritional supplements in bovine lameness—review of nutritional toxicities. In *Lameness in Cattle*, 3rd ed. P. R. Greenough, F. J. MacCallum, and A. D. Weaker (Eds.), Philadelphia: W. B. Saunders, pp. 145–161.

Statistics Canada. 2007. Cattle statistics. Catalog no. 23-012-XIE, vol 6, no. 2. http: //dsp-psd.pwgsc.gc.ca/collection_2007/statcan/23-012-X/23-012-XIE2007001.pdf (accessed October 30, 2007).

Stout, W. L. and D. E. Baker. 1981. Effect of differential adsorption of potassium and magnesium in soils on magnesium uptake by corn. *Soil Science Society of America Journal* 45: 996–997.

Stuedemann, J. A., S. R. Wilkinson, D. J. Williams et al. 1975. Long-term broiler litter fertilization of tall fescue grass pastures and health and performance of beef cows. In *Managing Livestock Wastes*, American Societal Agricultural English Publication Proceeding no. 275, St. Joseph, MI, pp. 265–268.

Tackeuchi, T., Y. Nakano, A. Aoki, S. Ohmori, and M. Kasuya. 1990. Elemental concentrations in hair of inhabitants of a cadmium contaminated area. In *Environmental Life Elements and Health*, T. Jian'an, P. J. Peterson, L. Ribang, and W. Wuyi (Eds.), Beijing: Science Press, pp. 253–259.

USDA. 1997. *Agricultural Statistics*. Washington, D.C.: U.S. Government Printing Office.

USDA. 2007. *National Agricultural Statistics Service*. http: //www.nass.usda.gov/QuickStats/index2.jsp (accessed October 30, 2007).

USEPA. 1973. Proposed criteria for water quality. Vol. I. Quality of water for livestock. *Environmental Reports of Current Developments* 4: 663–664.

Van Broekhoven, L. W., J. A. R. Davies, and J. H. Geurink. 1989. The metabolism of nitrate and praline in the rumen fluid of a cow and its effect on in vivo formation of *N*-nitrosoproline. *Netherlands Journal of Agriculture Science* 37: 157–163.

Volkmer, B. G., B. Ernst, J. Simon et al. 2005. Influence of nitrate levels in drinking water on urological malignancies: A community-based cohort study. *British Journal of Urology International* 95: 972–976.

Walker, D. J. and M. P. Bernal. 2008. The effects of olive mill waste compost and poultry manure on the availability and plant uptake of nutrients in a highly saline soil. *Bioresource Technology*, 99: 396–403.

Ward, M. H., S. D. Mark, K. P. Cantor, D.D. Weisenburger, A. Correa-Villasenor, and S. H. Zahm. 1996. Drinking water nitrate and the risk of non-Hodgkin's lymphoma. *Epidemiology* 7: 465–471.

Ward, M. H., T. M. deKok, P. Levallois et al. 2005. Workgroup report: Drinking-water nitrate and health—Recent findings and research needs. *Environmental Health Perspective* 131: 1607–1614.

Whalen, J. and C. Chang. 2001. Phosphorus accumulation in cultivated soils from long-term annual applications of cattle feedlot manure. *Journal of Environmental Quality* 30: 229–237.

Williams, P. E. V. 1995. Animal production and European pollution problems. *Animal Feed Science Technology* 53: 135–144.

Woodin, S. J. and A. M. Farmer. 1993. Impacts of sulphur and nitrogen deposition on sites and species of nature conservation importance in Great Britain. *Biology Conservation* 63: 23–30.

Wright, M. J. and K. L. Davison. 1964. Nitrate accumulation in crops and nitrate poisoning in animals. *Advances in Agronomy* 16: 197–247.

Zeegers, M. P., R. F. M. Selen, J. C. S. Kleinjans, R. A. Goldbohm, and P. A. van den Brandt. 2006. Nitrate intake does not influence bladder cancer risk: The Netherlands cohort study. *Environmental Health Perspective* 114: 1527–1531.

9 How Does Aerobic Rice Take Up Zinc from Low Zinc Soil? *Mechanisms, Trade-Offs, and Implications for Breeding*

Xiaopeng Gao, Thomas W. Kuyper, Fusuo Zhang, Chunqin Zou, and Ellis Hoffland

CONTENTS

9.1 INTRODUCTION

Scarcity of water is causing a shift from flooded (anaerobic) to aerobic conditions for rice production in North China, where many soils are zinc (Zn) deficient because of high pH and high carbonate content. Aerobic rice varieties are newly bred by crossing lowland with upland varieties and are grown as a dry field crop in irrigated but nonflooded and nonpuddled soils. The introduction of aerobic rice on low Zn soils places the problem of Zn deficiency in rice in a new perspective. The aim of our work is to understand the mechanisms of Zn mobilization in aerobic rice. This happens by using as a tool the genotypic variation in tolerance to Zn deficiency among aerobic rice. Genotypes of plants vary widely in tolerance to Zn deficiency. A mechanistic explanation, however, is still lacking.

The ability of a plant to take up Zn from the soils is considered to be the most determining factor of plant tolerance to Zn deficiency. Plant Zn uptake from low Zn soils can be increased both by Zn-mobilizing rhizosphere processes and by inoculation with arbuscular mycorrhizal fungi (AMF). Variation in responsiveness to inoculation with AMF among plant genotypes has been often documented, but the genetic basis remains poorly understood. We expect there is variation in mycorrhizal responsiveness based on Zn uptake among aerobic rice genotypes. A negative correlation is hypothesized between Zn uptake efficiency and mycorrhizal Zn responsiveness, which may indicate a trade-off between the two processes. Consequently, combining high Zn mobilization by root-induced chemical changes in rhizosphere and mycorrhizal effects seems impossible. The competition for carbon between production of low-molecular-weight organic acids in rhizosphere and for mycorrhizal fungi would cause the trade-off between both mechanisms for Zn acquisition. Implications for breeding Zn efficient aerobic rice are discussed.

9.2 RESEARCH BACKGROUND

Zinc deficiency has been reported in various parts of the world in annual crops (Cakmak 2002). About 30% of the world's soils are low in Zn. Low Zn soils can mainly be found in Australia, South Asia, and Central America (Alloway 2004). In China, more than 483 Mha of soils are Zn deficient, covering 50% of the total cultivated area (Lin and Li 1997). Low Zn soils are mainly found in North China, where the dominant soil type is calcareous with pH values higher than 7. The high pH and high calcium carbonate content reduce Zn availability due to the increased adsorptive capacity, the formation of hydrolyzed forms of Zn, and possible chemisorption on calcium carbonate. High bicarbonate concentrations (15–40 mM) strongly inhibit Zn uptake by rice roots and particularly transport to the shoots (Dogar and Hai 1980). In North China, the main crops affected by Zn deficiency are rice and maize (Takkar and Walker 1993). Worldwide, Zn deficiency affects over two billion people, and it is estimated that 14% of the total population in China is at risk of inadequate Zn intake (Chen 2004). Awareness is growing that Zn concentration in staple foods needs to be increased by agricultural approaches, and some global joint programs are being conducted with the aim of improving the Zn nutrition of humans.

Currently, a cultivation system shift in rice production is taking place in China caused by increasing water shortage. The traditional lowland system with flooded fields is being replaced by an aerobic system. Aerobic rice varieties are currently developed by crossing lowland with upland varieties and are grown as a dry field crop in irrigated but nonflooded and nonpuddled soil (Bouman et al. 2005). In China, breeders have produced aerobic rice varieties with an estimated yield potential of 6–7 Mg ha^{-1} that are now being pioneered by farmers on some 190,000 ha in North China (Wang et al. 2002). So far, the research has mainly focused on the water use and yield potential of the aerobic varieties. In practice, however, soils in North China are frequently Zn deficient. Investigations on the mechanisms of Zn acquisition and potential trade-offs between different mechanisms of Zn acquisition by new aerobic rice varieties are therefore needed for further genetic improvement of aerobic rice.

Research in the present chapter was conducted in the context of the Wageningen University program "From Natural Resources to Healthy People" (http://www.inref.wur.nl/), which investigates the fate of Zn in food chain and aims to develop agriculture- and/or food-based interventions to alleviate nutritional deficiencies of human consumers, especially those from the urban and rural regions in West Africa and China. In this chapter, we focus on processes involved in the transfer of Zn from soil to plants. More detailed information on this program is available in other chapters of this book.

9.3 ZINC (Zn) EFFICIENCY AMONG AEROBIC RICE VARIETIES

Genotypes of crop plants can vary widely in Zn efficiency, which is defined as the capacity of a genotype to grow well under Zn deficient conditions (Graham et al. 1992). Zinc efficiency can be expressed either by the relative shoot dry weight:

Zn efficiency = [(shoot dry weight at low Zn level) / (shoot dry weight at adequate Zn level)]

or by taking into account the biomass potential of genotypes;

Zn efficiency = [(shoot dry weight at low Zn level / experimental mean shoot dry weight at low Zn level) / (shoot dry weight at adequate Zn level / experimental mean shoot dry weight at adequate Zn level)]

By considering both formulas, we reported the existence of variation in Zn efficiency among some newly bred aerobic rice varieties (Gao et al. 2005). With 15 aerobic rice genotypes grown in a pot experiment, we observed a substantial Zn efficiency variation ranging from 50 to 94% (Figure 9.1). This variation offers opportunities for using breeding as a tool to resolve Zn deficiency problems in rice, and also suggests that in the newly developed aerobic rice production system, Zn deficiency problems can be alleviated by the development and use of genotypes with high tolerance to (efficiency in) low soil Zn conditions. Next to aerobic rice, substantial variation in Zn efficiency has been demonstrated for many crops including wheat (Cakmak et al. 2001), maize (Ramani and Kannan 1985), common bean (Hacisalihoglu et al. 2004), and lowland rice (Sakal et al. 1989). The precise mechanisms underlying Zn efficiency, however, are not yet clear. More than one mechanism is assumed to be responsible (Graham and Rengel 1993). Hacisalihoglu and Kochian (2003) thoroughly reviewed the possible factors, including (1) seed Zn content, (2) uptake of Zn from low Zn soil, (3) translocation of Zn in plants, (4) shoot Zn concentration, and (5) biochemical Zn utilization.

Within 15 aerobic rice genotypes, we found that variation in Zn efficiency is associated with Zn uptake and translocation from root to shoot (Figure 9.2). This variation indicates that the aerobic rice genotypes with a high tolerance to Zn deficiency are

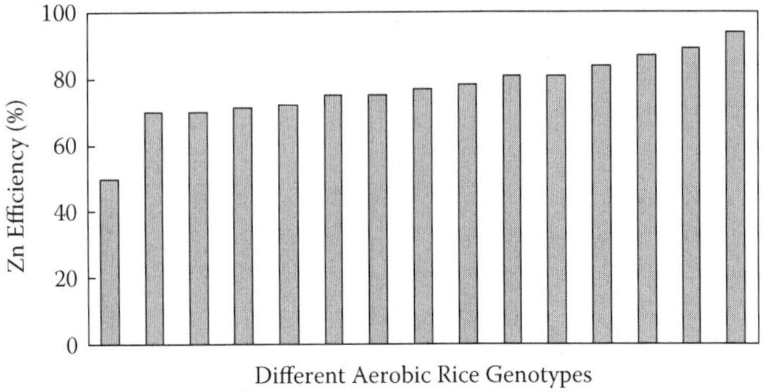

FIGURE 9.1 Variation in Zn efficiency among 15 aerobic rice genotypes. Zinc efficiency was calculated as the ratio of shoot dry weight under low Zn supply over that of an under-adequate supply.

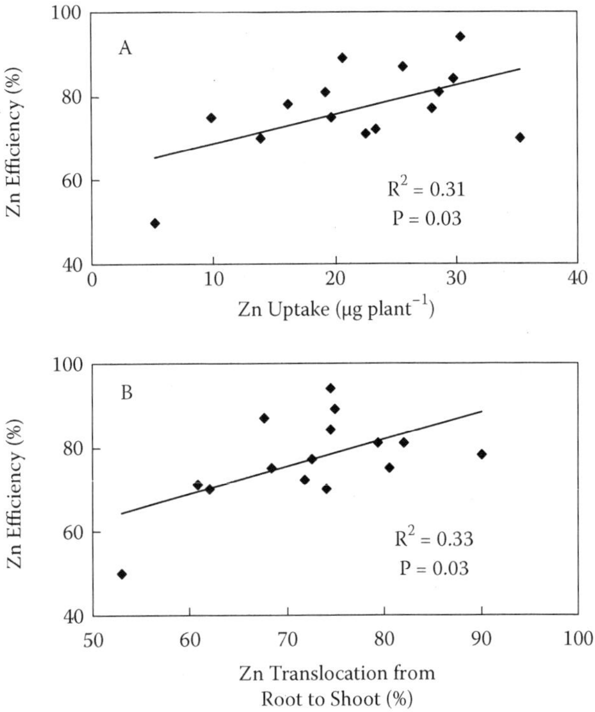

FIGURE 9.2 Correlation between Zn efficiency and Zn uptake (A) or Zn translocation from root to shoot (B) for 15 aerobic rice genotypes. (From Gao et al. 2005. *Plant and Soil* 278: 253–261. With permission.)

those that have a high capacity to take up Zn from the soil and transfer Zn to the shoot. Our results are in agreement with results obtained with wheat (Genc et al. 2006; Cakmak et al. 1997a), chickpea (Khan et al. 1998), and common bean (Hacisalihoglu et al. 2004). With Zn efficiency as the dependent variable and several parameters including Zn uptake, Zn translocation from root to shoot, seed Zn content, and shoot Zn concentration as independent variables, we did multiple stepwise regression analysis to determine the combination of factors that explained the variation in Zn efficiency statistically. Results showed that 62% of Zn efficiency could be explained for by Zn uptake (33%) and Zn translocation to the shoot (29%). Seed Zn content and shoot Zn concentration were not contributing variables. This relatively low percentage of explained variation in Zn efficiency (62%) confirms previous findings (e.g., on wheat by Cakmak et al. 2001) that tolerance to Zn deficiency is a complex trait in which many plant characteristics are involved.

We hypothesized that the unexplained variation in Zn efficiency (38%) might be related to differences in biochemical Zn utilization and Zn retranslocation from older into younger tissues in shoots. In several studies based on solution culture, Zn efficiency of different wheat genotypes positively correlated with activities of enzymes such as Cu/Zn SOD (superoxide dismutase) and carbonic anhydrase (Cakmak et al. 1997b; Hacisalihoglu et al. 2003). The efficient genotypes can maintain proper functioning of enzymes under low Zn conditions and thus the biochemical Zn utilization was considered as an important component of Zn efficiency in wheat. Also, in a study with flooded rice plants, Sasaki et al. (1998) found that the level of carbonic anhydrase mRNA decreased under Zn deficiency. Further investigations are required to quantify the role of biochemical Zn utilization in Zn efficiency of rice plants. Another possible mechanism that might be related to the unexplained 38% Zn efficiency is the retranslocation of Zn from older parts into young parts in shoot. In our study, we only examined the translocation capacity of Zn from root to shoot. In several studies, Zn retranslocation from old parts to young parts of the shoot was suggested as one possible mechanism affecting Zn efficiency, e.g., in common bean (Hacisalihoglu et al. 2004), wheat (Torun et al. 2000), and rice (Hajiboland et al. 2001).

More than one mechanism is often responsible for the level of Zn efficiency in a particular genotype. Genc et al. (2006) also showed the complexity of mechanisms in Zn efficiency among wheat genotypes. Similar to what we have found on aerobic rice, a number of different mechanisms were suggested to contribute to Zn efficiency in bread wheat. Genc et al. (2006) reported that Zn uptake was the most important determining factor, where Zn translocation from root to shoot had a relative small effect on Zn efficiency in bread wheat. These results lead to a generalization that the primary reason for variation in Zn efficiency among cereal genotypes revolves around the plants ability to take up Zn from the soil. Most studies leading to this generalization are based on soil culture experiments and they only examined early vegetative growth stage. In solution-culture-based experiments, variation in Zn efficiency may not be explained by variation in Zn uptake because of the absence of rhizosphere. Regarding the estimation of Zn efficiency based on grain yield rather than early vegetative growth, it is possible that additional mechanisms might be involved. Also, the relative importance of each mechanism might be different in different growth stages. For instance, an efficient barley genotype was able to remobilize greater amounts of Zn from vegetative to reproductive organs than the inefficient genotypes (Genc et al. 2004).

9.4 MECHANISMS OF Zn UPTAKE BY AEROBIC RICE

The major mechanisms increasing nutrient acquisition by plant roots are (1) the increase of the root surface area, (2) the involvement of mycorrhizas, and (3) root-induced chemical changes in the rhizosphere (Figure 9.3, see Marschner 1998). These mechanisms are specifically important for the immobile nutrients such as P and Zn. In the following text, we are going to discuss the contribution of each possible mechanism to plant Zn uptake, with an emphasis on aerobic rice. The potential trade offs between different mechanisms will also be discussed.

9.4.1 Role of Root Growth in Zn Uptake

Zinc bioavailability is limited by low mobility in the soil solution. Accordingly, root growth and surface area are important parameters determining Zn uptake. For aerobic rice, we found a positive correlation between root surface area and Zn uptake of 15 genotypes grown on a low Zn soil (Figure 9.4). On the one hand, the positive correlation highlights the importance of root growth on plant Zn acquisition. On the other hand, the relatively low explained percentage ($R^2 = 0.33$) suggests additional mechanisms other than root growth might play a major role in determining the capacity of aerobic rice to take up Zn from the soil. Root architecture and, more specifically, the distribution of root hairs determine the volume of soil from which Zn can be extracted. By studying root morphology of three wheat genotypes differing in Zn efficiency, Dong et al. (1995) found that more Zn-efficient genotypes had longer and thinner roots and a greater proportion of thinner roots in the early growth period. Better understanding of the regulation of root growth can help to develop better management strategies in agricultural production. For example, the localized

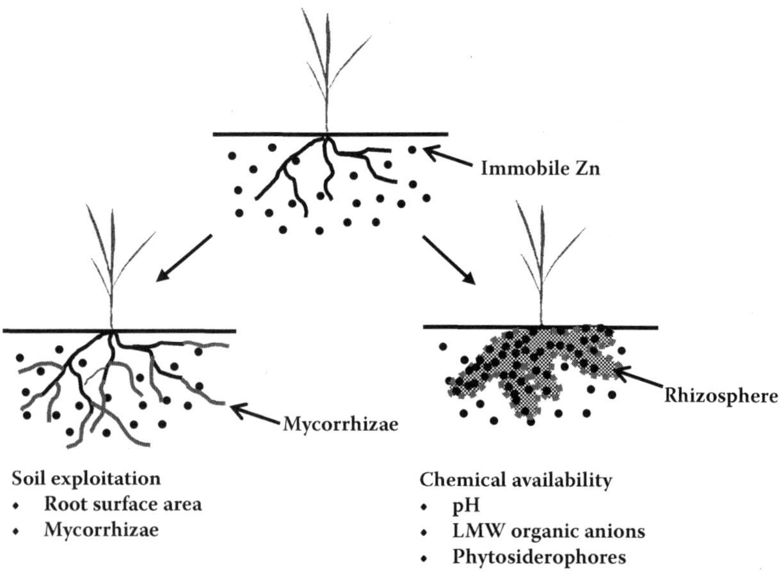

FIGURE 9.3 Hypothetical factors affecting Zn uptake by plants.

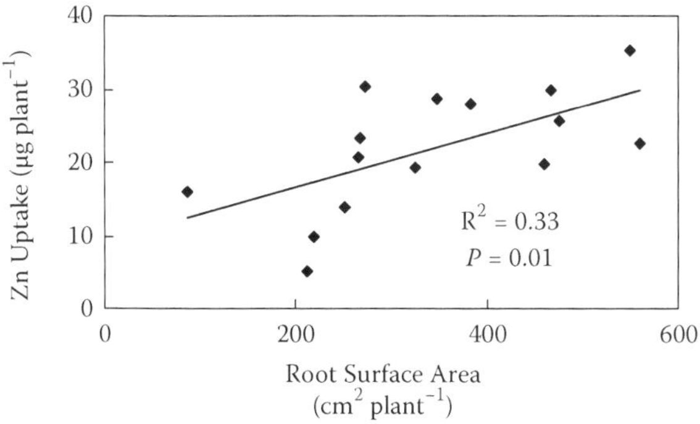

FIGURE 9.4 Correlation between Zn uptake and root surface area for 15 aerobic rice genotypes. (From Gao et al. 2005. *Plant and Soil* 278: 253–261. With permission.)

supply of macro- or micronutrients has been supposed to affect root development. A localized supply of NO_3^--stimulated lateral root elongation by 2–3 fold in the NO_3^--rich zone (Zhang et al. 1999). This knowledge offers the potential to increase plant Zn uptake on low Zn soils by localized fertilizer management techniques. If N fertilizer can be applied in a localized zone rather than homogeneously, the lateral root elongation might show a beneficial effect for plant Zn uptake.

9.4.2 Role of Arbuscular Mycorrhizae in Zn Uptake

Besides root growth, an increase of surface area for Zn acquisition can occur through the association between plant roots and mycorrhizal fungi. When the fungi colonize the roots, the external mycelium of arbuscular mycorrhizal fungi can access a greater volume of soil and also penetrate into smaller soil pores than the uncolonized root system alone, increasing the amount of Zn available to the plant. A large number of studies demonstrate the beneficial effect of arbuscular mycorrhizae fungi (AMF) on Zn uptake by the host plants, as reported for pigeon pea (Wellings et al. 1991), wheat (Khare et al. 1998; Ryan and Angus 2003) and wetland rice (Purakayastha and Chhonkar 2001). Upon colonization, the mycelium of the AMF increases the absorbing surface area of the symbiosis, and enhances exploration of a higher soil volume, and thereby increases nutrient uptake. With plants grown in compartmented boxes, the capacity of the external hyphae was estimated to account for 50% of the total Zn uptake in white clover (Li et al. 1991) and 25% in maize (Kothari et al. 1991). The effects are mostly reported for plants grown in soils with low Zn availability. Most studies were conducted under pot culture conditions, though several studies demonstrated the beneficial effects of AMF on plant Zn uptake under field conditions (e.g., on barley by Jakobsen [1983]).

With four aerobic rice genotypes grown in three different locations of Beijing, China, we did a preliminary field survey to gain insight in the natural colonization status of aerobic rice roots by AMF. The results showed very low mycorrhizal root colonization ranging from 0.1% to 2.7%, which might be caused by long-term phosphorus fertilization and tillage. This is in agreement with the findings of Hetrick et al. (1984), who reported that AMF showed little colonization and had no effect on winter wheat grown under field conditions. The lack of mycorrhizal colonization in the field might be due to low soil temperature in autumn. Under controlled conditions, however, levels of root colonization by AMF have been reported up to 70% of total root length for aerobic rice (Zhang et al. 2005) and 80% for wheat (Dodd and Jeffries 1989; Hetrick et al. 1992). In a field study by Jakobsen (1983), inoculation with AMF under field conditions improved the growth of barley and P uptake by mycorrhizal plants by 100% at harvest, compared with nonmycorrhizal plants. Also under field conditions, AMF inoculation at the nursery stage could increase growth, grain yield, and nutrient uptake (mainly P, Zn, and Cu) of wetland rice (Solaiman and Hirata 1997). These results indicate the feasibility of improving nutrient acquisition and growth of cereal crops on infertile soils by inoculation with AMF. As a rule, at high soil P content or high level of P-fertilization, root colonization by AMF decreases. The application of P fertilizer usually enhances P accumulation in plant shoot dry matter, while decreasing Zn accumulation, for example, in soybean (Lambert and Weidensaul 1991) or maize (Eivazi and Weir 1989). In mycorrhizal plants, however, both P and Zn contents increase in the shoot dry matter, indicating the importance of mycorrhizas in providing P and Zn to the host plants.

According to the field survey, results showed a very low percentage of mycorrhizal root colonization. The AMF colonization was also very low in the previously mentioned screening experiment in which we found substantial variation in Zn uptake among 15 aerobic rice genotypes. Thus, the observed variation in Zn uptake or tolerance to Zn deficiency was not likely related to a differential mycorrhizal effect. Combined with the relatively low contribution of root surface area to Zn uptake (33%), the root-induced chemical changes in rhizosphere might be involved in Zn acquisition by aerobic rice, as reported in lowland rice in a recent study (Wissuwa et al. 2006).

9.4.3 Role of Root-Induced Changes in Rhizosphere

Root-induced changes in the rhizosphere pH and release of low and high molecular weight exudates are of major importance for nutrient acquisition. In lowland rice, rhizosphere acidification can enhance the utilization of not only P (Kirk and Saleque 1995) but also Zn (Kirk and Bajita 1995). In addition, lowland rice grown in a nutrient solution experiment was found to respond similarly to P and Zn deficiency by increased citrate exudation (Hoffland et al. 2006). There is little information available regarding the root-induced chemical changes in the rhizosphere in this new production system because of the relative new concept of aerobic rice.

The rhizosphere of lowland rice growing in anaerobic soil can be acidified as a result of both (1) H^+ generated in Fe^{2+} oxidation by root-released O_2, and (2) H^+ release by roots to compensate for an imbalanced cation/anion uptake ratio (Begg et al. 1994). The form of nitrogen supply has the most prominent influence on cation/anion uptake ratio, and thus on rhizosphere pH (Marschner and Romheld 1983). When grown in anaerobic soil, rice roots dominantly take up N as NH_4^+, resulting in a release of H^+ by roots and a consequent decrease in rhizosphere pH. In the case of the newly developed aerobic rice cultivation, however, an increase in rhizosphere pH is expected because of the shift of the dominant N form from NH_4^+ to NO_3^-. This is expected to lead to an increased rhizosphere pH compared to the bulk soil and probably to reduced bioavailability of Zn to plants. Marschner et al. (1989) have showed in a Luvisol (pH 6.8) that the rhizosphere pH of bean plants was increased to 7.3 in nitrate-fed plants and decreased to 5.4 in ammonium-fed plants, while Zn concentrations in the plant increased from 34 to 49 mg kg^{-1} when ammonium was applied. In a study by Silber et al. (2004), the rhizosphere pH of rice flower (a new cut flower crop in Israel) supplied with N as a 3:1 NH_4/NO_3 ratio was 2 units lower than that of plants supplied with N as a 1:3 NH_4/NO_3 ratio, and as a consequence the leaf concentrations of Zn increased by a factor of 38% compared to the latter treatment. This expected shift in rhizosphere pH may partly explain the reduced Zn uptake by rice grown in aerobic fields compared to flooded fields (Gao et al. 2006).

A number of studies on root-induced chemical changes in the rhizosphere have been focused on the release of phytosiderophores (PS), a group of hydroxy- and amino-substituted iminocarboxylicacids. Graminaceous species (grasses, inclusive of cereals) respond to Fe deficiency by the release of PS (Marschner 1995). Besides Fe, phytosiderophores can also complex other micronutrients such as Zn, Mn, and Cu (Takagi et al. 1988; Treeby et al. 1989), indicating that phytosiderophore release can increase micronutrient availability in rhizosphere. Zhang et al. (1991) reported that Zn-deficient wheat released the same phytosiderophores as Fe-deficient wheat. Durum wheat genotypes that are sensitive to Zn deficiency exuded a smaller amount of PS than bread wheat genotypes that are tolerant to Zn deficiency (Walter et al. 1994; Rengel et al. 1998). Zinc efficiency in wheat was suggested to be causally related to PS-mediated enhanced mobilization of Zn from low Zn soils (Cakmak et al. 1996). In contrast, Gries et al. (1995) and Pedler et al. (2000) found that Zn deficiency didn't induce PS release both in barley and wheat cultivars. Furthermore, Erenoglu et al. (1996) found no relationship between the rate of phytosiderophore release and susceptibility of six bread wheat genotypes to Zn deficiency in a nutrient solution experiment. Pedler et al. (2000) found no differences in PS release in barley and wheat cultivars that have been reported to differ significantly in Zn efficiency. These conflicting results indicate that a final determination on the role of PS in Zn efficiency needs more detailed investigation. Probably phytosiderophores are not the only explanation, and their contribution to alleviating Zn deficiency is plant species or genotype-dependent. Although several papers showed the beneficial effect of PS on tolerance to Zn deficiency for wheat (Zhang et al. 1991), barley (Erenoglu et al. 2000), and sorghum and maize (Hopkins et al. 1998), we are not aware of any report on effects of phytosiderophore exudation on Zn nutrition in rice.

Rice plants have been reported to exude low molecular weight organic anions (LMWOAs), in which citrate and malate are the common identified LMWOAs. When grown hydroponically or in soil, malate and citrate exudation by rice roots was reported in several studies (Aulakh et al. 2001; Bhattacharyya et al. 2003; Hajiboland et al. 2005; Hoffland et al. 2006). In a recent study with six aerobic rice genotypes with different Zn uptake capacity, we found that genotypes with a higher Zn efficiency (and hence higher Zn uptake under Zn-limiting conditions) showed a stronger response to malate exudation than the genotypes with a lower Zn uptake (Gao et al. 2007). This observation was made both in a rhizotron experiment and in a nutrient solution experiment. To our knowledge, this is the only study reporting the release of malate in response to Zn deficiency of rice plants. Similarly, malate exudation increased as a response to bicarbonate stress, which is frequently associated with Zn deficiency on calcareous soils (Yang et al. 2003; Hajiboland et al. 2005). In several studies, citrate exudation was reported as a response to nutrient deficiency, including P and Zn (Hoffland et al. 2006), and Cu and Mn (Neumann and Römheld 2001). Further research is needed to understand why malate is released in the rhizosphere under some types of nutrient deficiencies and citrate at other times. These results indicate that the rhizosphere effect is one of the important processes to increase plant tolerance to Zn deficiency.

With a soil extraction experiment, Gao (2007) quantified the effect of malate exudation on Zn availability in soil solution. The increased malate concentrations in soil solution can increase Zn bioavailability. However, the estimated concentration of Zn in the rhizosphere for the efficient genotype was only 15% higher than that for the inefficient genotype, compared to the fivefold difference in plant Zn uptake. The small difference in estimated concentration of Zn in rhizosphere may be partly accounted for by the limitation of the rhizoton-lysimeter technique, which may vastly underestimate soil solution organic acid concentrations, especially close to the roots. On the other hand, this small difference also indicated the importance of investigations on other possible mechanisms, such as exudation of phytosiderophores. It was reported that Zn deficiency increases the secretion of phytosiderophores from wheat (Cakmak et al. 1994) and barley (Suzuki et al. 2006) roots into the rhizosphere. The detailed mechanism of the release of phytosiderophores by rice roots and its consequent effect on plant Zn uptake, however, has never been determined, even though rice was proven to produce small amounts of phytosiderophores under Fe deficiency (Zhang et al. 1998). Further investigations are necessary to quantify the contribution of rhizosphere effects on plant Zn acquisition. Introduction of models into the complex plant–soil system may be an interesting and plausible approach to link soil chemistry and plant physiology, as performed by Geelhoed et al. (1999), Kirk (2002), and Wissuwa (2003). Also, we only investigated the acquisition of Zn by aerobic rice at seedling stage. In view of various factors affecting or responsible for Zn uptake efficiency, more investigations on uptake over time are necessary.

9.4.4 Trade-Offs between Root Mycorrhizas–Organic Acids

Mycorrhizal fungi can elicit change in root architecture in plant species that are highly dependent upon mycorrhizal symbiosis (Hetrick 1991). Plants with a high degree of dependence on the AMF-root symbiosis tend to develop fewer root hairs in low-fertility soils. As a result, the plant root system is less able to obtain nutrients and thus creates a greater dependence of the plant on AMF-root symbiosis. In a study conducted by Itoh and Barber (1983), who compared the root hair development and P uptake of six plant species, the plant species with the most abundant root hairs were the ones that rely very little on mycorrhizal symbiosis for nutrient uptake. Russian thistle, which is a typically nonmycorrhizal plant species, had great root hair development (Miller 1988). In contrast, onions had few root hairs but are highly dependent on the mycorrhizal symbiosis (Stribley et al. 1980).

This trade-off between root development and mycorrhizal dependence can also be confirmed within different genotypes of aerobic rice. There was a significant negative correlation (Figure 9.5; $r = -0.81$, $P = 0.04$) between mycorrhizal Zn responsiveness and root surface area for six aerobic rice genotypes with different Zn-uptake efficiencies. Mycorrhizal Zn responsiveness was calculated as:

$$\text{MZnR} = [(\text{Zn uptake}_{+\text{AMF}} - \text{Zn uptake}_{-\text{AMF}})/\text{Zn uptake}_{-\text{AMF}}] \times 100$$

This negative correlation indicated that the Zn efficient genotypes (genotypes tolerant to Zn deficiency) tend to develop a higher root surface area, while showing a lower degree of dependence on the mycorrhizal symbiosis. We hypothesized this trade-off could be explained by the balance of metabolic investment by plants. In response to nutrient deficient conditions, plants may develop either higher root surface area (by increasing root fineness, root hair length, or number), or a greater dependence on mycorrhizal symbiosis, or reduced physiological changes (such as release of malate by aerobic rice roots). Each of these adaptations involves a different metabolic cost to the plant. The plant seems have the ability to select the one with minimal metabolic cost and maximum capacity of nutrient acquisition.

Similar to the trade-off between root surface area and mycorrhizas, we recently reported a significantly negative correlation between Zn uptake of six aerobic rice genotypes under nonmycorrhizal conditions and their MZnR (Gao et al. 2007). As a conclusion from a pot experiment with six aerobic rice genotypes growing on a low Zn soil, we reported that Zn uptake was significantly increased by inoculation with AMF, but only in genotypes with a low inherent Zn uptake when nonmycorrhizal (Gao et al. 2007). More specifically, Zn uptake of six aerobic rice genotypes under nonmycorrhizal conditions was significantly and negatively correlated to the mycorrhizal Zn responsiveness (MZnR) for mycorrhizal fungi (Figure 9.6). This negative correlation indicates that genotypes with high responsiveness to AMF colonization are inherently less efficient in taking up Zn from low Zn soils. Also, this negative correlation is not limited to aerobic rice and Zn.

There are some examples in the literature allowing generalization of the conclusion that there exists a trade-off between mycorrhizal responsiveness and nutrient efficiency. Cultivated oats with a lower P uptake and nonmycorrhizal were more

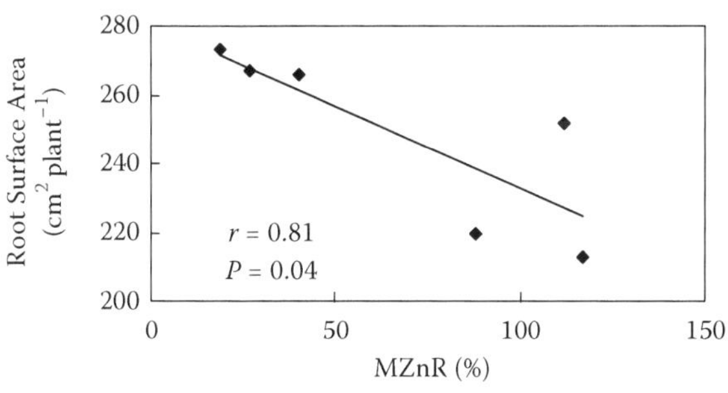

FIGURE 9.5 Correlation between root surface area when nonmycorrhizal and mycorrhizal Zn response of six rice genotypes inoculated with *Glomus mosseae*.

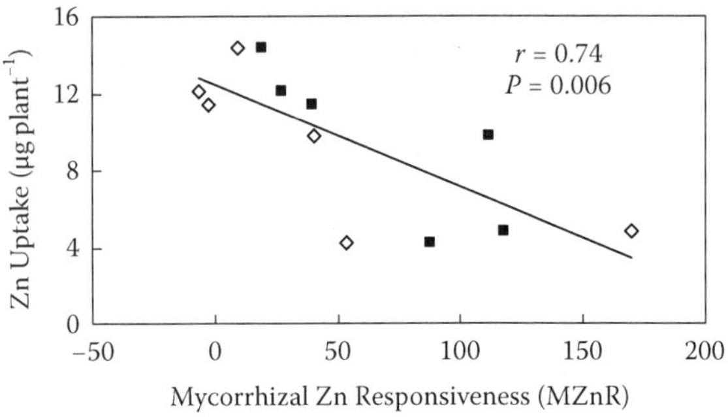

FIGURE 9.6 Correlation between Zn uptake when nonmycorrhizal and mycorrhizal Zn response of six aerobic rice genotypes for two AMF species (■ *G. mosseae*; ◊ *G. etunicatum*).

responsive to mycorrhizal inoculation than wild oats with a higher P uptake (Koide et al. 1988). A tomato genotype with a lower P uptake and nonmycorrhizal was also more responsive to AMF inoculation than a more efficient genotype with a higher P uptake (Bryla and Koide 1998). Baon et al. (1993) observed that barley genotypes with the lowest P uptake and nonmycorrhizal showed the largest mycorrhizal responsiveness. Kaeppler et al. (2000) reported a negative correlation between P uptake of inbred maize lines and mycorrhizal responsiveness under nutrient-poor conditions. In combination with our findings on aerobic rice and Zn, it seems that the negative correlation between mycorrhizal responsiveness and inherent nutrient uptake efficiency is a general rule in this plant–fungus symbiosis. This indicates that genotypes are either responsive to AMF, or have alternative mechanism to mobilize nutrients from soils. Plants that are poor mobilizers of nutrients on nutrient deficient soils will show a higher response to mycorrhizal fungi and benefit more from mycorrhizal inoculation.

In view of breeding, a combination of maximum mycorrhizal responsiveness and most efficient nutrient uptake capacity in one genotype seems unrealistic. To assess the best breeding strategy it would be important to determine the relative contribution of plant-involved mechanisms and mycorrhizal symbiosis to nutrient uptake efficiency (Smith et al. 1992). A dominant focus on mycorrhizal responsiveness could then lead to selection of genotypes that are less able to cope with low soil nutrient levels in a nonmycorrhizal condition. In the case of aerobic rice and Zn, we showed that inoculation with AMF did not significantly increase Zn uptake beyond levels that were found in genotypes with an inherently high uptake.

As we discussed in previous section, the capacity of releasing malate into rhizosphere might play a major role in determining Zn uptake by aerobic rice grown on low Zn soils. We also found a negative correlation between malate exudation rate and MZnR for the five selected genotypes (Figure 9.7). Although the correlation is nonsignificant, which might be caused by the number limitation of data points, it does suggest that malate production and mycorrhizas would be indeed incompatible and have to be traded off. We suggested that the competition for carbon in the production of low molecular weight organic acids and by mycorrhizal fungi might cause the trade off between both mechanisms for Zn acquisition. In response to Zn

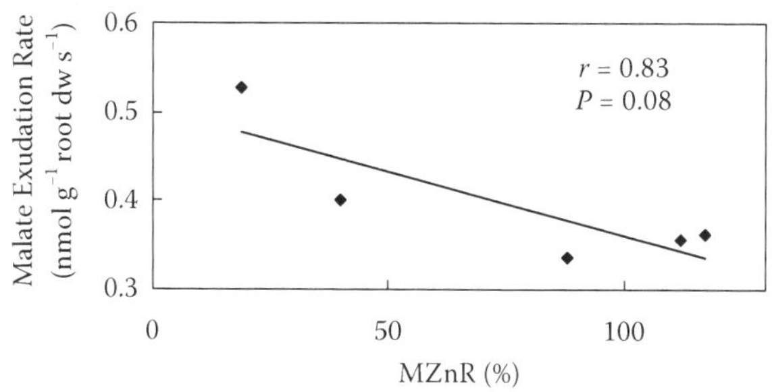

FIGURE 9.7 Correlation between malate exudation rates when nonmycorrhizal and mycorrhizal Zn response of five selected aerobic rice genotypes.

deficiency, roots of the efficient aerobic rice genotypes can develop the mechanisms to release malate into the rhizosphere and thus mobilize Zn from the soil. This process is a carbon-consuming process, and plants have to invest carbon to produce organic acids (Marschner 1995). As a result of the competition for metabolic investment, the efficient genotypes (with a higher tolerance to Zn deficiency and hence a higher Zn uptake) seem to invest more carbon to produce organic acids rather than to develop the mycorrhizal symbiosis. In contrast, the inefficient genotypes that show a greater dependence on the mycorrhizal symbiosis might invest less carbon for the production of organic acids.

9.5 CONCLUSIONS

It has long been the goal of agronomists to understand why one plant species or cultivar grows better on a particular soil with a low amount of available nutrients than another species or cultivar. We reported here a great genotypic variation in tolerance to Zn deficiency among different aerobic rice genotypes. We found that variation in tolerance was mainly associated with Zn uptake. Plant Zn uptake from low Zn soils can be increased by both Zn-mobilizing rhizosphere processes and inoculation with arbuscular mycorrhizal fungi. We demonstrated that mycorrhizal inoculation significantly increased Zn uptake, but only in genotypes with a low inherent Zn uptake. High Zn mobilization by a combination of chemical rhizosphere and mycorrhizal effects seems impossible and may not be feasible as a target for breeders. In a rhizotron experiment and a nutrient solution experiment, aerobic rice genotypes responded to Zn deficiency with increased root exudation of malate. Genotypes with a higher Zn uptake showed a stronger increase in malate exudation in response to Zn deficiency.

REFERENCES

Alloway, B. J. 2004. *Zinc in Soils and Crop Nutrition.* Brussels: International Zinc Association.

Aulakh, M. S., R. Wassmann, C. Bueno, J. Kreuzwieser, and H. Rennenberg. 2001. Characterization of root exudates at different growth stages of ten rice (*Oryza sativa* L.) cultivars. *Plant Biology* 3: 139–148.

Baon, J. P., S. E. Smith, and A. M. Alston. 1993. Mycorrhizal responses of barley cultivars differing in P efficiency. *Plant and Soil* 157: 97–105.

Begg, C. B. M., G. J. D. K., Kirk, A. F. Mackenzie, and H. U. Neue. 1994. Root-induced iron oxidation and pH changes in the lowland rice rhizosphere. *New Phytologist* 128: 469–477.

Bhattacharyya, P., S. C. Datta, and P. Dureja. 2003. Interrelationship of pH, organic acids, and phosphorus concentration in soil solution of rhizosphere and non-rhizosphere of wheat and rice crops. *Communications in Soil Science and Plant Analysis* 34: 231–245.

Bouman, B. A. M., S. Peng, A. R. Castañeda, and R. M. Visperas. 2005. Yield and water use of tropical aerobic rice systems in the Philippines. *Agricultural Water Management* 72: 87–105.

Bryla, D. R. and R. T. Koide 1998. Mycorrhizal response of two tomato genotypes relates to their ability to acquire and utilize phosphorus. *Annals of Botany* 82: 849–857.

Cakmak, I. 2002. Plant nutrition research: Priorities to meet human needs for food in sustainable ways. *Plant and Soil* 247: 3–24.

Cakmak, I., H. Ekiz, and A. Yilmaz et al. 1997a. Differential response of rye, triticale, bread and durum wheat to zinc deficiency in calcareous soils. *Plant and Soil* 188: 1–10.

Cakmak, I, K. Y. Gulut, H. Marschner, and R. D. Graham. 1994. Effect of zinc and iron deficiency on phytosiderophore release in wheat genotypes differing in zinc deficiency. *Journal of Plant Nutrition* 17: 1–17.

Cakmak, I., L. Ozturk, S. Eker, B. Torun, H. I. Kalfa, and A. Yilmaz. 1997b. Concentration of zinc and activity of copper/zinc-superoxide dismutase in leave of rye and wheat cultivars differing in sensitivity to zinc deficiency. *Journal of Plant Physiology* 151: 91–95.

Cakmak, O., L. Ozturk, S. Karanlik, H. Ozkan, Z. Kaya, and I. Cakmak. 2001. Tolerance of 65 durum wheat genotypes to zinc deficiency in a calcareous soil. *Journal of Plant Nutrition* 24: 1831–1847.

Cakmak, I., N. Sari, and H. Marschner et al. 1996. Phytosiderophore release in bread and durum wheat genotypes differing in zinc efficiency. *Plant and Soil* 180: 183–189.

Chen, C. M. 2004. *Ten-Year Tracking Nutritional Status in China.* Beijing: People's Medical Publisher (in Chinese).

Dodd, J. C. and P. Jeffries. 1989. Effect of fungicides on three vesicular–arbuscular mycorrhizal fungi associated with winter wheat (*Triticum aestivum* L.). *Biology and Fertility of Soils* 7: 120–128.

Dogar, M. A. and van T. Hai. 1980. Effect of P, N and HCO_{3-} levels in the nutrient solution on rate of Zn adsorption by rice roots and Zn content in plants. *Zeitschrift der Pflanzenphysiologie* 98: 203–212.

Dong, B., Z. Rengel, and E. D. Graham. 1995. Root morphology of wheat genotypes differing in zinc efficiency. *Journal of Plant Nutrition* 18: 2761–2773.

Eivazi, F. and C. C. Weir. 1989. Phosphorus and mycorrhizal interaction on uptake of P and trace elements by maize. *Fertilizer Research* 21: 19–22.

Erenoglu, B., I. Cakmak, and H. Marschner et al. 1996. Phytosiderophore release does not correlate well with zinc efficiency in different bread wheat genotypes. *Journal of Plant Nutrition* 19: 1569–1580.

Erenoglu, B., S. Eker, I. Cakmak, R. Derici, and V. Römheld. 2000. Effect of iron and zinc deficiency on release of phytosiderophores in barley cultivars differing in zinc efficiency. *Journal of Plant Nutrition* 23: 1645–1656.

Gao, X. P., T. W. Kuyper, C. Q. Zou, F. S. Zhang, and E. Hoffland. 2007. Mycorrhizal responsiveness of aerobic rice genotypes is negatively correlated with their zinc uptake when nonmycorrhizal. *Plant and Soil* 290: 283–291.

Gao, X. P., C. Q. Zou, X. Y. Fan, F. S. Zhang, and E. Hoffland. 2006. From flooded to aerobic conditions in rice cultivation: Consequences for zinc uptake. *Plant and Soil* 280: 41–47.

Gao, X. P., C. Q. Zou, F. S. Zhang, S. E. A. T. M. Van der Zee, and E. Hoffland. 2005. Tolerance to zinc deficiency in rice correlates with zinc uptake and translocation. *Plant and Soil* 278: 253–261.

Gao, X. P. 2007. Bioavailability of zinc to aerobic rice. PhD diss., Wageningen. ISBN 90-8504-646-2.

Geelhoed, J. S., W. H. Van Riemsdijk, and G. R. Findenegg. 1999. Simulation of the effect of citrate exudation from roots on the plant availability of phosphate adsorbed on goethite. *European Journal of Soil Science* 50: 379–390.

Genc, Y., G. K. McDonald, and R. D. Graham. 2004. Differential expression of zinc efficiency during the growing season of barley. *Plant and Soil* 263: 272–282.

Genc, Y., G. K. McDonald, and R. D. Graham. 2006. Contribution of different mechanisms to zinc efficiency in bread wheat during early vegetative stage. *Plant and Soil* 281: 353–367.

Graham, R. D. and Z. Rengel. 1993. Genotype variation in zinc uptake and utilization. In *Zinc in Soils and Plants*. Robson, A.D. (Ed.), Dordrecht: Kluwer Academic Publishers pp. 107–118.

Graham, R. D., J. S. Ascher, and S. C. Hynes. 1992. Selecting zinc-efficient cereal genotypes for soils of low zinc status. *Plant and Soil* 146: 241–250.

Gries, D., S. Brunn, D. E. Crowley, and D. R. Parker. 1995. Phytosiderophore release in relation to micronutrient metal deficiencies in barley. *Plant and Soil* 172: 299–308.

Hacisalihoglu, G. and L. V. Kochian. 2003. How do some plants tolerate low levels of soil zinc? Mechanisms of zinc efficiency in crop plants. *New Phytologist* 159: 341–350.

Hacisalihoglu, G., J. J. Hart, Y. H. Wang, I. Cakmak, and L. V. Kochian. 2003. Zinc efficiency is correlated with enhanced expression and activity of zinc-requiring enzymes in wheat. *Plant Physiology* 131: 595–602.

Hacisalihoglu, G., L. Ozturk, I. Cakmak, R. M. Welch, and L. Kochian. 2004. Genotypic variation in common bean in response to zinc deficiency in calcareous soil. *Plant and Soil* 259: 71–83.

Hajiboland, R., B. Singh, and V. Römheld. 2001. Retranslocation of Zn from leaves as important factor for zinc efficiency of rice genotypes. In *Plant Nutrition-Food security and sustainability of Agro-Ecosystems*. W.J. Horst, (Ed.), Dordrecht: Kluwer Academic Publishers, pp. 226–227.

Hajiboland, R., X. E. Yang, V. Romheld, and G. Neumann. 2005. Effect of bicarbonate on elongation and distribution of organic acids in root and root zone of Zn-efficient and Zn-inefficient rice (*Oryza sativa* L.) genotypes. *Environmental and Experimental Botany* 54: 163–173.

Hetrick, B. A. D., W. W. Bockus, and J. Bloom. 1984. The role of vesicular–arbuscular mycorrhizal fungi in the growth of Kansas winter wheat. *Canadian Journal of Botany* 62: 735–740.

Hetrick, B. A. D., G. W. T. Wilson, and T. S. Cox. 1992. Mycorrhizal dependence of modern wheat varieties, landraces, and ancestors. *Canadian Journal of Botany* 70: 2032–2040.

Hetrick, B. A. D. 1991. Mycorrhizas and root architecture. *Experientia* 47: 355–362.

Hoffland, E., C. Z. Wei, and M. Wissuwa. 2006. Citrate exudation by lowland rice (*Oryza sativa* L.) at zinc and phosphorus deficiency. *Plant and Soil* 283: 155–162.

Hopkins, B. G., D. A. Whiteney, R. E. Lamond, and V. D. Jolley. 1998. Phytosiderophore release by sorghum, wheat and corn under zinc deficiency. *Journal of Plant Nutrition* 21: 2623–2637.

Itoh, S. and S. A. Barber. 1983. Phosphorus uptake by six plant species as related to root hairs. *Agronomy Journal* 75: 457.

Jakobsen, I. 1983. Vesicular arbuscular mycorrhiza in field-grown crops. II. Effect of inoculation on growth and nutrient uptake in barley at two phosphorus levels in fumigated soil. *New Phytologist* 94: 595–604.

Kaeppler, S. M., J. L. Parke, S. M. Mueller, L. Senior, C. Stuber, and W. F. Tracy. 2000. Variation among maize inbred lines and detection of quantitative trait loci for growth at low phosphorus and responsiveness to Arbuscular mycorrhizal fungi. *Crop Science* 40: 358–364.

Khan, H. R., G. K. McDonald, and Z. Rengel. 1998. Chickpea genotypes differ in their sensitivity to Zn deficiency. *Plant and Soil* 198: 11–18.

Khare, A. K., A. K. Rawat, S. B. Dubey et al. 1998. Role of native vesicular-arbuscular mycorrhizal fungi in wheat-based cropping sequence for efficient use of phosphorus and zinc in black soils of Madhya Pradesh. *Journal of Agricultural Science* 68: 247–250.

Kirk, G. J. D. and J. B. Bajita. 1995. Root-induced iron oxidation, pH changes and zinc solubilization in the rhizosphere of lowland rice. *New Phytologist* 131: 129–137.

Kirk, G. J. D. and M. A. Saleque. 1995. Solubilization of phosphate by rice plants growing in reduced soil: Prediction of the amount solubilized and the resultant increase in uptake. *European Journal of Soil Science* 46: 247–255.

Kirk, G.J.D. 2002. Modelling root-induced solubilization of nutrients. *Plant and Soil* 255: 49–57.

Koide, R., M. Li, J. Lewis, and C. Irby. 1988. Role of mycorrhiza infection in the growth and reproduction of wild versus cultivated plants. I. Wild vs. cultivated oats. *Oecologia* 77: 537–543.

Kothari, S. K., H. Marschner, and V. Romheld. 1991. Contribution of VA mycorrhizal hyphae in acquisition of phosphorus and zinc by maize grown in a calcareous soil. *Plant and Soil* 131: 177–185.

Lambert, D. H. and T. C. Weidensaul. 1991. Element uptake by mycorrhizal soybean from sewage sludge treated soil. *Soil Science Society of America Journal* 55: 393–398.

Li, X. L., H. Marschner, and E. George. 1991. Acquisition of phosphorus and copper by VA-mycorrhizal hyphae and root to shoot transport in white clover. *Plant and Soil* 136: 49–57.

Lin, B. and J. Li. 1997. Problems and strategies of fertilizer in China. *Phosphorus Fertilizer and Complex Fertilizer* 23: 1–5 (in Chinese).

Marschner, H. and V. Romheld. 1983. In vivo measurement of root-induced pH changes at soil-root interface: Effect of plant species and nitrogen source. *Zeitschrift der Pflanzenphysiologie* 111: 241–251.

Marschner, H., M. Treeby, and V. Romheld. 1989. Root-induced changes in the rhizosphere: Importance for the mineral nutrition of plants. *Zeitschrift der Pflanzenernaehrung und Bodenkunde* 152: 197–204.

Marschner, H. 1995. *Mineral Nutrition of Higher Plants*. London: Academic Press.

Marschner, H. 1998. Role of root growth, arbuscular mycorrhizae, and root exudates for the efficiency in nutrient acquisition. *Field Crops Research* 56: 203–207.

Miller, R. M. 1988. The ecology of vesicular-arbuscular mycorrhizal in grass- and shrublands. In *The Ecophysiology of VA Mycorrhizal Plants*. G. R. Safir (Ed.), Boca Raton: CRC Press, pp. 135–170.

Neumann, G. and V. Römheld. 2001. The release of root exudates on rhizosphere microbial populations. In *The Rhizosphere: Biochemistry and Organic Substances at the Soil-Plant Interface,* R. Pinton, Z. Varanini, and P. Nannipieri (Eds.), New York: Marcel Dekker, pp. 95–141.

Pedler, J. F., D. R. Parker, and D. E. Crowley. 2000. Zinc deficiency-induced phytosiderophore release by the Triticaceae is not consistently expressed in solution culture. *Planta* 211: 120–126.

Purakayastha, T. J. and P. K. Chhonkar. 2001. Influence of vesicular-arbuscular mycorrhizal fungi (*Glomus etunicatum* L.) on mobilization of zinc in wetland rice (*Oryza sativa* L.). *Biology and Fertility of Soils* 33: 323–327.

Ramani, S. and S. Kannan. 1985. An examination of zinc uptake patterns by cultivars of sorghum and maize: Differences amongst hybrids and their parents. *Journal of Plant Nutrition* 8: 1199–1210.

Rengel, Z., V. Romheld, and H. Marschner. 1998. Uptake of zinc and iron by wheat genotypes differing in tolerance to zinc deficiency. *Journal of Plant Physiology* 152: 433–438.

Ryan, M. H. and J. F. Angus. 2003. Arbuscular mycorrhizae in wheat and field pea crops on a low P soil: Increased Zn uptake but no increase in P uptake or yield. *Plant and Soil* 250: 225–239.

Sakal, R., M. K. Verma, and A. P. Singh. 1989. Relative susceptibility of selected rice varieties to zinc deficiency in relation to utilisation of different pools of soil zinc from calcareous soil. *Annals of Agricultural Research* 10: 243–248.

Sasaki, H., T. Hirose, Y. Watanabe, and R. Ohsugi. 1998. Carbonic anhydrase activity and CO_2-transfer resistance in Zn-deficient rice leaves. *Plant Physiology* 118: 929–934.

Silber, A., L. Ben-Yones, and I. Dori. 2004. Rhizosphere pH as a result of nitrogen levels and NH4/NO3 ratio and its effect on zinc availability and on growth of rice flower (*Ozothamnus diosmifolius*). *Plant and Soil* 262: 205–213.

Smith, S. E., A. D. Robson, and L. K. Abbott. 1992. The involvement of mycorrhizas in assessment of genetically dependent efficiency of nutrient uptake and use. *Plant and Soil* 146: 169–179.

Solaiman, M. Z. and H. Hirata. 1997. Effect of arbuscular mycorrhizal fungi inoculation of rice seedlings at the nursery stage upon performance in the paddy field and greenhouse. *Plant and Soil* 191: 1–12.

Stribley, D. P., P. B. Sinker, and J. H. Rayner. 1980. Relation of internal phosphorus concentration and plant weight in plants infected by vesicular-arbuscular mycorrhiza. *New Phytologist* 86: 261–266.

Suzuki, M., M. Takahashi, and T. Tsukamoto et al. 2006. Biosynthesis and secretion of mugineic acid family phytosiderophores in zinc-deficient barley. *Plant Journal* 48: 85–97.

Takagi, S., S. Kawai, and M. H. Yu. 1988. Efficiency of iron extraction from soil by mugineic acid family phytosiderophores. *Journal of Plant Nutrition* 11: 643–651.

Takkar, P. N. and C. D. Walker. 1993. The distribution and correction of zinc deficiency. In *Zinc in Soils and Plants.* A.D. Robson (Ed.), Dordrecht: Kluwer Academic Publishers, pp. 151–166.

Torun, B., G. Bozbay, and I. Gultekin et al. 2000. Differences in shoot growth and zinc concentration of 164 bread wheat genotypes in a zinc-deficient calcareous soil. *Journal of Plant Nutrition* 23: 1251–1265.

Treeby, M., H. Marschner, and V. Römheld. 1989. Mobilization of iron and other micronutrient cations from a calcareous soil by plant-borne, microbial, and synthetic metal chelators. *Plant and Soil* 114: 217–226.

Walter, A., V. Römheld, H. Marschner, and S. Mori. 1994. Is the release of phytosiderophores in zinc-deficient wheat plants a response to impaired iron utilization? *Physiologia Plantarum* 92: 493–500.

Wang, H. Q., B. A. M. Bouman, D. Zhao, C. Wang, and P. E. Moya. 2002. Aerobic rice in northern China: Opportunities and challenges. In *Water-Wise Rice Production*. B. A. M. Bouman, H. Hengsdijk, B. Hardy, P. S. Bindraban, T. P. Tuong, and J. K. Ladha (Eds.), Proceedings of the International Workshop on Water-Wise Rice Production, 8–11 April 2002, Los Baños, Philippines, pp. 143–153.

Wellings, N. P., A. H. Wearing, and J. P. Thompson. 1991. Vesicular-arbuscular mycorrhizae (VAM) improve phosphorus and zinc nutrition and growth of pigeonpea in a Vertisol. *Australian Journal of Agricultural Research* 42: 835–845.

Wissuwa, M., A. M. Ismail, and S. Yanagihara. 2006. Effects of zinc deficiency on rice growth and genetic factors contributing to tolerance. *Plant Physiology* 142: 731–741.

Wissuwa, M. 2003. How do plants achieve tolerance to phosphorus deficiency? Small causes with big effects. *Plant Physiology* 133: 1947–1958.

Yang, X. E., R. Hajiboland, and V. Römheld. 2003. Bicarbonate had greater effects than high pH on inhibiting root growth of zinc-inefficient rice genotype. *Journal of Plant Nutrition* 26: 399–415.

Zhang, F. S., V. Römheld, and H. Marschner. 1991. Release of zinc mobilizing root exudates in different plant species as affected by zinc nutritional status. *Journal of Plant Nutrition* 14: 675–686.

Zhang, H., A. Jennings, W. B. Peter, and G. F. Brian. 1999. Dual pathways for regulation of root branching by nitrate. *Plant Biology* 96: 529–534.

Zhang, X. H., Y. G. Zhu, B. D. Chen, A. J. Lin, S. E. Smith, and F. A. Smith. 2005. Arbuscular mycorrhizal fungi contribute to resistance of upland rice to combined metal contamination in soil. *Journal of Plant Nutrition* 28: 2065–2077.

Zhang, X. K., F. S. Zhang, and D. R. Mao. 1998. Effect of iron plaque outside roots on nutrient uptake by rice (*Oryza sativa* L.): Zinc uptake by Fe-deficient rice. *Plant and Soil* 202: 33–39.

10 Breeding for Micronutrient Enriched Rice

Glenn B. Gregorio, Tin Htut, and Gloria S. Cabuslay

CONTENTS

10.1 INTRODUCTION

Rice is basically a starchy crop and one of the lowest in nutritional value compared to other staples. Postharvest processing of rice—especially polishing and then washing of the edible portion before cooking—causes the loss of more than 50% of the available nutrients. In the past, limited attention has been given to the improvement of rice's nutritional value. Breeding efforts in the 1970s to increase the protein content achieved modest results due to the large environmental impact, the complexity of the mode of inheritance, and the negative correlation to grain yield. Research priorities for improving grain quality that were recommended at International Rice Research Conferences in 1985 and 1990 did not focus on nutritive values. This lack of attention may have been due to the unsuccessful efforts in improving the protein value of rice and the small impact of quality protein maize (QPM) at that time. In many countries where the staple food is rice, per capita consumption is as high as 206 kg y^{-1}, with an average of 143 kg y^{-1} consumed in Southeast Asia. Rice consumption in sub-Saharan Africa is increasing yearly and will rise by 50% from 2005 to 2015 (FAO 2005). With the new paradigm shift from green revolution to nutritional revolution one must realize that any research effort to increase the nutritional value of rice could improve the heath of the rice-eating population. Among the popular cereals, rice is lowest in iron (Fe), with an average concentration of only 3–5 mg kg^{-1} after milling. New studies have proven the potential to exploit the genetic variation in rice grain micronutrient content without experiencing a negative correlation on yield, which is commonly observed with the improvement of protein content in many crops. The relationship between yield and grain mineral content may be positively correlated, particularly in mineral deficient soils where the plant may use its inherently high levels of Fe and zinc (Zn) to overcome the soil mineral deficiency. Hence, enhancing the essential micronutrient content in rice will be a win–win approach that can have a positive effect on human health and at the same time provide an agronomic advantage to the crop itself.

The success of any breeding program depends greatly on understanding the inheritance of plant traits, the nature of the combining ability among donor parents and recipient parent germplasm, and consequently developing appropriate breeding strategies. Therefore, it is essential to understand the combining ability of modern rice varieties and donor rice germplasm to establish breeding strategies to enhance grain mineral density for human nutrition.

10.2 GENETIC VARIABILITY OF MICRONUTRIENT CONTENT IN RICE GRAIN

Before 1990 there was very limited information on the micronutrient content in rice grain. In 1992 the International Rice Research Institute (IRRI) started to examine the effect of certain soil characteristics on the Fe content in the grain. Wide genetic variability among varieties was observed in regard to the Fe content in grain. This genetic variation provided hope for plant scientists to pursue the study and development of high Fe-containing rice for human health. In collaboration with the University of Adelaide and the International Food Policy Research Institute (IFPRI), the project was expanded to include Zn. As part of the collaboration a micronutrient profile among rice varieties was compiled into a database to aid plant scientists in selecting varieties for breeding for nutritional studies. Because the mineral analysis used inductively coupled plasma atomic emission spectrometry (ICP-AES) and could detect other essential minerals (e.g., boron [B], calcium [Ca], copper [Cu], potassium [K], magnesium [Mg], manganese [Mn], phosphorus [P], and sulfur [S]), these nutrients were included in the study, although the focus was more on Fe and Zn. This information could also provide guidance to plant scientists, human nutritionists, and economists in knowing the micronutrient profile of specific rice varieties consumed by a population.

The grain rice samples were grown in the same environmental conditions and carefully sampled to avoid possible contaminations. Amongst the initial 1259 samples analyzed (Table 10.1) Fe concentration in brown rice ranged from 4.59 to 27.26 mg kg^{-1} with a mean value of 15.16 mg kg^{-1}. For Zn, concentrations ranged from 10.63 to 68.75 mg kg^{-1} with a mean value of 30.22 mg kg^{-1}. These preliminary results then encouraged plant scientists to explore the micronutrient profile of other rice genotypes. Those genotypes that produced higher micronutrient content in the grain could be used as parents in the breeding program.

TABLE 10.1
Grain Mineral Density of 1,259 Rice Genotypes (Brown Rice) in mg kg^{-1}

Mineral	Minimum	Maximum	Average
B	0.5	25.4	5.3
Ca	37.0	264.0	120.5
Cu	0.9	29.7	5.9
Fe	4.6	27.3	15.2
K	700.0	6609.0	3558.5
Mg	172.0	2457.0	1473.9
Mn	6.9	144.2	29.9
P	870.0	6905.0	3963.3
S	488.0	2652.0	1363.7
Zn	10.6	68.6	30.2

Because the mineral content in grain is highly influenced by the environment, it is misleading to compare genotypes based on the grain mineral content data. Only those rice genotypes planted under similar growing conditions can be compared. The genotypes in Table 10.1 were grouped according to sets based on batches planted in similar growing conditions. Table 10.2 shows the Fe and Zn profiles for the rice varieties.

In variety set 1, most of the traditional varieties contained high Fe and Zn in the grain, whereas the modern-released varieties produced lower micronutrient content in the grain (data not shown). This particular observation can be attributed to the fact that direct selection for high micronutrient was not part of the previous rice breeding program. The low Fe and Zn content in the improved rice content genotypes can be supported by the results in sets 6 and 7 of Table 10.2, where the promising varieties for release and the IRRI breeding lines showed lower Fe and Zn content compared to the other sets. The popular varieties and donor parents in international breeding programs in variety set 5 have low Fe and Zn content, and they will produce low micronutrient progenies. Adding some high micronutrient parents in the rice breeding program with direct selection for high micronutrient traits will improve the nutritional value of the next generation of varieties. However, breeding of an additional trait (e.g., high micronutrient content) will require more expense and time for plant breeders within any breeding program. The good news is that the new plant types (NPT) in variety set 8 demonstrated that new breeds of high yielding rice varieties with parents from tropical japonicas (set 3) have elevated Fe and Zn content in their brown rice. This observation confirms that using new donor parents will improve the micronutrient content in their progenies.

10.3 GENETIC COMBINING ABILITY ANALYSIS OF MICRONUTRIENT-DENSE TRAITS

Understanding the combining ability of donor parents is necessary to understand the ability to transfer the desirable trait and to help identify the best breeding combina-

TABLE 10.2
Brown Rice Fe and Zn Content of Rice Varieties Grown under Similar Growing Conditions in Eight Different Sets

	Variety Set	Number of Samples	Fe (mg kg^{-1}) Mean ± SE (Range)	Zn (mg kg^{-1}) Mean ± SE (Range)
1	Traditional and released varieties	140	13.2 ± 2.9 (7.8–24.4)	24.2 ± 4.6 (13.5–41.6)
2	Aromatic rice	51	14.6 ± 3.2 (10.8–23.2)	31.9 ± 6.0 (23–50)
3	Tropical japonicas	250	12.9 ± 1.5 (8.7–23.9)	26.3 ± 3.8 (15.0–40.1)
4	Wild rice and derivatives	21	15.6 ± 2.3 (11.8–21)	37.9 ± 8.6 (23–52)
5	Popular varieties and donor parents	199	13.0 ± 2.5 (7.7–19.2)	25.7 ± 4.6 (15.3-37.3)
6	Promising lines	83	8.8 ± 1.3 (6.3–14.5)	25.4 ± 4.2 (17.0–38.0)
7	IRRI breeding lines	350	10.7 ± 1.6 (7.5–16.8)	25.0 ± 7.6 (15.9–58.4)
8	New plant types	44	16.7 ± 2.1 (11.5–24)	29.6 ± 3.2 (23–36)

tions to effectively incorporate the trait of interest into the recipient genotype. A nine-parent diallel population was developed by crossing nine parental rice varieties in all possible combinations. The parental rice varieties involved were Azucena, Basmati 370, BG-300, IR64, IR68144, IR72, IR74, and Madhukar. Genotypes included Xua Bue Nuo, Azucena, Basmati, IR68144, and Xua Bue Nuo. They were selected as donor rice varieties for their higher grain Fe and Zn density according to preliminary germplasm screening conducted at the IRRI. The 72 F_1s and 9 parental genotypes were evaluated at the IRRI experimental farm with three replications under similar growing conditions to minimize environmental effects. Dehulled brown rice was analyzed for micronutrient density using ICP-AES. The combining ability analysis was performed according to Griffing's (1956) model 1, method 3.

Results showed highly significant variations among 72 crosses for grain Fe, Zn, Ca, Mg, Cu, Mn, P, S, and K concentrations (Table 10.3). When variation among the crosses was partitioned into general combining ability (GCA), specific combining ability (SCA), and reciprocal sums of squares, highly significant GCA, SCA, and reciprocal mean squares for all grain mineral densities (GMDs) were observed. The GCA is the overall performance of a particular parent variety to transfer the desirable traits to its progenies, irrespective of the other genotype being crossed. The GCA value is significant if the additive genes (fixable) are controlling the trait. In contrast, SCA is the individual performance of a particular cross in expressing the trait. The SCA will show how some specific parental combinations will perform better than other combinations, even though they may have the same levels of mineral content in the grain. A high SCA value is significant, if the nonadditive genes (fixable genes) are controlling the trait of interest. In the next discussion, more detailed genetic interpretations will be presented to build the foundation in establishing the breeding strategy to enrich rice with micronutrients.

TABLE 10.3

Combining Ability Analysis of Variance for Rice Grain Mineral Concentration of a Nine-Parent Diallel Population Based on Griffing Method

Source	df	Mean Square								
		Fe	Zn	Ca	Mg	Cu	Mn	P	S	K
Rep	2	1.53 ns	8.2 ns	352.0	15,468.5	0.08 ns	10.2 ns	9,595 ns	3,273 ns	328,692 ns
Crosses	71	18.13	188.3	324.1	90,821.0	1.60	94.8	1,030,021	70,143	281,390
GCA	8	97.53	941.7	2,101.1	429,487.6	7.80	627.4	5,461,346	294,795	1,511,549
SCA	27	16.55	190.3	130.0	99,656.2	0.87	38.7	963,051	85,219	238,042
Reciprocal	36	1.68	19.3	74.9	8,935.3	0.77	18.5	95,509	8,913	40,532
Error	142	1.03	3.6	25.8	4,393.0	0.46	5.4	7,024	4,334	23,128

Note: GCA = general combining ability, SCA = specific combining ability, df = degrees of freedom, ns = not significant; other mean squares are statistically significant at $P < 0.05$.

The GCA effects of the parental rice varieties were both positive and negative for all GMDs. Parental rice varieties (Azucena, Basmati 370, IR68144, Madhukar, and Xua Bue Nuo) had a positive GCA effect for most GMDs. This observation suggested that donor parental rice varieties could transfer the high GMD to its progenies. All commercial high-yielding rice varieties (BG-300, IR64, IR72, and IR74) showed negative GCA for GMD, suggesting that these parental rice varieties in general would not contribute higher mineral density traits to the next generation. However, because of the highly significant SCA mean square, some particular crosses between donor parents and commercial rice varieties are expected to be promising breeding materials. Among donor parents, Xua Bue Nuo showed the largest GCA effect on grain Fe density, followed by Madhukar and Azucena. The new highly mineral-dense line IR68144 showed a larger GCA than Basmati 370. Generally, the size of the GCA effect of a parent varied depending on the mineral element. For example, Xua Bue Nuo showed the lowest GCA effect for grain Zn concentration, whereas Madhukar showed the largest GCA effect, and Azucena the second-largest GCA effect. However, for grain Ca density, Xua Bue Nuo showed the third largest GCA, whereas Azucena showed the largest negative GCA effect. Similar phenomena were also true for other parents and mineral elements. Therefore, the association of GCA effects of the parents on nine mineral nutrients needs to be discussed.

Generally, high correlation coefficients among GCA effects were observed on some mineral densities. However, the GCA effects of the parents on certain mineral densities showed a low correlation and were statistically nonsignificant. For example, the GCA effects of the parents on grain Fe and Zn content, showed low correlation coefficients with Ca density. Similarly, the GCA effects of the parental rice varieties on grain Mn and other grain mineral content showed low correlation coefficients. This observation suggested that parental selection to breed for enhanced GMD based on the GCA could not be applied for enhancing overall mineral concentration of the rice grain. One reason may be that the parental rice varieties were originally selected only for high Fe and Zn content. Therefore, caution should be taken in selecting parental rice varieties based on only a few mineral nutrients so as not to upset the biological ratio of nutritionally important grain minerals in future rice varieties. Moreover, it may be suggested that breeding for enhancing grain mineral elements should aim at optimizing the mineral elements important for human nutrition, or for using selection strategies for multiple traits, which is needed for enhancing GMD. Using the ICP-AES is highly recommended for simultaneously analyzing many minerals.

A differential SCA effect of crosses across mineral element was observed. For grain Fe, Zn, Mg, P, S, and K, the combinations of IR68144 × Madhukar, Azucena × Basmati 370, Basmati 370 × Xua Bue Nuo, and Azucena × Madhukar showed large SCA effects. However, these crosses did not show a large SCA effect on grain Ca. Therefore, it was necessary to examine the correlations among SCA effects on grain minerals. Highly significant and large positive correlation coefficients among SCA effects of the crosses on grain Fe, Zn, Mg, P, K, and S content were observed. However, significant negative correlation coefficients were observed between the SCA effect on grain Ca and other minerals, e.g., Mg, Mn, P, S, and Zn. Similarly, the SCA effect on grain Mn showed a negative correlation effect on grain Cu, Fe, K, Mg, Mn,

P, S, and Zn. Therefore, selection of particular crosses for some GMDs would have a negative effect on other grain minerals. The selection based on SCA should be made on the most nutritionally-limited grain minerals on the condition that negative consequences on other nutrients would not alter either the human or plant nutritional status. Alternatively, selection should be based on all traits involving GMD. In fact, correlation among GCA effects of the parents and SCA effects of the crosses on the nine mineral densities could be considered as genotypic correlation. This relationship can still be modified by the environment or a cancellation between GCA and SCA effects.

It was clear that the phenotypic correlation among the mean density of each mineral nutrient based on F_1s and reciprocal F_1s showed no negative relationship, suggesting some cancellation of GCA and SCA or environmental effect. According to SCA values, there were interesting crosses between commercial rice varieties and traditional donor rice varieties, whose SCA effects were relatively large. However, selection based on SCA alone cannot guarantee higher GMD because the SCA was composed of dominance and some type of epistasis (unpredictable/nonfixable) components. Therefore, primary selection should be made based on GCA, and then on SCA for better parental combination coupled with their mean performance. There were significant reciprocal effects on GMD traits, however, this was relatively smaller than the GCA and SCA effects. These significant reciprocal effects imply that a proper choice of male and female parents must be considered to obtain better combinations for high micronutrient content in the grain. Some cross combinations may not produce high micronutrient content but the reciprocal cross combination where the reverse male and female parents were made will produce high micronutrient content. This could be due to maternal effect or because the nonnuclear chromosomal effects were expressed. Moreover, this suggests the importance of whether the donor parent used would be the male or female in the breeding program.

10.4 GENETIC COMPONENT OF HIGH IRON IN THE GRAIN

A separate 10 × 10 parent diallel cross to determine genetic component analysis (Hayman 1954) was undertaken for high Fe in the grain using four traditional high-Fe rice varieties (Azucena, Basmati 370, Xua Bue Nuo, and Tong lang Mo Mi), three advanced lines (IR61608, PP2462-11, and AT5-15), and three IRRI released varieties (IR36, IR64, and IR72). Results show highly significant differences between the cross combinations and their F_1 progenies. This observation clearly indicates a genetic effect on grain Fe concentration, suggesting that selection among F_1 progenies is possible. The genetic analysis of variance revealed the presence of additive gene action (fixable genes) in addition to a significant nonadditive genetic variance (nonfixable genes or unpredictable genes). Environmental effects are present but smaller than the genetic effects. Narrow-sense heritability of the trait was found to be moderately low (43%). The large difference between estimates of narrow-sense and broad-sense heritability (88%) further confirmed the effect of both the additive and the nonadditive type gene action controlling the high Fe concentration.

Analysis of combining ability indicated a high and significant GCA, and this suggests that some parents may be combined with a range of varieties to produce a high Fe concentration in its progenies. Specific combining ability is also present, indicating that specific combinations between the parents (e.g., Azucena × Basmati 370) produces increasing Fe concentration in progenies. The presence of reciprocal effects suggests the importance of the use of the parent as female or male. These results suggest that high-iron donors such as Basmati 370 and Xua Bue Nuo should be used as females in the crossing program to obtain progenies with higher Fe. However, other donor parents maybe useful as males.

Grain Fe analysis of selected F_1 crosses showed very high Fe content (Table 10.4), suggesting the high potential of these crosses to produce higher Fe and Zn content than their donor parents. High-Fe donors Milagrosa (17 mg Fe kg^{-1}) and Banjaiman (21.3 mg Fe kg^{-1}) were able to transfer their desirable traits to their progenies, and some crosses had an even higher Fe than the highest parent. The initial breeding results provided optimism that developing rice varieties with enhanced micronutrients in the grain is a practical strategy to improve human health in rice consuming populations.

10.5 BREEDING STRATEGIES TO ENHANCE MICRONUTRIENT CONTENT IN RICE

For rice consumers, micronutrient-enriched rice is an affordable source of Fe and Zn. The high consumption of rice makes developing rice with high amounts of desirable micronutrients (e.g., Fe and Zn), a high priority in any breeding program. These new breeding programs are expensive, and require more precise strategies. A better understanding of the genetics and inheritance of the trait will allow the exploitation of new donor parents for the improvement of micronutrient concentration in rice.

The genetics of micronutrient concentration was found to be a very complex trait. Although genetic variations for micronutrient content in the grain exist in breeding populations, the desired variance is not readily transferable. This is due to the moderately low heritability coupled with high environmental influence on the trait.

TABLE 10.4
Iron (Fe) Concentrations of Promising Breeding Populations for Iron-Dense Rice

F2	Parentage	Fe (mg kg^{-1})
IR71927	Pulot G/Milagrosa	18.5
IR71928	Fuji/Milagrosa	19.3
IR71929	Hayayuki/Milagrosa	23.2
IR71930	IR62446/Milagrosa	17.4
IR71931	Jimbyeo/Milagrosa	19.0
IR71932	Koshikihari/Milagrosa	19.1
IR72482	Banjaiman/IR62446	19.5
IR72483	Banjaiman/IR61356	17.0
IR72486	Banjaiman/Koshikihari	17.3
	Banjaiman	21.3
	Milagrosa	17.0

The nonadditive genetic variation (non fixable or unpredictable genes) was more predominant than additive genetic variation (fixable genes). Because the nonadditive genes were more controlling of the high micronutrient traits, selection in the early generation of the population may result in the loss of the potential progenies possessing the enhanced micronutrient as the desirable traits are not yet expressed at the early generation. As the filial generation is advanced, the nonadditive gene effects are dissipated, and the additive gene effects will be more pronounced. Selection for the trait should be delayed until later generations, such as the fifth filial generation (F_5), when the progenies are more homozygous and the nonadditive gene effect is no longer expressed. Whereas, selection for such agronomic traits as yielding ability, early maturity, and resistance to pest and deceases can be made at the early filial generations without selection yet for the high-Fe trait. Bulk breeding or single seed descent (SSD) up to F_5 may be the appropriate breeding method, followed by the pedigree method at F_6, where a micronutrient analysis will again be undertaken. Environmental effects also played a significant role in the inherit capability of the rice grain to accumulate nutrients. To minimize the environmental effects, the variability of each genotype must be maximized. Thus, selection under optimum growing conditions with sufficient soil nitrogen and potassium will maximize the genetic differences between genotypes (Gregorio et al. 2000).

Another important consideration is the choice of donor parent as the male or female parent in the hybridization. The significant reciprocal effects in the diallel analysis suggest that some high micronutrient donors should be used as females in the hybridization program to achieve higher macronutrient concentrations in the progenies, rather than using them as males. The high heterosis for macronutrient concentrations suggests the potential of hybrid rice technology to enhance macronutrient content in the grain, especially by selecting high-micronutrient-containing parents, and using them in the breeding program.

10.6 CONCLUSIONS

The GCA effects of the parents on certain mineral concentrations showed low correlations, suggesting that breeding for enhancing grain mineral elements should aim for the optimization of desirable micronutrients to avoid the imbalance of the different minerals in the grain. This optimization breeding needs a selection strategy for multiple micronutrient traits to enhance grain mineral concentrations. Caution must, however, be exerted for selecting donor rice varieties by using only one or two micronutrients and disregarding the other nutrients, so that the biological ratio of nutritionally-important grain minerals are not disrupted in future rice varieties. Importantly, breeding strategies to enhance the micronutrient concentrations in the rice grain requires collaboration among the different disciplines to attain the desired nutrition levels in rice.

ACKNOWLEDGMENTS

This work was financially supported by grants from DANIDA of Denmark and the Asian Development Bank coordinated by Dr. Howdy Bouis at the International Food

Policy Research Institute (IFPRI). We wish to thank Dr. Robin Graham at University of Adelaide for his advice and assistance in the micronutrient analysis.

REFERENCES

FAO (Food and Agriculture Organization of the UN). 2005. *FAO–STAT Data Base*. Rome, Italy: FAO.

Gregorio, G. B., D. Senadhira, T. Htut, and R. D. Graham. 2000. Breeding for trace mineral density in rice. *Food Nutrition Bulletin* 21: 382–386.

Griffing, B. 1956. Concepts of general and specific combining ability in relation to diallel crossing systems. *Australian Journal of Biological Sciences* 9: 463–493.

Hayman, B. I. 1954. Theory and analysis of diallel crosses. *Genetics* 39: 789–809.

11 Biofortification in a Food Chain Approach for Rice in China

Maja Slingerland, Fusuo Zhang, Tjeerd-Jan Stomph, Xiaopeng Gao, Jianfen Liang, and Wen Jiang

CONTENTS

11.1 INTRODUCTION

Worldwide micronutrient deficiencies are responsible for many health problems in people, especially in developing countries. In the past, iodine, iron (Fe), and vitamin A deficiencies received major attention. Zinc (Zn) deficiency as a micronutrient problem has long been ignored as data and, as a result, interpretation was insufficient, but recently Zn deficiency data were revisited in a metastudy (IZiNCG 2004). According to the available data on Zn requirements and food intake, the proportion of the population at risk for low Zn intake, and hence Zn-deficiency-related health problems, is more than 25% in South and Southeast Asia and sub-Saharan Africa, where diets are largely plant based (IZiNCG 2004). Looking from the side of natural resources, 8 million ha of soils in Southeast Asia suffer from Zn deficiency, which leads to a high risk of Zn deficiency in crops (Alloway 2004), and in people living in these locations. These data show that risk of Zn deficiency in people is high, which has resulted in placing Zn deficiency high on the research and development agenda.

In China, diets are largely based on plant-derived foods (with rice being one of the most important food components), especially for the poorer segment of the rural population. The daily Zn supply from rice depends on Zn concentration and the molar ratio of phytic acid to Zn in the total diet, because phytic acid inhibits the use and absorption of Zn by the human body. Preferably, the phytic acid to Zn molar ratio should be below 15 to guarantee an ample supply of Zn (WHO 2002). Looking at China in detail, many people, especially young children, suffer from Zn deficiency problems (Chen et al. 1985). In 1990 the average daily dietary Zn supply in seven provinces varied between 9.0 and 13.3 mg per capita for urban and between 11.4 and 14.8 mg per capita for rural households (Chen et al. 2004), with large differences between provinces and income groups. In the same year an average of 9% of children in urban areas and 41% of rural children were found to be nutritionally in the category of moderate stunting (Z scores between −2 and −3); in rural areas, whereas another 19% of children were found in the category of severe stunting (Z scores < −3) (Chen et al. 2004). Stunting is indicated as a lower height for age than a given standard, and it is accepted as an indicator of Zn deficiency in young children (IZiNCG 2004). Anthropometric measurements further revealed that stunting in urban children under six decreased from 9% in 1995 to 4% in 1998, and to 3% in 2000. Stunting also decreased in rural children under six years of age from 39% in 1995 to 22% in 1998, and to 20% in 2000. Stunting appears to be related to income level. Among children from poor rural households the average prevalence of stunting in children below six was 30% in both 1998 and 2000.

China has large areas where Zn deficiency in soils hampers crop production (Silanpäa 1990; Alloway 2004). In total about one-third of Chinese soils are Zn deficient, predominantly in northern and central China (Alloway 2004). This situation is particularly observed on calcareous and/or loess soils, soils along the Yellow River and Yangtze, and soils used for rice production in northern China (Takkar and Walker 1993). Zinc deficiency in soils used for rice production may lead to lower rice yields and to a lower Zn supply from rice for people living in these regions. Zinc availability from soils to rice crops depends partly on cultivation methods. Currently, severe competition for water between urban, industrial, and agricultural uses has induced a large shift from flooded to aerobic rice cultivation (Bouman et al. 2005). The new cultivation system required the development of new aerobic rice cultivars by crossing lowland with upland varieties (Wang Huaqui, personal communication) to be grown as a dry field crop in irrigated but nonflooded and nonpuddled soils (Bouman et al. 2005). As a result, potential effects on Zn uptake by rice need to be investigated.

There are a range of successful targeted interventions to address micronutrient deficiencies, such as distribution of iodized salt and iron-fortified wheat flour, and administration of vitamin A supplements to vulnerable groups in society. Most proposed interventions require major infrastructure and logistics for production and distribution of fortified products and medicines. For example, fortification needs centralized food processing factories and an importation of the fortificants. In some areas others require major human resources for increasing awareness and training on changing dietary habits. In developing countries where the enabling environment for such interventions is essentially absent, biofortification is proposed as an alternative and more sustainable intervention (Bouis et al. 2000). Biofortification aims to develop micronutrient-rich staple food crops such as rice through plant breeding (Gregorio et al. 2000). Once the traits for micronutrient density are available, they can be stacked with any other desired traits into improved varieties that can be delivered to farmers through the existing seed or tuber distribution system. Depending on the system, farmers will continue to buy elite seeds or tubers, or produce their own planting materials. In both cases the intervention does not require additional infrastructure, logistics, or extension work, as farm households do not need training to change their cultivation methods or their food habits.

Focusing on biofortification the Global Challenge Program *HarvestPlus* was initiated by the Consultative Group on International Agricultural Research, coordinated by the International Center for Tropical Agriculture (CIAT) and the International Food Policy Research Institute (IFPRI). The program was cofunded by a number of donors, such as the Bill Gates Foundation, United States Agency for International (USAID), and Danida. One of the objectives of *HarvestPlus* is to mitigate Zn deficiency problems in rice-consuming people in China by developing Zn-enriched rice cultivars. This program will develop new genotypes that have a high potential to accumulate larger quantities of essential micronutrients, such as Zn in their edible parts, that is, rice seeds. The expression of this potential in the rice seeds depends, however, on genotype–environment–management interactions. The results in terms of micronutrient-rich foods depend on the fate of the micronutrients and antinutritional factors, such as phytic acid, during processing of the seeds into food. During breeding and during food processing, interactions among micronutrients also require attention. To improve our understanding of interactions between the different processes in the rice food chain in China, a joint research program has been initiated in 2002 between China Agricultural University, Chinese Academy of Agricultural Sciences (CAAS), Chinese Centre for Disease Control and Prevention in Nanjing, and Wageningen University (WU) (Slingerland et al. 2006). With a food-chain approach (Figure 11.1), the program aimed to enhance Zn supply from rice, capturing both the downstream and upstream aspects of biofortification. The program is called "From Natural Resources to Healthy People," and includes research on soil-plant interactions, plant physiology, food processing, and nutrition; the research focuses on Zn and phytic acid. A program on vegetable genomics ran parallel as a partnership between Wageningen University and CAAS (Bonnema et al. 2006). In this program food consumption patterns and the potential role of dietary components in meeting micronutrient requirements were investigated (Ma et al. 2005). In the same program genetic approaches were explored in *Arabidopsis* species as model plants to gain insight into breeding options to increase micronutrients, reduce phytic acid, or to increase the molar ratio of phytic acid to Zn (Bentsink et al. 2003; Vreugdenhil et al. 2004).

This chapter reports on the results of the joint programs, and aims to explore avenues to increase supply of bioavailable Zn to people from rice crops produced in China. We will consider the Chinese context (see Figure 11.1, the outer ring) explicitly by investigating the potential on Zn-deficient soils, the shift in the cultivation system, and the particulars of Chinese dietary habits.

11.2 MATERIALS AND METHODS

The food chain approach (Figure 11.1) considers different levels of interventions along the food production and consumption chain that can either reinforce or counteract each other. Within the program the potential contribution of each intervention was investigated at each level. The approach is applied to guide research aiming at increasing Zn supply from rice-based foods to Zn-deficient people in China.

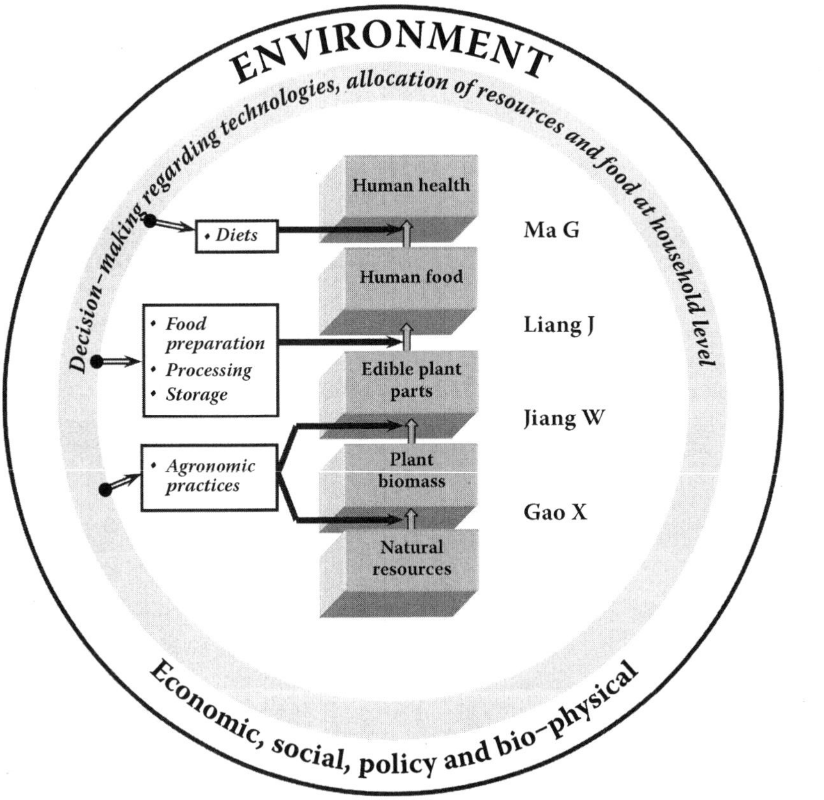

FIGURE 11.1 Food chain approach indicating the position of four PhD researchers (Gao X., Jiang W., Liang J., and Ma G.), and the areas of research from natural resources (plant genotypes and soil quality) to rice biomass, rice grain, rice-based food, and daily diets. (Adapted from Slingerland et al. 2006. *Netherlands Journal of Agricultural Sciences* 53(3/4): 253–279. With permission).

11.2.1 Zinc (Zn) and Phytic Acid Analyses

Within the program the following procedures for Zn and phytic acid determination were used, and led to comparable results. For Zn analysis, samples of roots, shoots, or grains were washed in deionized water. Dry matter was determined by oven drying samples at 75–80°C for 48–72 hours. Ground samples were digested either in a mixture of HNO_3 and $HClO_4$ (4:1, v/v) according to the method of Jackson (1973). A microwave digestion system was used with nitric acid, hydrogen peroxide and hydrofluoric acid, as described by D'Ilio et al. (2002) for other experiments (Liang et al. 2008); Gao et al. (2005, 2006), Ma et al. (2005, 2007a), Jiang et al. (2007a, 2008a), and Liang et al. (2007a). Zinc in plant digests was analyzed using ICP or flame atomic absorption spectrophotometers. The analyses were validated using certified Zn standard reference material obtained from Wageningen Evaluating Programmes for Analytical Laboratory (WEPAL, Wageningen University, The Netherlands) or by commercial standards from the National Center for Standard Substance, Beijing.

Ma et al. (2005) and Liang et al. (2008) measured phytic acid by the anion-exchange method (Harland and Oberleas 1986). Liang et al. (2007a) measured phytic acid, also by the China National Standard Analysis Method (GB/T 17406-1998) based on extraction, separation on anion exchange resin, and spectrophotometric detection of reaction product with ferric chloride ($FeCl_3$) and sulfosalicylic acid. The concentration unit is used throughout the text for both solutions and dry matter.

11.2.2 Criteria of Zn Deficiency

Soil Zn availability is measured as DTPA extractable Zn. When the concentration was less than 0.5 mg Zn kg^{-1} soil, we consider this as Zn deficient. Different parameters are used to establish Zn deficiency in people, including the measurement of dietary Zn supply, calculation of dietary phytic acid to Zn molar ratios in relation to type of diet, measurement of hair (Chen et al. 1985) and nail Zn concentrations, anthropometric measures in children below 5 years of age, and Zn blood plasma levels (Chen et al. 1985). In this paper the anthropometric measure stunting (low height for age) is used as an indicator for Zn deficiency. Supportive evidence is provided by average dietary Zn supply expressed as the Zn concentration (mg Zn kg^{-1}) in the food, and by phytic acid to Zn molar ratio as indicator for the bioavailability of Zn. Zinc absorption is expected to be as low as 15% when phytic acid to Zn ratios are over 15, 30% when molar ratios are from 5 to 15, and 50% when molar ratios are below 5 (IZiNCG 2004). As the research program aims to increase the supply of bioavailable Zn through improvements in the rice food chain, the desired outcome may be higher Zn concentrations and/or lower phytic-acid to Zn ratios in the rice grain and in rice-based foods.

11.3 RESULTS

Data on human consumption are presented to set the targets for improved Zn delivery from rice-based foods. The contribution of rice to daily intake of Zn and of phytic acid was considered. Options for improving Zn intake through the following: (1) soil

Zn delivery and plant Zn uptake, (2) within plant Zn allocation to edible parts, and (3) rice processing effects on Zn and phytic acid.

11.3.1 Contribution of Rice to Daily Intake of Zn and Phytic Acid from Chinese Diets

Ma et al. (2007b) analyzed the data from the 2002 China Nationwide Nutrition and Health Survey and concluded that Chinese generally consume cereal-based diets Their median daily dietary intake of phytic acid and Zn was 1186 and 10.6 mg, respectively, which is equal to an average molar ratio of 11 (Figure 11.2). For Zn, this intake is low, as the revised recommended dietary allowances (IZiNCG 2004) for unrefined cereal-based diets for adults are 10–19 mg d^{-1} depending on gender, physiological stage (pregnancy in women), and body weight. The average value is, however, not so dramatic, because low phytic acid to Zn molar ratios are to expected from an unrefined cereal diet.

From a 24-hour study people on average consumed 238 g white rice per day and all dishes had phytic acid to Zn molar ratios below 15 (Ma et al. 2005). When the data from the nationwide survey were stratified distinguishing between urban and rural populations, Ma et al. (2007b) found only 8% of all urban and 29% of all rural had food intake with molar ratios over 15. Further stratification according to four development levels showed that in rural populations 45% of the subjects in the poorest development level had diets with molar ratios over 15. In urban areas phytic acid intake was almost half of that in rural areas, whereas Zn intake was about equal, explaining the difference in molar ratios (Figure 11.2).

Rice dishes contributed 25 to 33.8% to daily Zn and 52% to daily phytic acid supply (Ma et al. 2007a). Wheat, another staple crop, was responsible for supplying an additional 29% of dietary phytic acid. In total, Ma et al. (2005) analyzed 60 commonly consumed Chinese dishes showing that other dietary components, such as

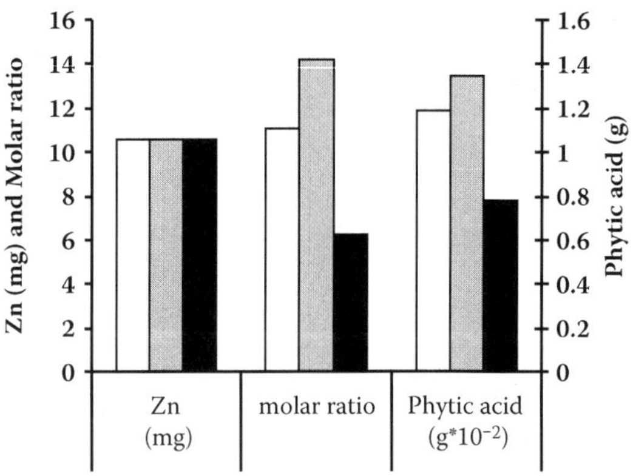

FIGURE 11.2 Zinc and phytic acid composition of Chinese diets consumed by individuals over a 24-hour period. Values are medians of the total population (white bars), for rural (gray bars), and for urban (black bars) populations. (After Ma et al. 2007b.) *European Journal of Clinical Nutrition* 61: 368–374; data from China Nutrition and Health Study. 2002.)

soybean products, contain similar content of Zn to rice but large amounts of phytic acid. As a result, the molar ratio of phytic acid to Zn ranged between 35 and 68 in fourteen different soybean products. Eating soybean and rice products together may also inhibit Zn uptake from rice because of the high phytic acid in the soybean products. But, given the small amounts of soybean products consumed, soybean contribution to phytic acid was not large per se.

There is a large variation between rice varieties and rice cultivation practices in China, leading to rice grains of different quality in diets. Grains of 56 of the most commonly cultivated rice varieties have been collected in the main rice cultivation areas in China from research stations and farmers' fields (Liang et al. 2007a) to evaluate the phytic acid and Zn concentrations in brown rice. The variation between lowest and highest Zn concentrations appeared to be about 300% times, whereas the variation in phytic acid concentration was only about 60% (Table 11.1). Classifying the kernels according to shape showed higher average Zn and phytic acid concentrations with increasing kernel length (Table 11.2). If rice was not polished before consumption, the phytic acid to Zn molar ratio was above 15, which would seriously hamper Zn absorption. In this study genotype, growth conditions, and cultivation methods were not included as explanatory variables, and should be researched in more detail.

The importance of rice for the intake of both Zn and phytic acid supports the suggestion by Liang et al. (2008) and Ma et al. (2005) to establish rice breeding targets for increased Zn and decreased phytic acid concentrations in rice grains. For Chinese consumers the quality of white rice is of more interest. Zinc and phytic acid concentrations in white rice are much lower than in brown rice (Table 11.1), depending largely on milling time (Liang et al. 2008). The breeding targets should account for these differences.

11.3.2 Uptake of Zn as Influenced by Cultivation Practices

Due to water scarcity in North China, farmers are partly shifting from flooded to aerobic rice cultivation. The effect of this shift in water regime on rice yield and Zn

TABLE 11.1
Average Zn and Phytic Acid (PA) Concentrations in Brown Rice for 56 Rice Genotypes Collected from Main Rice-Growing Areas in China from the Harvest of 2002 and in White Rice from Four Commercial Rice Samples Purchased in Supermarkets

Micronutrient	Mean (n = 56) brown rice[a]	Range (n = 56) brown rice[a]	Range (n = 4) white rice[b]
Zn (mg kg^{-1})	22.8	12.9–38.2	10.9–17.6
PA (g kg^{-1})	9.6	7.2–11.9	0.6–1.8
PA–Zn molar ratio	42.9	27.0–66.6	3.0–11.1

[a] Data from Liang et al. 2007. *Journal of the Science of Food and Agriculture* 87: 540–510.
[b] Data from Ma et al. 2005. *Journal of Agriculture Food and Chemistry* 53: 10285–10290.

TABLE 11.2
Phytic Acid and Zn Concentrations in Grain (Brown Rice) and Calculated Phytic Acid to Zn Molar Ratios for 56 Collected Rice Genotype Stratified by Kernel Size[a] According to the Industrial Standard of the Ministry of Agriculture of China[b]

Kernel Shape	Length (mm)	Phytic Acid ($g\ kg^{-1}$)	Zn ($mg\ kg^{-1}$)	Phytic Acid–Zn Molar Ratio
Short-grain (n = 20)	5.1 ± 0.2	9.2 ± 1.2	20.3 ± 4.0 a	44.2
Medium-grain (n = 17)	6.1 ± 0.3	9.7 ± 0.9	22.8 ± 3.8 b	41.5
Long-grain (n = 19)	7.0 ± 0.3	9.9 ± 1.4	24.6 ± 3.4 c	39.3

Note: The mean Zn levels differ significantly ($p < 0.05$). The Zn content is also correlated with kernel length ($p < 0.01$).

[a] Data from Liang et al. 2008. *Journal of Cereal Science* 48(1): 83–91.

[b] Data from CHISA. 2002.

content has been investigated in a field experiment with four aerobic rice cultivars on a calcareous alluvial soil (Cambisoil) (ISSS-ISRIC-FAO, 1998) in a Northern suburb of Beijing (Gao et al. 2006). The soil had a loamy texture, pH (H_2O) 8.0, organic matter 21.4 g kg^{-1}, and DTPA extractable Zn of 2.0 mg kg^{-1} soil. Treatments included Zn fertilizer (0 and 23 kg ha^{-1} Zn applied as $ZnSO_4$ three days prior to sowing) and the aerobic rice genotype. These treatments were tested in fields that were either subjected to flooded or aerobic cultivation practices in the preceding years. Only the averaged dry matter (DM) and Zn concentrations for the four aerobic rice cultivars (Handao 297, 277, 72, and 89B271-17hun) are shown in Figure 11.3. At maturity, the averaged shoot DM was similar between aerobic and flooded cultivation practices or between fertilizer applications. However, shoot Zn concentrations were significantly lower under aerobic than under flooded cultivation conditions ($P < 0.0001$) and both responded positively to Zn application ($P < 0.0005$). Grain DM yields under aerobic cultivation condition were significantly lower than those under flooded conditions ($P < 0.0001$) and neither yield responded to Zn application. Grain Zn concentrations were not affected by the two cultivation practices ($P < 0.79$). The conclusion is that aerobic conditions generally led to less grain and lower total Zn in grains per hectare, even in soils that were not classified as Zn deficient, based upon their DTPA extractable Zn concentration. The effects are expected to be even more pronounced in Zn deficient soils.

To investigate genotypic differences in early stage of Zn uptake and efficiency, Gao et al. (2005) also conducted a pot experiment under greenhouse conditions with air temperature ranging from 30 ± 3°C at day to 23 ± 3°C at night and for natural day length in Hefei of Anhui Province. The treatments included 15 aerobic and 8 lowland (flooded) rice genotypes and two Zn levels (0 and 5 mg Zn kg^{-1} of soil applied as $ZnSO_4{\cdot}7H_2O$). In pots, Zn-deficient clay soil from Shou city in Anhui was used with soil pH (H_2O) of 6.5, organic matter of 1.7%, and the DTPA-extractable Zn concentration of 0.32 mg kg^{-1}. The experiment was conducted in a completely randomized design with three replicates. Plants were harvested 28 days after germination. There

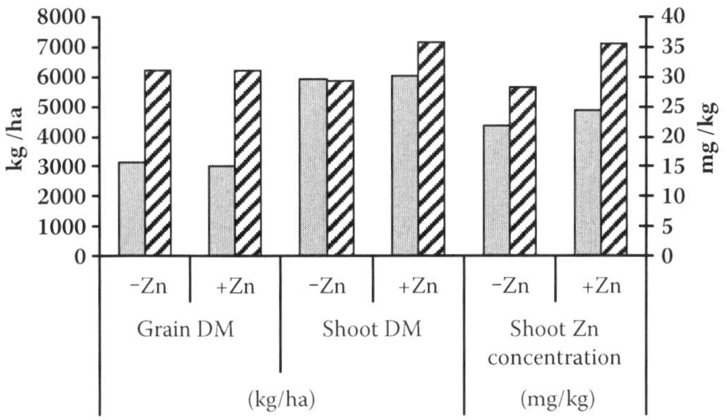

FIGURE 11.3 Average grain and shoot DM yield and shoot Zn concentrations at maturity for four aerobic rice cultivars grown under flooded (hatched bars) and aerobic (filled bars) cultivation practices and without Zn (0 kg ha^{-1}) or with Zn (23 kg ha^{-1} as $ZnSO_4$). (From Gao et al. 2006. *Plant and Soil* 280: 41–47.)

was a large genotypic difference in prevalence and severity of Zn deficiency symptoms, both among aerobic and among lowland genotypes, and between the averages of the aerobic and lowland genotypes. Zinc deficiency resulted in a decrease in shoot and root dry weight for most genotypes. In the treatments with Zn fertilizer application, all genotypes grew well and showed no symptoms of Zn deficiency.

Overall, 53% of the observed variation in tolerance to Zn deficiency (expressed as ratio of shoot dry weight at Zn deficiency over that at adequate Zn supply) was associated with Zn uptake and translocation from roots to shoots (Gao et al. 2005). Genotypic variation of Zn uptake could be explained only for 32% by root surface area. Uptake of Zn may, therefore, increase due to root elongation, e.g., in response to localized NO_3^- supply (Zhang et al. 1999). Further research showed that tolerance to low Zn availability was related to the capacity of plants to exude low-molecular weight organic anions, with citrate being the most effective one in mobilizing Zn (Hoffland et al. 2006). This observation concurs with earlier work by Zhang et al. (1989, 1991), and demonstrates the importance of studying root systems in explaining Zn uptake capacity of plants.

Jiang et al. (2008a) conducted further screening experiments on 16 genotypes of aerobic rice in a greenhouse pot experiment and on 14 genotypes in a field experiment in Anhui province in China. The aim of these experiments was to screen for genotypes that attain relatively high grain yield, high grain Zn concentration, or both. The soil in these experiments was identical: a Shajiang black soil (Vertisols, ISSS-ISRIC-FAO 1998) with pH 6.8 and a concentration of DTPA-extractable Zn of 0.3-0.4 mg kg^{-1}. Treatments were genotypes and Zn application rates from $ZnSO_4$ sources in the field at either 0 or 22.5 kg ha^{-1} Zn, and in the pots at either 0 or 10 mg Zn kg^{-1} soil. In both experiments plants were harvested 30 days after flowering to determine dry weight and Zn concentrations at grain maturity.

The effect of Zn application on grain Zn concentration appeared to be genotypic dependent (Figure 11.4). The same finding has also been observed from the experiments reported by Gao et al. (2006). The overall conclusion is that the maximum

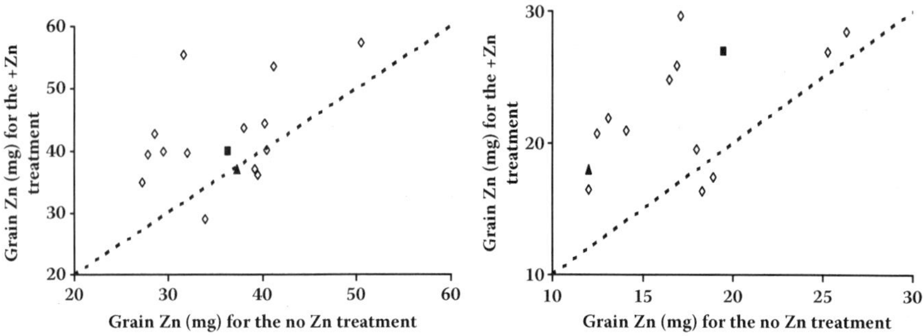

FIGURE 11.4 Observed grain Zn concentrations for two sets of rice cultivars, grown in a pot experiment (a) or in the field (b) and either with (y-axis) or without (x-axis) additional Zn. Each data point represents three replications of a single genotype. The dotted line represents the 1:1 line. Filled squares are for genotype Handoa 502, filled triangles are for genotype Baxiludao (cf. Table 11.4). (Data from Jiang et al. 2007b. *Physiology and modeling of zinc allocation in aerobic rice*, PhD diss., Wageningen University.)

observed increase in Zn concentration in the grain was roughly 75% in the pot experiment and 70% in the field experiment after an application of 23 kg Zn ha^{-1} from $ZnSO_4$. The data (Figure 11.4) indicate that some cultivars may have a greater potential to reach appreciable Zn levels in the brown rice grown in the pot experiment than rice grown under field conditions. A much higher rate of Zn fertilization may be needed or other management practices to improve total Zn uptake under field conditions compared to those obtained in the pot experiment. As observed in both field experiments (Gao et al. 2006; Jiang et al. 2008a), these Zn applications were not accompanied by any yield effect. Hence, it may not make economic sense to introduce this Zn application technology to the farmers unless Zn-enriched rice can be sold for a higher price.

In the same field and pot experiments described above, the application of Zn leads to a large increase in total Zn uptake, but to a relatively low increase in grain Zn concentration. In order to target breeding it seems most relevant to identify those processes from plant uptake to final grain loading of Zn that determine a poor grain Zn loading response despite an enhanced plant Zn uptake after fertilization. Jiang et al. (2007, 2008a) further investigated the cultivars Handao-502 and Baxiludao 28 days after sowing in a pot experiment. Results in Table 11.3 show that under low Zn conditions without Zn fertilizer application, Handao-502 yielded higher root and shoot biomass per pot than Baxiludao, as well as contained higher Zn concentrations in plant tissues. Handao-502 is, however, more efficient in growing in low Zn circumstances and able to take up more Zn from low Zn soils. Under a higher Zn supply Handao-502 still yielded higher root and shoot biomass per pot, but had lower Zn concentrations in the plant tissues than Baxiludao. As we are ultimately interested in increasing Zn concentration in plant tissues, especially in rice grain, the early growth data indicate that Baxiludao is most promising. In contrast, the grain Zn data from the field experiment (Figure 11.4) lead us to a different conclusion. Handao-502 reaches much higher grain Zn concentrations than Baxiludao. These data show that

TABLE 11.3
Root and Shoot DM Yield and Zn Concentrations in Root and Shoot of Two Aerobic Rice Genotypes 28 Days after Germination under Two Zn Fertilizer Treatments (No Zn versus 5 Mg Zn per kg Soil)

Rice Genotype	Treatment (mg Zn kg^{-1})	Root DM (g per pot)	Root Zn Concentration (mg kg^{-1})	Shoot DM (g per pot)	Shoot Zn Concentration (mg)
Baxiludao	0	0.20	22.3	1.61	10.8
Handao 502	0	0.26	24.0	1.74	15.7
Baxiludao	5	0.21	222	1.98	116
Handao 502	5	0.33	130	2.15	95

Source: Adapted from Gao et al. 2005. *Plant and Soil* 278: 253–261.

early season screening (i.e., 28 days after sowing) for Zn concentration in plant tissue has little potential to be used as an indicator of Zn concentration in grains.

11.3.3 Distribution of Zn over Plant Organs: Potential for Zn Loading of Grains

Zinc that is taken up will be distributed to different plant organs, and the amount of Zn mass being allocated or reallocated to the grain is of special importance for achieving the goal of our program. Prior to flowering, the flow of Zn within the plant includes at least the following three steps: root uptake, root to shoot transfer, and within shoot allocation to vegetative plant parts. After flowering the additional processes consist of direct transport of newly taken-up Zn to the panicle, to the grains, and the retranslocation of earlier allocated Zn from vegetative plant parts to the grain (Clemens et al. 2002). An insight into plant physiology is necessary for designing interventions to influence these (re)allocation processes.

Jiang et al. (2008b) conducted two experiments to explore plant growth and the maximum Zn concentration in plant organs. In a "sand culture" experiment solutions with increasing Zn concentrations between 0.15 and 2250 μmol l^{-1} were applied to plants growing in quartz sand. In the solution culture experiment Zn was applied to plants of the genotypes Handao-502 and Baxiludao growing in a nutrient solution. In this experiment Zn was applied every three days in incremental quantities, based on plant growth rates targeting plant Zn concentrations of 10 to 200 mg kg^{-1} during their entire growth. In both experiments plants were harvested at different growth stages: 45 days after emergence, panicle initiation, flowering, 15 days after flowering, and 30 days after flowering (maturity).

Increasing Zn application up to 750 μmol l^{-1} in the "sand culture" led to a 15-fold increase in Zn uptake by plants, whereas grain Zn accumulation was only enhanced by a factor of 1.5–2. Higher Zn application rates led to decreased grain and total plant DM yield, presumably because of toxicity.

Table 11.4 shows the result for Handao-502 and Baxiludao genotypes in the solution culture experiment. The data indicate that Zn concentrations dropped from

TABLE 11.4
Zinc Concentrations (mg Zn kg^{-1} DM) of Different Organs and Tissues of Rice in a Nutrient Solution Experiment, with Either Sufficient or Excess Plant Zinc Uptake

	Zn Supply Levels				
		Excess			
	Sufficient	150 mg kg^{-1}		200 mg kg^{-1}	
	Both Cultivars	Handao-502	Baxiludao	Handao-502	Baxiludao
Grain (brown rice)	20	45	40	50	50
Endosperm	20	50	40	55[a]	50[a]
Bran	60	120	90	120[a]	100[a]
Glumes	25	60	60	65	65
Rachis	25	120	100	250	130
Leaf blades	25	50	40	100	60
Stem[b]	25	300	300	450	400
Root	60	200	250	350	400
Shoot	25	120	120	180	180

[a] Calculated.

[b] Without sheath.

Source: Adapted from Jiang et al. 2008b. *Annals of Applied Biology* (DOI 10.1111/j.1744-7348.2008.00243.x)

stem to rachis (in particular, in Baxiludao), from rachis to bran, and from bran to endosperm. Consequently, we conclude that the enhanced grain Zn uptake is limited by Zn allocation within the plant rather than by root uptake. Moreover, genotypes may differ in rachis Zn loading, which indicates that further research is needed on the mechanisms of Zn loading to rachis, bran, and endosperm.

Timing and location of Zn fertilization are two important aspects potentially influencing Zn allocation within plants. Jiang et al. (2007) conducted another hydroponic experiment with Handao-502 and Baxiludao genotypes under greenhouse conditions. During the entire growth cycle of the plants sufficient (20 mg kg^{-1}) and excessive (200 mg kg^{-1}) Zn concentrations were attained in the plants by applying Zn as $ZnSO_4$ proportional to plant growth. In this regard, when a plant reaches a greater biomass, the amount of Zn supply was increased to maintain the same preset concentration to avoid the tissue dilution effect ^{65}Zn was applied to nutrient solution either at flowering or 15 days after flowering as a 50 μL $^{65}ZnSO_4$ solution containing 148 kBq ^{65}Zn. One week after application half of the plants were harvested and their tissues were analyzed for both unlabeled and labeled Zn content. The remaining plants were washed and placed in a fresh nutrient solution without radioisotope Zn but with the original concentration of unlabeled Zn up to harvest. In the second experiment ^{65}Zn was applied to the tip (5-cm in length) of the flag leaf or to the lowest senescent leaf as a 5 mL $^{65}ZnSO_4$ solution containing 450 kBq ^{65}Zn and a leaf surface wetting agent. Plants were harvested at maturity and individual organs were analyzed for radioactivity.

For root Zn application most of the Zn accumulated in the grains originated from concurrent uptake by roots after flowering and not from Zn remobilization from leaves (Figure 11.5). Moreover, this figure also shows that timing is important. Root application 15 days postflowering led to a larger amount of radioactive Zn in the stem and a smaller amount in the grain than when application occurred at flowering itself. Foliar Zn application had no effect on Zn loading of grains. At maturity 45–50% of absorbed ^{65}Zn had nevertheless left the treated leaf to be reallocated to other vegetative organs. Genes responsible for efficient root Zn uptake capacity after flowering should, therefore, be identified and included in biofortification breeding objectives as potentially an effective way to enhance Zn concentration in rice. In addition, the grain Zn loading ability needs to be improved.

11.3.4 Distribution of Zn within the Grain: Options for Food Processing

Although people in China tend to mainly consume white rice and noodles, rice in the Chinese market is also available as paddy, brown rice, parboiled rice, and other rice-based products. Interventions impacting micronutrient supply from rice to people depend largely on the quality of the harvested rice grain, which is the basis of all rice products. Hence, it is important to know in which fractions of the rice grain Zn and phytic acid can be found, and how hulling, polishing, and other processing methods impact their respective concentrations and the molar ratio in the end product.

Phytic acid and Zn are not evenly distributed within the brown rice grain. Phytic acid is mostly found in the outer layers and the germ, while Zn is also present in endosperm (Liang et al. 2008). As polishing (milling) removes the outer layers of

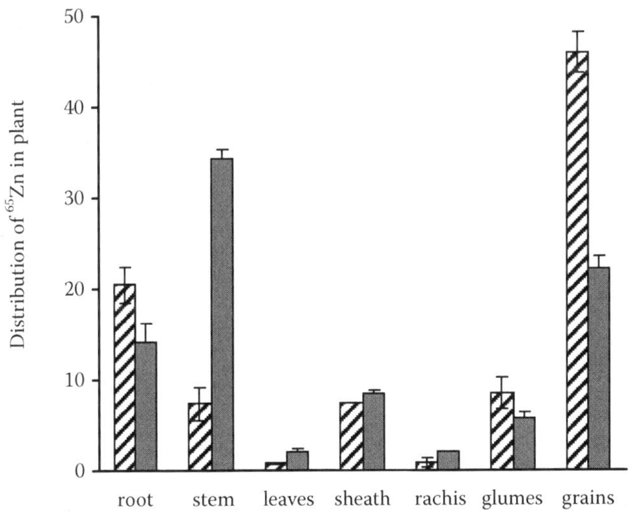

FIGURE 11.5 Distribution of Zn in plant organs after roots were supplied with Zn^{65} at flowering (hatched bars) and at 15 days postflowering (filled bars) for rice cultivar Handao-502. (Adapted from Jiang et al. 2007a. *Annals of Applied Biology* 150: 383–391.)

the grain, including the germ, the process removes relatively more of the phytic acid than of the Zn. Therefore, the phytic acid to Zn molar ratio in white rice is lower than in brown rice. The extent of this effect varies with the shape of the rice kernel, because the molar ratio between surface and content differs between round and long-shaped kernels. Liang et al. (2008) explored how milling time influenced Zn concentrations and the phytic acid to Zn molar ratio in white rice. Phytic acid concentration decreases considerably with increasing milling time, whereas Zn concentration changes at much lower rate. Hence, the phytic acid to Zn molar ratio, therefore, gradually decreases (Figure 11.6). Long kernels initially have higher phytic acid and Zn concentrations, and lower phytic acid to Zn molar ratios than short kernels (Table 11.2).

The phytic acid to Zn molar ratio in the short kernels (Beijing-37) decreased faster than in the medium length kernels (Zhangyou-752), especially during the milling time period of 30–120 sec (Figure 11.6). Based on phytic acid to Zn molar ratios, the ranking between genotypes during these milling times differs from the initial ranking, whereas the values between genotypes hardly differ any more at milling times over 180 s. At a milling time of 120 sec the phytic acid to Zn molar ratio is below 15, and hence absorption of Zn from diet should increase to around 30% (IZiNCG 2004). At this milling time 65–75% of the initial phytic acid is removed, depending on genotype, 40% of the initial grain Zn is removed, and 9–13% of grain mass is lost during the process. To improve Zn absorption, more phytic acid needs to be removed. At a milling time of 300 s, 99% of phytic acid is removed, as well as 50% of grain Zn and 20–25% of total mass. Polishing to acceptable molar ratios will be achieved at a cost of reducing grain Zn concentration and grain mass. Therefore, breeding for low phytic acid content in brown rice and short round kernels may lead to an acceptable phytic acid to Zn molar ratio without the need to increase milling time, while breeding simultaneously increases the fraction of Zn retained in the white rice and also decreases mass loss (Liang et al. 2008).

Phytic acid can also be removed or inactivated by other processing methods, such as adding phytase, soaking, and fermenting (Liang 2007). When these methods

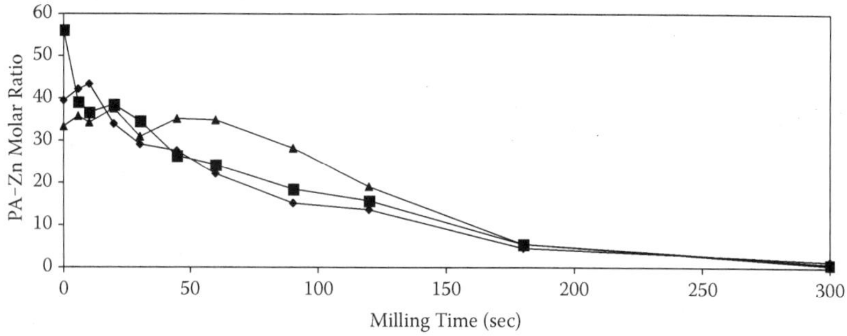

FIGURE 11.6 Decrease of phytic acid to Zn molar ratio with increasing milling time, with short grain variety Bejing-37 (diamond shape), medium grain variety Zhangyou-752 (triangle) and long grain variety Ganwanxian-30 (block) rice genotypes. (Data from Liang et al. 2008. *Journal of Cereal Science* 48(1): 83–91.)

are applied to brown rice or after a short milling time, a higher Zn concentration can be maintained in the end product.

11.4 DISCUSSION

11.4.1 Soil–Plant Relations

The goal of this study was to enhance Zn content in Chinese diets using a food chain approach (Figure 11.1). Gao et al. (2005 and 2006) concluded that Zn uptake and plant growth are seriously inhibited on calcareous or high pH soils. The problem is enhanced when rice is cultivated under aerobic conditions. Zinc fertilization at sowing is needed to guarantee normal plant growth and development. Cakmak et al. (1999) have shown that Zn fertilizer on Zn-deficient soils can likewise greatly enhance growth, development, and yield in wheat, although early growth was more improved in rice experiments by Gao et al. (2005, 2006) and Jiang et al. (2008a, 2007b). Breeding and selection for rice genotypes that have a good Zn uptake are essential. Breeding targets may therefore include genotypes that have special mechanisms to increase Zn uptake. Other mechanisms revolve around the potential for releasing low molecular weight organic anions or other phytosiderophores (Zhang et al. 1989, 1991; Hoffland et al. 2006), or the potential for elongating roots (Zhang et al. 1999).

Recently, Shimizu et al. (2004) detected that phosphorus-deficiency-induced root elongation and its QTL in rice may therefore be relevant when breeding for rice genotypes with improved Zn uptake (Zhang et al. 1999; Gao et al. 2005). A third mechanism could be related to maintaining high root Zn uptake capacity after flowering, as Zn uptake is important during flowering for ensuring an ample good loading in grains (Jiang et al. 2007). In general, cereals gradually reduce root activity during plant senescence (Woolhouse 1978), which accompanies grain filling, although the experiment showed that there are plants that can maintain root uptake capacity.

Genotype by environment interactions and fertilizer effects on phytic acid in rice need also to be investigated. Liu et al. (2005) collected 72 cultivars or breeding lines from China, cultivated them under identical circumstances, and found that grain phytic acid varied between 1.1 and 0.6%. Seeding and cultivating a selection of these grains in four locations with wide difference in ecological conditions showed a major environmental effect (70% of the variance was from the environment, with phytic acid ranging from 0.69 to 0.97%). The cultivar effect and the cultivar by environment interaction were also significant. The interaction effect did not compromise the main environmental effect.

Phosphate plays a major role in productivity and leads to phytic acid formation in grains. Based on research in sorghum (Traore 2006), we suggest that late uptake of massive amounts of P should be avoided to limit high levels of phytic acid in rice grains, and P fertilization should be geared towards early plant P uptake and reduced availability of P for plant uptake during grain filling.

11.4.2 Allocation and Reallocation Processes

A distinction must be made between the following processes for discussing findings on plant allocation and translocation of Zn: (1) rate of root uptake, (2) rate of root to shoot allocation, (3) rate of sequestration in none edible above ground plant parts, (4) rate of re-translocation from nonedible plant parts, (5) rate of transport from stem to nongrain panicle parts such as rachis, rachilea, and glumes (hereafter referred to as rachis, etc.), (6) the rate of transport from rachis, etc., to the outer layers of the rice grain (hereafter referred to as rice bran), and (7) rate of transport from the bran to the endosperm (Clemens et al. 2002). The Zn concentrations of different organs and tissues were studied for a range of Zn supply levels (Jiang et al. 2007a, 2008b). In Table 11.4 the most limiting transfers can be derived from the difference in Zn concentrations between two connected tissues, with larger differences indicating a stronger limitation to transfer.

The processes mentioned by Clemens et al. (2002) will be discussed individually in reference to the numbered points following:

1. It is clear from the field and pot experiments (Gao et al. 2005, 2006; Jiang 2008a,b) that enhanced Zn uptake through fertility management does not translate into an increased grain Zn loading. Plant Zn uptake per se, therefore, does not seem the most limiting factor to rate of grain Zn loading.
2. The root-to-shoot transfer does not seem to limit the loading of aboveground plant parts in either of the two studied cultivars similarly to that observed at early growth stages in a field study with these and other cultivars (Gao et al. 2005; see Table 11.3).
3. With increasing plant Zn uptake, levels of additional Zn are mainly sequestered in roots and stems, while leaf tissue Zn concentrations are maintained at levels below 100 mg Zn kg^{-1} dry matter. This observation may not be surprising as tissue Zn concentrations above this value are reported to be toxic (Marschner 1995). In fact, when Zn supply levels were enhanced to extreme levels, leaves were found to accumulate more Zn, and dry matter accumulation was reduced, indicating toxicity may have occurred (Jiang et al. 2008b).
4. With the use of radioisotope (^{65}Zn) (Jiang et al. 2007) established that at flowering Zn taken up from roots at flowering was readily allocated to the grains, while Zn in leaves was translocated to other vegetative plant parts. These effects were observed when sufficient or surplus Zn was supplied via nutrient solution. In general, translocation of Zn tends to take place as tissue senescence sets in. Given the role of Zn in protein synthesis, it is logical that massive translocation of biologically active Zn from a tissue either will be the cause of the senescence or can only occur once senescence begins to other causes. The rate of grain Zn loading was observed to be different among cultivars in the field as evidenced by differences in final grain Zn concentrations in Figure 11.4. This observation has also been established in much wider screening in rice germplasm (Gregorio et al. 2000).

5. The transfer rate of Zn from stems to the rachis and other nongrain parts differed between two studied cultivars, and in CV Baxiladao it was clearly more limited, and potentially reduced final grain Zn concentration in this cultivar (Table 11.4).
6. The transfer of Zn between rachis and the outer grain tissues (bran) seemed to be limiting grain Zn loading in the cultivar that had higher rachis Zn levels (Handao 502).
7. The transfer of Zn between outer grain tissues (bran) and the endosperm limited endosperm Zn loading, as tissue concentrations dropped by a factor of two to three between bran and endosperm. The final amount of Zn loaded in the endosperm is the most important outcome, as white rice only contains the endosperm.

The studied cultivars did not differ much in Zn levels in stem or bran, but differed to a larger extent in Zn levels in the rachis, and thus showed different points at which the tissue Zn concentrations between stem and bran were most reduced. The underlying cause has not been identified yet. The transfer from xylem to phloem may in both cases be the reason for the observed drop in tissue Zn concentrations but the Zn transfer may take place both in the rachis and in the grains. This latter transfer is in contrasts with wheat, where xylem to phloem transfer always takes place outside the grain due to an anatomical xylem discontinuity (Zee and O'Brien 1970). In rice such a discontinuity does not exist (Zee 1971), and it seems therefore unlikely that all xylem-to-phloem transfer of Zn would take place in tissues prior to the grain. The observed drop in Zn concentration between bran and endosperm and the observed drop in concentration between stem and rachis may therefore both be partly related to the xylem to phloem transfer limitations of Zn. The data from Liang et al. (2007a, 2008) suggest that breeding could focus on enhancing the rate of Zn transfer from bran to endosperm, and from stem to bran. Once these transfers of Zn were successfully enhanced, then research is needed to evaluate ways to secure Zn uptake for the increased concentrations of Zn in the grains.

11.4.3 Breeding

An inventory with 1138 genotypes was conducted by the International Rice Research Institute (IRRI) in the Philippines, and showed that the range of Zn concentrations in brown rice was 13.5–58.4 $\mu g\ g^{-1}$ (Gregorio 2002). This data offered a perspective for breeding and selection of high Zn rice. High grain Zn concentrations were observed in rice from different environments, and there was also some evidence that genotype by environment interactions may play a role, as well as a high grain mineral density could be combined with high yield. Next, a complete diallel population was developed using 9 rice cultivars with varying grain mineral densities (Gregorio and Htut 2003). A significant specific combining ability (SCA) was produced, indicating that combinations of parents could be selected that contained high mineral densities in the grains of their progenies. A highly significant positive correlation among specific combining ability effect was observed on grain Fe, Zn, Mg, P, K, and S, suggesting that breeding for high Zn and Fe might go together with breeding for high P, which is undesirable because 65 to 80% of P is present as phytic acid in mature seeds (Raboy 2003). Significant negative correlations were observed between grain Ca and Zn and Fe concentrations, respectively, and between grain Mn and Fe and Zn concentration, respectively. These results show that enhanced overall grain mineral concentrations may not be easy to breed for, and focusing on a single mineral could lead to decrease in another nutrient.

In general, conventional breeding processes are slow and expensive, especially when a trait involves many genes. The use of biotechnological tools, such as marker-assisted selection (MAS), could significantly increase the pace and reduce the costs of breeding. For MAS, knowledge of the genetic basis of Zn accumulation in seed is needed. Much of this information is collected in model plants, such as *Arabidopsis thaliana* (Bentsink et al. 2003; Vreugdenhil et al. 2004; Ghandilyan et al. 2006), and the information can hopefully be applied to staple crops. Rather than breeding for a specific outcome, such as increased grain Zn density, the use of Quantitative Trait Locus (QTL) analyses can provide information on the different processes affecting mineral accumulation, such as genes encoding for metal uptake, transport, sequestration mechanisms, and underlying processes such as chelator synthesis.

In *Arabidopsis thaliana,* QTL analyses for seed cationic mineral content were identified, and showed that some were involved in controlling the accumulation of more than one mineral (Vreugdenhil et al. 2004). Bentsink et al. (2003) found one major QTL explaining most of the variation in seed phytic acid concentration in *Arabidopsis*, which was at the same time responsible for high P levels in seeds. Breeding for low phytic acid would, therefore, always coincide with low P. The comparison of the results showed that the loci affecting Zn accumulation were largely different from those controlling phytic acid concentration of seeds (Ghandilyan et al. 2006). If the same relationship applies to cereal crops, breeding for genotypes with low phytic acid and high Zn concentrations in their seeds should be possible. This possibility is, however, in contrast with comments resulting from the rice breeding experiment by Gregorio and Htut (2003). Converting the information in molecular markers, advanced breeding tools can be used in selecting for the presence or absence of specific alleles of genes known to play a role in seed mineral accumulation (Ghandilyan et al. 2006). Our research on most limiting steps can be effectively linked with such genetic research, and would suggest not to "waste" time on genes involved in Zn uptake but rather to look for QTL in the identification of genes affecting the processes responsible for the transfer of Zn from stem to rachis to bran to endosperm. A combination of genetic and physiological research on *Arabidopsis thaliana* and rice could help to find relevant markers on the basis of an understanding of underlying processing leading to grain Zn loading.

Liang et al. (2007) suggested that genetic differences in accumulation of Zn and phytic acid and their distribution over the grain might exist, and hence could be targets for breeders. Ren et al. (2007) explored phytic acid P, total P, and mineral density in different rice grain parts of a low phytate mutant and its parent Xieqingzao. In brown rice, the phytic acid P in the mutant was about 60% of that in the parent, whereas the P content in both was similar, which contrasts with the already, mentioned data in *Arabidopsis* (Bentsink et al. 2003), and with the fixed phytate-P ratios suggested by Raboy (2003). Distribution effects of phytic acid P were also found.

Firstly, a significant higher percentage of phytic acid-P was deposited in the bran part of the mutant than that of the parent. Secondly, more Zn seemed to be diverted from the hull to the endosperm in the mutant than in the parent, leading to a higher proportion of total Zn retained after milling. Subsequently, Zn content in white rice was about 50% higher in the mutant than in the parent. The low phytic acid mutant of rice showed the potential to increase both content and bioavailability of Zn.

For normal *japonica* rice (Liu et al. 2005) and for low phytic acid mutants (Ren et al. 2007), interactions between genotypes and environment on grain phytic acid concentrations have been reported, but in either case the environment has not been described in terms of soil pH, P, and Zn contents or any other factor potentially responsible for the interaction. For Zn, uptake interactions between genotype and environment have been explored for aerobic and lowland rice genotypes (Gao et al. 2005, 2006). The mentioned genotype by environment interactions and the need for maintained Zn uptake rates during grain filling (Jiang et al. 2007) imply that next to breeding efforts, a proper understanding of gene expression under different agro-ecologies and fertilizer application rates is needed.

11.4.4 Dietary Composition and Food Processing

To attain diets with high amounts of Zn and to have phytic acid to Zn molar ratios that are favorable for Zn uptake by humans, a careful mix of interventions is needed. Important sources of phytic acid, such as soybean products, are present in Chinese daily diets but the low quantities consumed give no reason for concern. The recent change to diets with more meat is also promising because meat contains relatively high amounts of Zn and no phytic acid. Thus, meat can positively influence the phytic acid to Zn ratio of the diet as a whole. In China, where rice is a major dietary component, especially for the poor, it was found to provide half of the phytic acid intake and about one-third of the Zn. Current milling time does take away phytic acid, but the reported contribution of rice to phytic acid in the diet is for polished rice, hence more phytic acid needs to be removed. Liang et al. (2007b, see Figure 11.6) showed that enhanced milling (up to 300 sec) is needed to remove 99% of phytic acid. This length of time would obviously lead to considerable loss of both mass and Zn. Therefore, an increase in milling time does not seem to provide a solution.

Another option to enhance mineral supply from rice-based foods is to use brown rice or rice that has been milled for a very short time (which maintains more minerals in the end product) in combination with food processing; methods such as soaking (Lestienne et al. 2005), fermenting, germinating, and applying phytase are needed to remove or inactivate phytic acid. Such combinations may lead to an end product with higher nutritive value. Studies to explore these options have also been conducted within the research program and will be reported by Liang (2007). Obviously, these options would imply a long trajectory of introduction and potential behavioral change. In this respect the recent introduction of Fe-fortified soybean sauce in six counties in China is well received (see *China Daily* newspaper, July 14, 15, 2007, p. 3). Cofortification with Zn might be an option if the product is targeted at Fe- and Zn-deficient people.

11.5 CONCLUSIONS

The use of Zn fertilizers at the seedling stage is needed to guarantee normal plant growth and development in the Zn-deficient growing conditions. Breeding and selecting for cultivars with enhanced Zn uptake under low Zn conditions could contribute to normal growth, but not to increased Zn concentration in rice seeds. Furthermore, Zn uptake during flowering may be essential for high-grain Zn loading; thus, prolonged active root Zn uptake may be a positive trait. A major contribution to enhanced-grain Zn concentrations can be expected from increased Zn transfer from stem to rachis to bran to endosperm, an aspect that has had little attention so far. Breeding for low phytic acid, higher Zn concentrations, and improving distribution of Zn and phytic acid over the rice kernel, is desirable and possible. Combining research on QTLs in *Arabidopsis* and physiology of seed Zn loading in rice offers a potential to increase breeding pace and precision.

Although no silver bullet has been found, the food chain approach has shown to be useful in advancing the biofortification approach by providing breeding programs with objectives that address the shifting of rice production to aerobic systems, the current processing methods of rice, and the proportion of rice in a mixed diet. In addition, this food chain approach provides options for improving fertilizer application methods, for breeding targets related to Zn loading to grains, and for processing for enhanced mineral bioavailability from brown rice grains.

ACKNOWLEDGMENTS

We are grateful for the funding support from "From Natural Resources to Healthy People" program by the Interdisciplinary Research and Education Fund of Wageningen University. Staff and students from China Agricultural University and CAAS in Beijing and the Centre for Disease Control and Prevention in Nanjing and Wageningen University are acknowledged for being research partners in this program.

REFERENCES

Alloway, B. J. 2004. *Zn in Soils and Crop Nutrition*. Brussels: International Zn Association.

Bentsink, L., K. Yuan, M. Koorneef, and D. Vreugdenhil. 2003. The genetics of phytate and phosphate accumulation in seeds and leaves of *Arabidopsis thaliana* using natural variation. *Theoretical and Applied Genetics* 106: 1234–1243.

Bonnema, A. B., Z. Lin, L. Qu, and E. Jacobsen. 2006. Development of institutional collaboration between Wageningen University and the Chinese Academy of Agricultural Sciences. *Netherlands Journal of Agricultural Sciences* 53(3/4): 369–386.

Bouis, H. E., R. D. Graham, and R. M. Welch. 2000. The Consultative Group on International Agricultural Research (CGIAR) Micronutrient Project: Justification and objectives. *Food Nutrition Bulletin* 21: 374–381.

Bouman, B. A. M., S. Peng, A. R. Castaneda, and R. M. Visperas. 2005. Yield and water use of tropical aerobic rice systems in the Philippines. *Agriculture Water Management*. 72: 87–105.

Cakmak, I., M. Kalayci, H. Ekiz, H. J. Braun, Y. Kilinc, and A. Yilmaz. 1999. Zinc deficiency as a practical problem in plant and human nutrition in Turkey: A NATO-science for stability project. *Field Crops Research* 60: 175–188.

Chen, X. C., T. A. Yin, J. S. He, Q. Y. Ma, Z. M. Han, and L. X. Li. 1985. Low levels of zinc in hair and blood. Pica, anorexia, and poor growth in Chinese preschool children. *American Journal of Clinical Nutrition* 42: 694–700.

Chen, C., W. He, and Z. Fu. 2004. *Ten Year Tracking Nutritional Status in China 1990–2000.* Peoples Medical Publishing House, ISBN 7-117-06367-X/R-6368, pp. 326.

Clemens, S., M. G. Palmgren, and U. Krämer. 2002. A long way ahead: understanding and engineering plant metal accumulation. *Trends in Plant Science* 17(7): 309–315.

D'Ilio, S., M. Alessandrelli, R. Cresti, G. Forte, and S. Caroli. 2002. Arsenic content of various types of rice as determined by plasma based techniques. *Microchemical Journal* 73: 195–201.

Gao, X. P., C. Q. Zou, X. Y. Fan, F. S. Zhang, and E. Hoffland. 2006. From flooded to aerobic conditions in rice cultivation: Consequences for zinc uptake. *Plant and Soil* 280: 41–47.

Gao, X. P., C. Q. Zou, F. S. Zhang, S. E. A. T. M. van der Zee, and E. Hoffland. 2005. Tolerance to zinc deficiency in rice correlates with zinc uptake. *Plant and Soil* 278: 253–261.

Ghandilyan, A., D. Vreugdenhil, and G. M. G. Aarts. 2006. Progress in the genetic understanding of plant iron and zinc nutrition. *Physiologia Plantarum* 126: 407–417.

Gregorio, G. B. and H. H. Dharmawansa Senadhira, and R.D. Graham. 2000. Breeding for trace mineral density in rice. *Food Nutrition Bulletin.* 21: 382–386.

Gregorio, G. B. 2002. Progress in breeding for trace minerals in staple crops. In *Plant breeding: a new tool for fighting micronutrient malnutrition. Journal of Nutrition* 132: 500S–502S.

Gregorio, G. and T. Htut. 2003. Micronutrient dense rice: Developing breeding tools at IRRI. In *Rice Science: Innovations and Impact for Livelihood*, Proceedings of the international rice research conference, Beijing, China, 16–19 September 2002, T. W. Mew, D. S. Brar, S. Peng, D. Dawe and B. Hardy (Eds.), Manila, Philippines: IRRI, pp. 371–378.

Harland, B. F. and D. Oberleas. 1986. Anion exchange method for determination of phytates in foods: Collaborative study. *Association of Official Analytical Chemists* 69: 667–670.

Hoffland, E., C. Z. Wei, and M. Wissuwa. 2006. Organic anion exudation by lowland rice (*Oryza sativa* L.) at zinc and phosphorus deficiency. *Plant and Soil* 283: 155–162.

ISSS, ISRIC, and FAO. 1998. *World Reference Base for Soil Resources.* World soil resources reports 84. Rome: FAO.

IZiNCG (International Zinc Nutrition Consultative Group). 2004. Assessment of the risk of zinc deficiency in populations and options for its control. C. Hotz and K. H. Brown (Eds.), *Food and Nutrition Bulletin* 25: S91–S202.

Jackson, M. L. 1973. *Soil Chemical Analysis.* New Delhi: Prentice Hall of India.

Jiang, W., M. Zhao, and T. J. Stomph. 2005. Effects of zinc application on plant growth and zinc partitioning in aerobic rice. In *Plant Nutrition for Food Security, Human Health, and Environmental Protection*, XV IPNC 2005, C. J. Li et al. (Eds.), Beijing, China, Tsinghua University Press, pp. 418–419.

Jiang, W., P. C. Struik, J. Linga, H. van Keulen, Z. Ming, and T. J. Stomph. 2007. Uptake and distribution of root-applied or foliar-applied ^{65}Zn after flowering in aerobic rice. *Annals of Applied Biology* 150: 383–391.

Jiang, W., P.C. Struik, M. Zhao, H. van Keulen, T.Q. Fan, and T.J. Stomph. 2008a. Indices to screen for grain yield and grain zinc concentration in aerobic rice at different soil Zn levels. *Netherlands Journal of Agricultural Sciences* 55(2): 181–197.

Jiang, W., P.C. Struik, H. van Keulen, M. Zhao, L.N. Jin, and T.J. Stomph. 2008b. Does increased Zn uptake enhance grain Zn mass fraction in rice? *Annals of Applied Biology* DOI 10.111/j.1744-7348.2008.00243.x (published on line 23-Apri-2008).

Lestienne, I., C. Mouquet-Rivier, C. Icard-Verniere, I. Rochette, and S. Treche. 2005. The effects of soaking of whole, dehulled and ground millet and soybean seeds on phytate degradation and Phy/Fe and Phy/Zn molar ratios. *International Journal of Food Science and Technology* 40: 391–399.

Liang, J. F. 2007. *Iron, zinc and phytic acid in rice from China: wet and dry processing towards improved mineral bioavailability.* PhD diss., Wageningen University.

Liang, J. F., B. Z. Han, L. Han., M. J. R. Nout, and R. J. Hamer. 2007. Iron, zinc and phytic acid content of selected rice varieties from China. *Journal of the Science of Food and Agriculture* 87: 540–510.

Liang, J.F., Z. Li, K. Tsuji, K. Nakano, M.J.R. Nout, and R.J. Hamer. 2008. Milling characteristics and distribution of phytic acid and zinc in long, medium and short grain rice. *Journal of Cereal Science* 48(1): 83-91.

Liu, Z., F. Cheng, and G. Zhang. 2005. Grain phytic acid content in japonica rice as affected by cultivar and environment and its relation to protein content. *Food Chemistry* 89: 49–52.

Ma, G. S., Y. Jin, J. Piao, F. Kok, G. Bonnema, and E. Jacobsen. 2005. Phytate, calcium, iron and zinc contents and their molar ratios in foods commonly consumed in China. *Journal of Agriculture Food and Chemistry* 53: 10285–10290.

Ma, G, Y. Li, Y. Jin, F. Zhai, F. J. Kok, and X. Yang. 2007a. Phytate intake and molar ratios of phytate to zinc, iron and calcium in diets of people in China. *European Journal of Clinical Nutrition* 61: 368–374.

Ma, G., Y. L, F. Zhai, F. J. Kok, E. Jacobsen, and X. Yang. 2007b. Iron and zinc deficiencies in China: What is a feasible and cost effective strategy? *Public Health and Nutrition*: doi: 10.1017/S13689800070001085. Published online by Cambridge University Press 26-Sep-2007.

Marschner, H. 1995. *Mineral Nutrition of Higher Plants*, 2nd ed. London: Academic Press.

Raboy, V. 2003. Molecules of interest *myo*-Inositol-1.2.3.4.5.6-hexa*kis*phosphate. *Phytochemistry* 64: 1033–1043.

Ren, X. L., Q. L. Liu, H. W. Fu, D. X. Wu, and Q. Y. Shu. 2007. Density of nutrient elements in rice grains of a low phytate mutant. *Food Chemistry* 102: 1400–1406.

Shimizu, A., S. Yanagihara, S. Kawasaki, and H. Ikehashi. 2004. Phosphorus deficiency-induced root elongation and its QTL in rice (Oryza sativa L.). *Theoretical and Applied Genetics* 109, 1361–1368.

Sillanpää, M. 1990. *Micronutrient Assessment at the Country Level: An International Study.* FAO Soils Bulletin No. 63, Rome: Food and Agricultural Organisation.

Slingerland, M., K. Traore, P. Kayode, and C. E. S. Mitchikpe. 2006. How to fight iron deficiency malnutrition in West Africa? A food chain approach. *Netherlands Journal of Agricultural Sciences* 53(3/4): 253–279.

Takkar, P. N. and C. D. Walker. 1993. The distribution and correction of zinc deficiency. In *Zinc in Soils and Plants.* A. D. Robson (Ed.), 1993. Dordrecht: Kluwer Academic Publishers, pp. 151–165.

Traore, K. 2006. Can sorghum grain quality be improved by applying inorganic Zn and P fertilisers and organic amendments? Chapter 6, 87–102. In *Effects of soil amendments and drought on zinc husbandry and grain quality in Sahelian sorghum.* PhD diss., Wageningen University.

Vreugdenhil, D., M. G. M. Aarts, M. Koornneef, H. Nelissen, and W. H. O. Ernst. 2004. Natural variation and QTL analysis for cationic mineral content in seeds of *Arabidopsis thaliana. Plant, Cell and Environment* 27(7): 828–839.

Woolhouse, H. W. 1978. Senescence processes in the life cycle of flowering plants. *Bioscience* 28(1): 25–31.

World Health Organization (WHO), Food and Agriculture Organization (FAO), International Atomic Energy Association (IAEA). 2002. *Trace Elements in Human Health and Nutrition*, 2nd ed. Geneva: World Health Organization.

Zee, S. Y. 1971. Vascular tissue and transfer cell distribution in the rice spikelet. *Australian Journal of Biological Sciences* 25: 411–414.

Zee, S. Y. and T. P. O'Brien. 1970. A special type of tracheary element associated with "xylem discontinuity" in the floral axis of wheat. *Australian Journal of Biological Sciences* 23: 783–791.

Zhang, F. S., V. Römheld, and H. Marschner. 1989. Effect of zinc deficiency in wheat on the release of zinc and iron mobilizing root exudates. *Zeitschrift der Pflanzenernaehrung und Bodenkunde*. 152: 205–210.

Zhang, F. S., V. Römheld, and H. Marschner. 1991. Release of zinc mobilizing root exudates in different plant species as affected by zinc nutritional status. *Journal of Plant Nutrition* 14: 675–686.

Zhang, H. M., A. Jennings, P. W. Barlow, and B. G. Forde. 1999. Dual pathways for regulation of root branching by nitrate. *Plant Biology* 96(11): 6529–6534.

12 Expression of Pea Seed Ferritin cDNA in Indian Mustard

Nutritional Value and Oxidative Stress Tolerance of the Transformants

Nadella Nirupa, Majeti Narasimha Vara Prasad, and Pulugurtha Bharadwaja Kirti

CONTENTS

12.1 INTRODUCTION

To meet the nutritional needs of a rapidly growing world population (estimated to reach 8 billion by 2030), 50% more food grains must be produced. Mineral and vitamin deficiencies affect a greater population than protein energy malnutrition (Bouis 2003). Deficiencies of micronutrients such as iron (Fe), zinc (Zn), and vitamin A afflict over three billion people (more than 50% of the world's population). Most are women, infants, and children in the developing world. Deficiencies of micronutrients result in increased morbidity and mortality rates, decreased worker productivity, stagnated national development, permanent impairment of cognitive development in infants and children, and large economic costs and suffering to those societies affected (Welch 2005). This global crisis in nutritional health is the result of dysfunctional food systems that do not consistently supply enough of these essential nutrients to meet the nutritional requirements of high-risk groups such as pregnant and lactating women.

12.2 STRATEGIES FOR ERADICATING DEFICIENCIES AND IMPROVING HEALTH

A number of strategies have been adopted by various organizations (e.g., World Health Organization [WHO]) to improve nutrition. Fortifying foods has been conventional to ensure easy access to a diversified healthy diet, including vitamins and minerals, among the rapidly growing population. Recently, the trend of fortification has been emerging with the aim of achieving optimal health and reducing the chronic disease burden (WHO 2004). An important problem in several food fortification programs is the lack of bioavailability of compounds used to combat deficiencies. For example, common widespread vehicles for commercial Fe fortification are cereal flours (UNICEF 1999). Bioavailability of Fe, selection of the vehicle, the balance of inhibitors and enhancers, and total Fe intake are the factors determining efficiency of fortification. The other most common strategy is the utilization of plants.

12.2.1 Plants' Role in Eradicating Deficiencies

Human existence entails that crops provide at least 50 nutrients including vitamins, minerals, trace elements, amino acids, and essential fatty acids in adequate levels. Poor agricultural systems and shrinking arable land curtail man's ability to meet human nutritional needs in the developing world. Changes in agricultural policies and systems must be made that will ensure consistent and adequate supplies of all essential nutrients to all people. The queries and concerns that should be addressed when these technologies are used to improve the micronutrient status of populations involve staple food crops (Welch 2005). Many successful efforts to increase agriculture productivity were undertaken to meet demanding human requirements during the 20th century. Improving the nutrient density of crops without affecting the environment will be a large challenge in the future due to declining resources such as water and arable land. Moreover, there will be a need to improve nutrient density of crops without affecting the environment. Biofortification is the process of improving the nutritional value of staple food crops by the development of genotypes, either conventionally or by genetic engineering. Conventional breeding, a method to harness genes in plants by either sexual crossing or in vitro techniques, such as cell fusion, have produced a diverse array of differentiated crops, which emphasize output traits like grain quality. For example, Lam et al. (2006) investigated the quantitative trait loci (QTLs) for phytate in rice grain and their relationship with Fe. In contrast, plant genetic engineering is the ability to specifically introduce genetic changes into a plant. Unlike conventional ways, transgenics may further increase the nutrient density of the foods by complementing strategies that reduce antinutrient substances (e.g., phytic acid) from inhibiting micronutrient bioavailability (Lucca et al. 2006; Holm et al. 2002). Plant biotechnology, therefore, should considerably supplement plant breeding.

Improving the biotic and abiotic stress tolerance of economically important plants would serve as an important criterion to aid in both human and plant nutrition. Environmental stress associated with oxidative damage at the cellular level is the most important limiting factor in plant productivity. Overexpression of genes that combat stress and improve nutritional quality of foods is a major concern. Some agronomic traits (e.g., heat, frost, and drought resistance, as well as salinity and heavy metal tolerances) have already been targeted for improvement (Pilon-Smits et al. 1999; Hegedüs et al. 2002; Rodrigues et al. 2006). The other principal aspect of plant nutrition to consider for better productivity is the availability of essential minerals like Fe, boron (B), copper (Cu), Zn, manganese (Mn), etc. Optimizing the delivery of these minerals ensures an increase in efficiency, and quality of the plant, and also may advance human nutrition by increasing the mineral content to meet human requirements. Recent research has shown that these trace minerals also assist plants to resist disease and other environmental stresses (Bouis 2002).

Nevertheless, large-scale production of biofortified crops is a challenging output as significant research needs to be compiled on safety risk assessments for the required regulatory approvals. Some of the major concerns regarding genetic engineering are unknown effects and impact on the environment due to transgene escape, crossing species barriers, biodiversity, and long-term effects on human health as new proteins acting as allergens, metabolites, and potential toxins are released due to ectopic expression of genes (Singh et al. 2006). Bioengineered crops with genes that sequester minerals efficiently under different environmental conditions may result in undesirable metabolic consequences such as heavy metal accumulation. Hence, the study on plant–metal interactions in such plants is important for the safety of the environment.

Optimized seed Fe content is a major biotechnological challenge identified by the WHO, and it is therefore crucial to understand the underlying mechanisms. Deficiencies of this micronutrient cause a number of severe health problems that globally affects about two billion people and can result in anemia in women and irreversible impairment of children's learning capability (WHO 2001). Functional endpoints of Fe deficiency include immunity, oxidative stress, neurocognitive function, and development. Many programs have aimed at supplementing food with additives but infrastructure, economics, local habits, and availability of centralized facilities and health care professionals can limit their utility. It is reasonable to attempt to use all approaches available to improve Fe nutrition. The bioavailability of Fe is fairly low (nearly 10%) in vegetable foods. Vegetable-based diets are the primary food source in most developing countries. Improving the plant's Fe content through existing technologies might serve to accomplish the task of removing Fe deficiency. Expression of genes that code for trace element binding protein, storage proteins, and/or increased expression of proteins that are responsible for trace element uptake into plants (Lucca et al. 2002, 2006) may aid in increasing the micronutrient (e.g., Fe) content in the plant.

Dietary ferritin, a protein cage around an iron mineral, was until recently an underestimated source of bioavailable Fe. Ferritin was employed to fortify rice, wheat, and lettuce with bioavailable Fe by various groups (Goto et al. 1999; Goto and Yoshihara 2001; Drakkaki et al. 2000) through stable accumulation of the ferritin subunit in the seeds and vegetative parts. Recent evidence also suggests that ferritin may aid in the plant's Fe availability and combat oxidative stress resulting from biotic and abiotic stress (Goto 2001; Deak et al. 1999). In the light of these findings, ferritin was overexpressed in *B. juncea,* an important oil crop in the Indian subcontinent. Mustard is used globally in salad dressings, pickles, and processed meat products. The ferritin transgenics generated were assessed for the heavy metal-induced oxidative tolerance. Overexpression of pea seed ferritin in *Brassica* under the control of a strong constitutive promoter CaMV35S would be appropriate to enhance the Fe content of the vegetative parts.

Exposure to excessive heavy metals severely affects growth and productivity of the plants. A wide range of cellular responses are associated with the elevated levels of heavy metals. Iron, an essential nutrient for plants, has a pivotal and dual role in free radical chemistry. High reactivity of this transition metal with oxygen species produces hazardous hydroxyl radicals. Cadmium, a nonessential heavy metal, disturbs the redox control of the cell when not detoxified, and it is also correlated with disturbances in the uptake and distribution of macro-and micronutrients in plants (Schutzendubel et al. 2002). Ferritins are a class of ubiquitous multimeric proteins able to store Fe in a soluble and nontoxic form. Ferritin, by regulating the intracellular free Fe, controls iron-catalyzed generation of reactive oxygen species (Rama Kumar and Prasad 1999; Fourcroy et al. 2004). Effects of Fe and oxidative stress on ferritin expression have been extensively studied in animals, but not in plants (Cozzi et al. 1990; Balla et al. 1994). Ferritin can also sequester Fe and nonferrous metals, reducing the availability of metals that catalyze cell damage (Sczekan and Joshi 1989). Besides enzymatic scavenging, control of the concentrations of metals (known for their prime role in oxygen activation and enzyme inactivation) by sequestering them could form an important complementary way for preventing toxic metal effects.

Indian mustard has been identified as a high biomass-producing plant with the capacity to accumulate various heavy metals at higher concentrations in plant cells (Kumar et al. 1995; Salt et al. 1995). Although all *Brassicas* accumulate metals, *B. juncea* shows a strong ability to accumulate and translocate Cd, Cr, Cu, Ni, Pb, and Zn to the shoots (Palmer et al. 2001). However, higher metal concentrations will result in structural changes in roots, stems, and leaves and alter physiological, and morphological characters. *B. juncea* is amenable to genetic transformation (Barfield and Pua 1991), and it is a plant species ideal for a wide range of experimental topics. Genetic engineering has already been used successfully to enhance metal tolerance and accumulation in *Brassica* (Pilon-Smits et al. 1999). Knowledge of the metal tolerance in transgenic plants overexpressed with genes like ferritin, having the potential to sequester metals, would provide optimistic insights for developing plant genotypes with high genetic ability to adapt to toxic soil conditions.

12.2.1.1 Importance of Ferritin

Ferritin is a large protein (12-nm diameter, 480,000 Da) with a large cavity (256 nm^3) for the Fe mineral that is created by the spontaneous assembly of 24 ferritin polypeptides folded into four-helix bundles bound to each other by hydrogen and salt (ionic) bonds (Theil and Briat 2004). It is comprised of H and L subunits, where H subunit has an active ferroxidase site for rapid Fe oxidation, and it occurs in multiple forms in humans. The L subunit plays a role in protein stability and Fe mineralization. However, plant and bacterial ferritins have only a single type of subunit, which probably fulfills both oxidation and mineralization (Rucker et al. 1996). Plant ferritins are synthesized as precursor proteins with a unique N-terminal sequence that is composed of two domains (Ragland et al. 1990; Van Wuytswinkel and Briat 1995). These N-terminal domains are not present in the mammalian or other ferritins. The N-terminal transit peptide is the first domain that contains 40–50 residues, and is presumed to facilitate transport of the ferritin precursor to plastids (Briat et al. 1995). The second domain, the extension peptide, is a part of the mature protein (Laulhere et al. 1988) whose function is still unclear.

Plant ferritin genes have been obtained from many different plant species that include *Lens esculenta* (Crichton 1978), *Glycine max* cell suspensions, cotyledon (Sczekan and Joshi 1987), *Pisum sativum* seed (Van Wuytswinkel and Briat 1995), *Vigna ungiculata* (Wardrop et al. 1999), *Pisum sativum* (Lobreaux et al. 1992), *Phaseolus vulgaris* (Spence et al. 1991), maize (Lobreaux et al. 1992), *Medicago truncatula* (Gyorgyey et al. 2000), and *Medicago sativa* (Deak et al. 1999). Conservation of ferritin protein sequence, folding, and tertiary and quaternary structure among plants and animals is very high (Ragland et al. 1990). Studies of plant ferritins have revealed several important differences in the structure, localization, and regulation of plant ferritins as compared to animal ferritins (Connolly et al. 2002). Iron-regulated expression of animal ferritin is controlled primarily at the level of translation, while in plants, it is mainly at the transcriptional level. Plant ferritin is a developmentally-regulated protein that is detected only during specific stages of the plant life cycle (Lobreaux and Briat 1991). Ferritin subunits increase in seeds during their maturation, and are detected in dry seeds. During germination, ferritins are degraded, and the pattern observed by immunodetection is reminiscent of free-radical degradation of ferritin during iron exchanges in vitro. Plant ferritins are usually the products of a small gene family with differential regulation and all plant ferritin genes so far reported are single-copy genes (*Zea mays*, Fobis-Loisy et al. 1995; *Vigna unguiculata*, Wardrop et al. 1999; *Arabidopsis*, Petit et al. 2001; *Glycine max,* Goto and Yoshihara 2001). The bulk of Fe in leaves is found within the chloroplasts, and Fe ferritin represents more than 90% of the Fe found in a pea embryo axis (Marentes and Grusak 1998).

12.3 ROLE OF FERRITIN

The role of ferritin is to concentrate Fe in the cells to an effective level that matches cellular need (Goto and Yoshihara 2001). The concentration of Fe leads to Fe homeostasis. When Fe concentrations are very high, ferritin also has a protective function by sequestering the Fe inside the protein (Rama Kumar and Prasad 1999; Fourcroy et al. 2004), thus performing a detoxification function. The possible role of ferritin acting against biotic stresses has also been reported (Deak et al. 1999; Hegediis et al. 2002; Dellagi et al. 2005). Recent reports indicate the potential role of ferritin as a protector of the genome (Surguladze et al. 2005). Ferritin and soybean meals were evaluated as dietary treatment for Fe deficiency in rats, where a full recovery from anemia and an increase in tissue Fe levels was apparent after a 28-day treatment (Beard et al. 1996; Lucca et al. 2002; Murray-Kolb et al. 2003). Because ferritin Fe is separated from the iron-binding (chelating) components in food by its protein coat, it is less sensitive to chelators such as phytates, whose concentrations depend on soil and climatic conditions and their stability in solution (Theil 2000). The protein and/or Fe clusters help the meals survive digestion. Existing reports corroborate that Fe is transiently stored in ferritins and used for the accumulation of iron-containing proteins. The role of ferritin as a transient Fe buffer has been documented in the developmental processes of plants (Buchanan-Wollaston and Ainsworth 1997; Strozycki et al. 2003). Seed formation studies have indicated that ferritins are also key proteins in long-term Fe storage (Lobreaux and Briat 1991; Marentes and Grusak 1998).

Presence of free Fe in the living cells is a major detrimental factor producing ROS (reactive oxygen species) through Fe-mediated Fenton reactions. Uncomplexed Fe (III) is reduced to Fe (II) by superoxide and further generate toxic radicals (OH^-). Control of the concentration of free Fe in the cells could be a possible way to reduce oxidative damage. Experimental evidence from humans and animals demonstrated the upregulation of ferritin during oxidative stress. Further manipulation of Fe-centered metabolism through genetic engineering of various plants may provide new tools for providing a better insight of the interrelated pathways of Fe uptake and accumulation, as well as for investigating the role of ferritin during oxidative stress (Sczekan and Joshi 1989; Rama Kumar and Prasad 1999), and its role in contributing to solving the global problems of Fe deficiency (Prasad and Nirupa 2007).

12.3.1 Expression of Pea Seed Ferritin cDNA in Indian Mustard

Transgenic mustard (*Brassica juncea*) plants with high Fe content in leaves have been produced by transferring pea seed ferritin cDNA (a generous gift from Prof. Jean François Briat, Laboratoire de Biochimie et Physiologie Moléculaire des Plantes, CNRS, INRA, Montpellier, France) under the control of a CaMV35S promoter through genetic transformation via *Agrobacterium tumefaciens* (Figure 12.1). The transmission of transgenes to T_1 generation was analyzed by seed germination on hygromycin containing germination medium (Figure 12.2). The identification of plants containing the ferritin gene was verified via PCR and Southern blot analysis (Figures 12.3 and 12.4). The expressed protein in transformants was visualized by the immunohistochemical method (Figure 12.5). The proper assembly of ferritin protein into 24-subunit quaternary structures is essential for exhibiting a functional activity (Goto et al. 1999). The molecular weight of the native protein varies from 440 kDA to 540 kDA in mammals and plants, respectively. The assembly of pea seed ferritin in transgenic *B. juncea* was monitored on the 5% nondenaturing polyacrylamide gels stained with Perl's Prussian stain (Laulhere et al. 1988; Rama Kumar and Prasad 2000). The native pea seed ferritin, with molecular weight of 540 kDa in the transformed plants migrated slower than the horse spleen ferritin whose molecular weight is 480 kDa, indicating the correct assembly into native 24

FIGURE 12.1 (a) Well rooted putative transgenic plantlets; (b) and (c) Acclimatization of transgenic *Brassica juncea*; and (d) Transgenic *Brassica* plants at the end of life cycle with siliqua.

FIGURE 12.2 T_1 *Brassica juncea* seedlings (from the seeds of T_0 plants) transformed with ferritin, germinated on half strength MS basal medium with 8% agar as a gelling agent containing 30 mg L^{-1} hygromycin. Seeds from each putative primary transformant (T_1) and the control were cultured on half strength basal medium with 30 mg L^{-1} hygromycin. Seeds from primary transformants germinated and grew normally with green color and produced secondary and tertiary leaves. Sensitive seedlings were bleached and stunted. Resistant seedlings were transferred to soil for future studies.

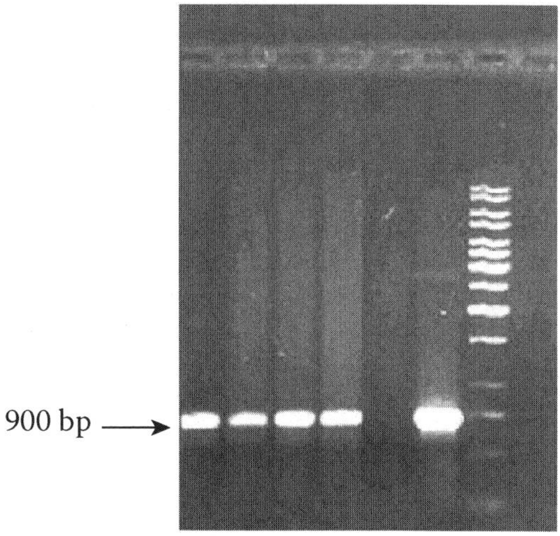

FIGURE 12.3 PCR analysis showing amplification of 900 bp fragment of ferritin fragment. Lane 1–4: Four individual transformants of *B. juncea*. Lane 5: Negative control (nontransformed *B. juncea*). Lane 6: Positive control (pCAMBIA 1300 ferritin). Lane 7: Mwt.marker 1Kb ladder. Total genomic DNA was extracted from the potted plants using a CTAB (cetyl trimethyl ammonium bromide) protocol as per Doyle and Doyle (1990). Polymerase chain reaction was carried out using ferritin cDNA primers in initial screening of transformants. The reaction conditions were 94°C for 3 min, 30 cycles of 1 min at 94°C, 45s at 56°C, and 45s at 72°C.

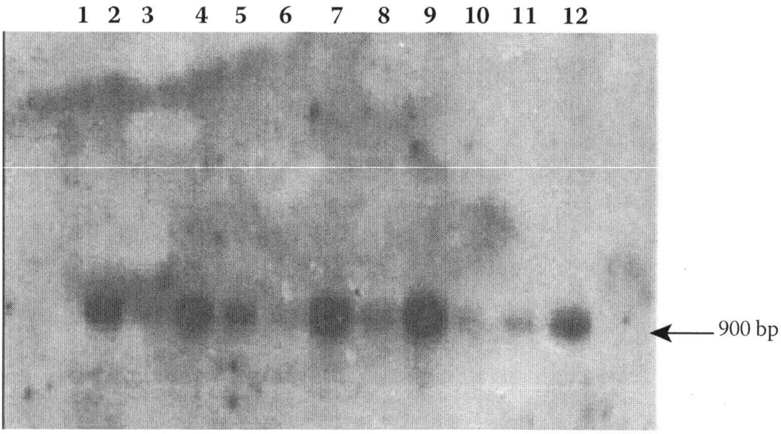

FIGURE 12.4 Southern blot hybridization pattern of DNA from T_1-transformed and nontransformed *B. juncea* plants. Genomic DNAs (15 μg) were digested with *EcoRI* (*Escherichia coli*) R-stain restriction enzyme and probed with [^{32}P]-labeled ferritin fragment. The release of a 900bp pea seed ferritin fragment from the genome of the transgenic plants upon digestion of the genomic DNA with *EcoRI* is indicated with an arrow. λDNA *EcoRI /Hind* III (*Haemophilus influenzae*) double digest was used as a size marker. Lane 1: Negative control (nontransformed plant). Lane 2–5, 7–12: Transformed *B. juncea*. (Tfr 2.1, 3.1, 4.1, 5.1, 7.1, 8.1, 9.1, 10.1, 11.1, and 12.1).

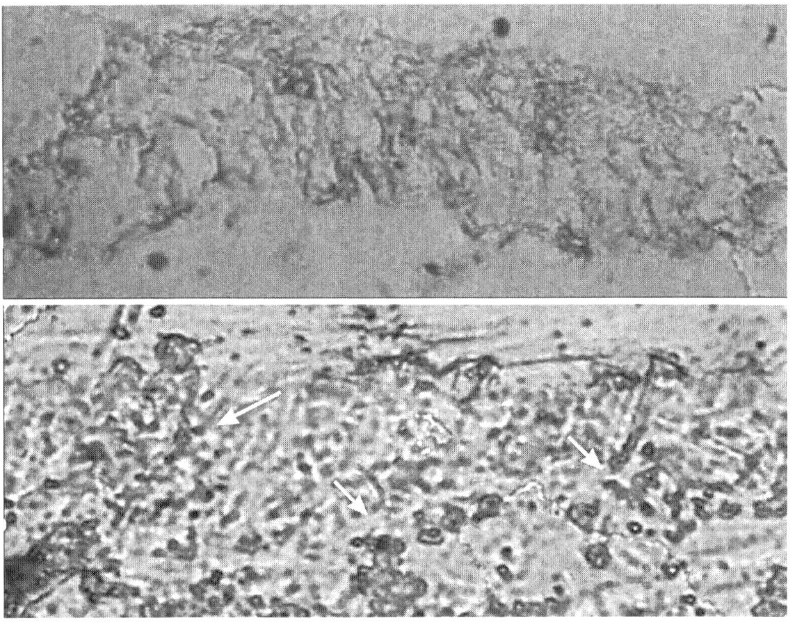

FIGURE 12.5 Immunohistochemical localization of ferritin in (top) control leaf and (bottom) ferritin transformed *B. juncea* leaf (Tfr 2.1). The sites where the ferritin immunohistochemical staining is comparatively more distinct are indicated. Freshly cut leaves were fixed in 4% paraformaldenhyde in 50mM phosphate buffer (pH 7.2). The tissues were washed in the same buffer and dehydrated in graded series of ethanol dilutions, followed by incubation in graded series of xylene dilutions, and infiltrated and embedded in paraffin (Paraplast, St. Louis, Missouri). Samples were sectioned into 10 μm slices and mounted on slides. Mounted sections were deparaffinized and ferritin was immunodetected with an antiferritin primary antibody in combination with an alkaline phosphatase staining. Digital images were taken from a Leica DMR microscope at 40X magnification.

subunit form (Figure 12.6). There was no corresponding band in the control sample lane due to the developmental regulation of ferritin in normal leaves.

The bulk of Fe in leaves was found within the chloroplasts, and Fe ferritin represents more than 90% of the Fe found in a pea embryo axis (Marentes and Grusak 1998). The constitutive expression of the transgene in leaves and seeds resulted in an enhanced iron accumulation particularly in leaves (data not shown). *B. juncea* is widely cultivated as an oilseed crop and its leaves are consumed as leafy vegetable in northern parts of India. Overexpression of pea seed ferritin in *B. juncea* apart from conferring enhanced nutritional significance, may contribute to increased crop yields, because the storage of Fe from the soil by ferritin helps alleviate the problem of absorption of Fe present as water insoluble compounds in the soil (Goto et al. 1998; Goto and Yoshihara 2001). Apart from these functions, the complex integrated mechanisms of Fe uptake and compartmentalization can be studied with the help of ferritin transgenics, and also the other possible roles of ferritin such as sequestration of divalent cations (Sczekan and Joshi 1989; Rama Kumar and Prasad 1999), its possible roles as a protective antioxidant against reactive oxygen species generated due

FIGURE 12.6 Detection of native pea seed ferritin expressed in T_1 *B. juncea* leaves on a 5% nondenaturating polyacrylamide gel by staining in 2% potassium ferricyanide in 50% ethanol and 7% acetic acid. Lane 1: horse spleen ferritin. Lane 2: nontransformant plant. Lane 3–6: transformed *B. juncea* (Tfr 2.1, 3.1, 4.1, 7.1). Native pea seed ferritin in transformed plants was detected in poly acrylamide gel by separating the total protein extracts after precipitating with 50 mM $MgCl_2$ by 5% nondenaturating PAGE according to Laemmli (1970). Ferritin was stained by soaking the gel in 2% (w/v) potassium ferricyanide in 50% ethanol and 7% acetic acid for 4 hours. Horse spleen ferritin (Sigma, USA) was loaded as a positive control.

to biotic and abiotic stress (Deak et al. 1999), and its use as a defense agent guarding against plant pathogens.

12.4 FERRITIN TRANSGENICS—OXIDATIVE STRESS

Plant stresses, such as ozone or ethylene treatment, Fe overload, or impaired photosynthesis induce ferritin accumulation in chloroplasts (Van Wyutswinkel and Briat 1995). However, heavy metal induced oxidative stress was not investigated in ferritin overexpressors. Cadmium (Cd) is a nonessential element and is considered to be highly toxic to humans and plants. Cadmium is more readily taken up by plants than other metals such as lead (Pb). The concentration of Cd in foods is related to its level in the soil and its bioavailability (da Silva et al. 2005). Cadmium acts at the cellular level, and how plants defend themselves against this pollutant is receiving increasing attention. Studies have shown that Cd, which is not a transition metal, may cause oxidative stress (Schutzendubel et al. 2001). Ferritin is also capable of binding cations such as aluminum, beryllium, Cd, and Zn apart from Fe in the mineral core (Sczekan and Joshi 1989; Rama Kumar and Prasad 1999; Polanams et al. 2005). Most of the experimental data on Cd toxicity leaves a dearth of information on the specifics of essential (Cu, Fe, Zn) and nonessential metals (e.g., Pb, mercury) (Rauser 2000). There have not been many studies designed specifically to address the effect of micronutrient status on toxicity from exposure to nonessential metals (Peraza et al. 1998). It is generally believed that Cd uptake by plants represents opportunistic transport via cation channels for Ca and Mg, or via a carrier for other divalent cations such as Zn, Cu, or Fe (Welch and Norvell 1999). Cadmium-induced biological effects could result from its interaction with –SH groups of functional enzymes and/or due to an ionic imbalance as a result of altered translocation of ions such as Fe^{2+}, Zn^{2+}, Cu^{2+}, or Mn^{2+} and/or metal ion substitution. Hence, Cd was selected along with Fe for determining the metal accumulation and tolerance. Taken together, all these observations clearly need to be considered with regard to the role of ferritin in defending the plant against heavy metal stress. Because *B. juncea* is a hyper-accumulator and ferritin is also reported to accumulate Cd (Sczekan and Joshi 1989), the tolerance capacity of the ferritin transformants against Cd, Fe, and Fe + Cd has been examined (Nirupa et al. 2007). Two-week-old wild type and T_1 transgenic mustard seedlings grown in agar medium at varying concentrations of Cd, Fe, and different combinations were evaluated for metal-induced oxidative stress. Lower TBARS and cellular superoxide content were observed in the transgenic seedlings than in wild-type plants, even at higher concentrations of Cd and combinations of Cd and Fe. The concentrations of Cd in plant tissues of the transgenics grown in the greenhouse were below the detection limit (data not shown). As ferritin stores free Fe, the Fe (II) interaction with H_2O_2 to form hydroxyl radicals can be prevented and thus protect cells against damage from oxidative stress. A similar mechanism might be acting in the ferritin overexpressors of *B. juncea* and, thus, a decrease in the cellular level of superoxide radicals in both roots and shoots of the transformants were observed in all the concentrations of Cd and Fe treatments. Iron-treated ferritin overproducing lines might have resulted in restricted cellular damage by inhibiting the Fenton reaction in the Cd treatments.

Ferritin protects plant cells from oxidative damage induced by a wide range of stresses (Table 12.1). Van Wyutswinkel and Briat (1995) indicated a protective role of ferritin in restricting oxidative stress by a methylviogen toxicity test. Methylviogen, an inducer of oxidative stress, requires free Fe for its activity and is antagonized by Fe chelators. Transgenic tobacco plants, which express alfalfa ferritin in vegetative tissues either in its processed form in chloroplasts or in the cytoplasmic nonprocessed form, retained photosynthetic function upon free-radical toxicity generated by excess Fe or paraquat (Pq) treatment, and exhibited tolerance to necrotic damage caused by viral (tobacco necrosis virus) and fungal (*Alternaria alternata, Botrytis cinerea*) infections (Deak et al. 1999). Transgenic tobacco overexpressing ferritin described by Hegedüs et al. (2002) showed significantly higher chlorophyll fluorescence (Fv/Fm) values, a lower rate of MDA (malondialdehayde) formation, increased antioxidant activities (catalase: CAT; peroxidase: POD; glutathione peroxidase: GPX) at low temperatures, where as wild type caused a marked decrease in the Fv/Fm ratios, indicating adaptation towards low temperature (Hegedüs et al. 2002).

Fourcroy et al. (2004) indicated that the coregulation of APX and ferritin gene expression by excess Fe could provide a synergistic way to protect plants from ROS formation because APX scavenges H_2O_2, whereas ferritin stores excess free Fe, and thus prevents the generation of the highly toxic hydroxyl radical. The *C. canadensis* ferritin2 genes were upregulated by Pq in both susceptible and resistant horseweed *Conyza canadensis* biotypes plants. The improved expression level was possibly associated with defense reactions in the plants after Pq treatment (Soós et al. 2006). Oxidative stress and ABA mediate ferritin induction in *Atfer2, Zmfer2,* whereas only oxidative stress operates in *Atfer 1, Atfer 3, Zmfer* genes (Petit et al. 2001). Iron mediates oxidative stress in *Vigna mungo* (L.) Hepper, as revealed by upregulation of activities of antioxidative

TABLE 12.1
Selected Reports of Biotic and Abiotic Stress Tolerance of Ferritin Transgenics

Plant/Gene Used	Type of Stress Tolerance	Mechanism	Reference
Nicotiana tabacum			
Soy bean ferritin cDNA P6 (chloroplastic ferritin), C5 (cytoplasmic ferritin)	Methyl viogen Abiotic	Oxidative stress tolerance by sequestering free iron. Unaffected total protein and MDA concentrations in ferritin transgenics.	Van Wyutswinkel and Briat 1995
Alfa Alfa ferritin cDNA	Iron excess or paraquat treatment. Viral (tobacco necrosis virus) and fungal (*Alternaria alternata, Botrytis cinerea*) infections.	Oxidative stress tolerance by sequestering intracellular Fe. Normal photosynthetic function and chlorophyll content. Paraquat transfers electrons from PSI to oxygen, a reduced photon capture by chloroplasts may end up in a reduced ROS generation and then result in an apparent resistance to oxidative stress.	Deak et al. 1999
	Low temperature and high irradiance.	Higher Fv/Fm values than control plants, and lower rate of MDA formation. Increased antioxidant enzyme activity.	Hegedüs et al. 2002
Arabidopsis thaliana			
Soybean ferritin P6 (chloroplastic ferritin), C5 (cytoplasmic ferritin)	Photoinhibition ozone fumigation.	No protection conferred by ferritin transgenics. Increased ferritin mRNA abundance. Very low protein accumulation. Reduced photochemistry and transport of electrons to PSI.	Murgia et al. 2001

enzymes and reduction of the activity in the presence of antioxidants and free- radical scavengers (Rama Kumar and Prasad 1999, 2000). However, Murgia et al. (2001) showed that tobacco plants overexpressing a soybean ferritin gene (and accumulated ferritin) were not more protected from oxidative stress than the control plants against photoinhibition. The ferritin gene was expressed differently in distinct potato tissues as a consequence of infection *P. infestans* (Mata et al. 2001). Dellagi et al. (2005) reported that ferritin accumulation during infection of Arabidopsis by *E. chrysanthemi* is a basal defense mechanism that is mainly activated by bacterial siderophores.

12.5 CONCLUSIONS

A range of possible cellular means may be involved in the detoxification of heavy metals and thus tolerance to metal stress. Along with this mechanism, our results support the evidence for an involvement of ferritin in heavy metal detoxification and tolerance. Further insight into the detailed process needs to be elucidated for a comprehensive overview, as it is possible that more than one mechanism may be involved in reducing the toxicity and imparting tolerance to a particular metal.

ACKNOWLEDGMENTS

The authors are thankful to Prof. Jean François Briat at Laboratoire de Biochimie et Physiologie Moléculaire des Plantes, CNRS, INRA, in France for the generous gift of many reprints and pea seed ferritin cDNA (to M.N.V. Prasad) that helped us to conduct research on phytoferritins. Part of this work has been carried out with financial support from the Department of Science and Technology (A38199 to PBK), Government of India, New Delhi.

REFERENCES

Balla, G., H. S. Jacob, and J. Balla et al. 1992. Ferritin: A cytoprotective antioxidant strategem of endothelium. *Journal of Biological Chemistry* 267: 18148–18153.

Barfield, D. G. and E. C. Pua. 1991. Gene transfer in plants of *Brassica juncea* using *Agrobacterium tumefaciens* mediated transformation. *Plant Cell Reports* 10: 308–314.

Beard, J. L., J. W. Burton, and E. C. Theil. 1996. Purified ferritin and soybean meal can be sources of iron for treating iron deficiency in rats. *Journal of Nutrition* 126: 154–160.

Bouis, H. E. 2003. Micronutrient fortification of plants through plant breeding: Can it improve nutrition in man at low cost? *Proceedings of the Nutrition Society* 62: 402–411.

Bouis, H. E. 2002. Plant breeding: A new tool for fighting micronutrient malnutrition. *Journal of Nutrition* 132: 491–494.

Briat, J. F., A. M. Laboure, J. P. Laulhere et al. 1995. Molecular and cellular biology of plant ferritins. In *Iron Nutrition in Soil and Plants*, J. Abadia (Ed.), Dordrecht: Kluwer Academic Publishers pp. 265–275.

Buchanan-Wollaston, V. and C. Ainsworth. 1997. Leaf senescence in *Brassica napus*: Cloning of senescence related genes by subtractive hybridisation. *Plant Molecular Biology* 33: 821–834.

Connolly, E. L., J. P. Fett, and M. L. Guerinot. 2002. Expression of the IRT1 metal transporter is controlled by metals at the levels of transcript and protein accumulation. *Plant Cell* 14: 1347–1357.

Cozzi, A., P. Santambrogio, S. Levi, and P. Arosio. 1990. Iron detoxifying activity of ferritin: Effects of H and L human apoferritins on lipid peroxidation in vitro. *FEBS Letters* 277: 19–122.

Crichton, R. R., Y. Ponze-Ortiz, and H. B. Stuhrman. 1978. Isolation and characterization phytoferritin from pea (*Pisum sativum*) and lentil (*Lens esculenta*). *Biochemical Journal* 171: 349–356.

da Silva, A. L. O., S. Jacob do Couto, and J. J. C. Moreira et al. 2005. Dietary intake and health effects of selected toxic elements. *Brazilian Journal of Plant Physiology* 17: 79–93.

Deak, M., G. V. Horváth, and S. Davletova et al. 1999. Plants ectopically expressing the iron binding protein, ferritins are tolerant to oxidative damage and pathogens. *Nature Biotechnology* 17: 192–196.

Dellagi, A., M. Rigault, and D. Segond et al. 2005. Siderophore-mediated upregulation of Arabidopsis ferritin expression in response to *Erwinia chrysanthemi* infection. *Plant Journal* 43: 262–272.

Doyle, J. J. and J. L. Doyle. 1990. Isolation of plant DNA from fresh tissue. *Focus* 12: 13–15.

Fobis-Loisy, I., K. Loridon, S. Lobreaux, M. Lebrun, and J. F. Briat. 1995. Structure and differential expression of two maize ferritin genes in response to iron and abscisic acid. *European Journal of Biochemistry* 231: 609–619.

Fourcroy, P., G. Vansuyt, S. Kushnir, D. Inze, and J. F. Briat. 2004. Iron-regulated expression of a cytosolic ascorbate peroxidase encoded by the APX1 gene in *Arabidopsis* seedlings. *Plant Physiology* 134: 605–613.

Goto F, T. Yoshihara, and H. Saiki. 1998. Iron accumulation in tobacco plants expressing soybean ferritin gene. *Transgenic Research* 7: 173–180.

Goto, F. and T. Yoshihara. 2001. Improvement of micronutrient contents by genetic engineering development of high iron content crops. *Plant Biotechnology* 18: 7–15.

Goto F. 2001. Genetic improvement of iron content and stress adaptation in plants using ferritin gene. *Biotechnology and Genetic Engineering Reviews* 18: 351–371.

Goto, F., T. Yoshihara, N. Shigemoto, S. Toki, and F. Takaiwa. 1999. Iron fortification of rice seed by the soybean ferritin gene. *Nature Biotechnology* 17: 282–286.

Gyorgyey, J., D. Vaubert, and J. I. Jimenez-Zurdo et al. 2000. Analysis of *Medicago truncatula* nodule expressed sequence tags. *Molecular Plant Microbial Interactions* 13: 62–71.

Hegedüs, A., S. Erdei, T. Janda, J. Szalai, D. Dudits, and G. Horváth. 2002. Effects of low temperature stress on ferritin or aldose reductase overexpressing transgenic tobacco plants. *Acta Biologica Szegediensis* 46: 97–98.

Holm, P. B., K. N. Kristiansen, and H. B. Pedersen. 2002. Transgenic approaches in commonly consumed cereals to improve iron and zinc content and bioavailability. *Journal of Nutrition* 132: 514–516.

Kumar, P. B., A. N. Dushenkov, V. H. Motto, and I. Raskin. 1995. Phytoextraction the use of plants to remove heavy metals from soils. *Environmental Science and Technology* 29: 1232–1238.

Lam, B. H., J. C. R. Stangoulis, R. M. Welch, and R. D. Graham. 2006. QTLs for phytate in rice grain and their relationship with iron. In *Proceedings of the 13th Australasian Plant Breeding Conference,* 18–21 April, 120 Christchurch: N.Z., p. 120.

Laulhere, J. P., A. M. Lescure, and J. F. Briat. 1988. Purification and characterization of ferritin from maize, pea and soybean seeds. *Journal of Biological Chemistry* 26: 10289–10294.

Lobreaux, S. and J. F. Briat. 1991. Ferritin accumulation and degradation in different organs of pea (*Pisum sativum*) during development. *Biochemical Journal* 274: 601–606.

Lobreaux, S., S. J. Yewdall, J. F. Briat, and P. M. Harrison. 1992. Amino-acid sequence and predicted three-dimensional structure of pea seed (*Pisum sativum*) ferritin. *Biochemical Journal* 288: 931–939.

Lucca, P., R. F. Hurrell, and I. Potrykus. 2002. Fighting iron deficiency anemia with iron rich rice. *Journal of American College of Nutrition* 21: 184–190.

Lucca, P., S. Poletti, and C. Sautter. 2006. Genetic engineering approaches to enrich rice with iron and vitamin A. *Physiologia Plantarum* 126: 291–303.

Mata, C. G., L. Lamattina, and R. O. Cassia. 2001. Involvement of iron and ferritin in the potato-*Phytophera infestans* interaction. *European Journal of Plant Pathology* 107: 557–562.

Marentes, E. and M. A. Grusak. 1998. Iron transport and storage within the seed coat and embryo of developing seeds of pea (*Pisum sativum* L.). *Seed Science Research* 8: 367–375.

Murgia, I., J. F. Briat, D. Tarantino, and C. Soave. 2001. Plant ferritin accumulates in response to photoinhibition but its ectopic overexpression does not protect against photoinhibition. *Plant Physiology and Biochemistry* 39: 797–805.

Murray-Kolb, L. E., R. Welch, E. C. Theil, and J. L. Beard. 2003. Women with low iron stores absorb iron from soybeans. *American Journal of Clinical Nutrition* 77: 180–184.

Nirupa, N., M. N. V Prasad, and P. B. Kirti. 2007. Expression of pea seed ferritin cDNA in indian mustard: nutritional value and oxidative stress tolerance of the transformants. In *Biogeochemistry of Trace Elements: Environmental Protection, Remediation and Human Health*, Y. Zhu, N. Lepp, and R. Naidu (Eds.), Beijing: Tsinghua University Press, pp. 285–286.

Palmer, C. E., S. Warwick, and W. Keller. 2001. *Brassicaceae* (cruciferae) family—Plant biotechnology and phytoremediation. *International Journal of Phytoremediation* 3: 245–287.

Peraza, M. A., F. A. Fierro, D. S. Barber, E. Casarez, and L. T. Rael. 1998. Effects of micronutrients on metal toxicity. *Environmental Health Perspectives* 106: 203–216.

Petit, J. M., O. Van Wuytswinkel, J. F. Briat, and S. Lobreaux. 2001. Characterization of an iron-dependent regulatory sequence involved in the transcriptional control of *AtFer1* and *ZmFer1* plant ferritin genes by iron. *Journal of Biological Chemistry* 276: 5584–5590.

Pilon-Smits, E. P. M., S. Hwang, and C. M. Lytle et al. 1999. Overexpression of ATP sulfurylase in Indian mustard leads to increased selenate uptake, reduction, and tolerance. *Plant Physiology* 119: 123–132.

Polanams, J., A. D. Ray, and R. K. Watt. 2005. Nano iron phosphate, iron arsenate, iron vandate and, iron molybdate minerals synthesized within the protein cage of ferritin. *Inorganic Chemistry* 44: 3203–3209.

Prasad, M. N. V. and N. Nirupa. 2007. Phytoferritins—Implications for human health and Nutrition. *The Asian and Australasian Journal of Plant Science and Biotechnology* 1: 1–9.

Ragland, M., J. F. Briat, J. Gagnon, J. P. Laulhere, O. Massenet., and E. C. Theil. 1990. Evidence for conservation of ferritin sequences among plants and animals and for transit peptide in soybean. *Journal of Biological Chemistry* 265: 18339–18344.

Rama Kumar, T. and M. N. V. Prasad. 1999. Ferritin induction by iron mediated oxidative stress and ABA in *Vigna mungo* (L.) Hepper seedlings: Role of antioxidants and free radical scavengers. *Journal of Plant Physiology* 155: 652–655.

Rama Kumar, T. and M. N. V. Prasad. 2000. Partial purification and characterization of ferritin from *Vigna mungo* (Blackgram) seeds. *Journal of Plant Biology* 27: 241–246.

Rauser, W. E. 2000. The role of thiols in plants under metal stress. In *Sulfur Nutrition and Sulfur Assimilation in Higher Plants*, C. Brunold, H. Rennenberg, L. J. De Kok, I. Stulen, and J. C. Davidian (Eds.), Bern: Paul Haupt, pp. 169–183.

Rodrigues, S. M., M. O. Andrade, A. P. S. Gomes, F. M. DaMatta, M. C. Baracat-Pereira., and E. P. B. Fontes. 2006. Arabidopsis and tobacco plants ectopically expressing the soybean antiquitin-like ALDH7 gene display enhanced tolerance to drought, salinity, and oxidative stress. *Journal of Experimental Botany* 57: 1909–1918.

Rucker P., F. M. Torti., and S. V. Torti. 1996. Role of H and L subunits in mouse ferritin. *Journal of Biological Chemistry* 271: 33352–33357.

Salt, D. E., R. C. Prince, I. J. Pickering, and I. Raskin. 1995. Mechanisms of cadmium mobility and accumulation in Indian mustard. *Plant Physiology* 109: 1427–1433.

Schutzendubel, A., P. Schwanz, T. Teichmann et al. 2001. Cadmium-induced changes in antioxidative systems, hydrogen peroxide content, and differentiation in scots pine roots. *Plant Physiology* 127: 887–898.

Sczekan, S. R. and J. G. Joshi. 1987. Isolation and characterization of ferritin from soybean (*Glycine max*). *Journal of Biological Chemistry* 262: 13780–13788.

Sczekan, S. R. and J. G. Joshi. 1989. Metal binding properties of phytoferritin and synthetic iron cores. *Biochimica et Biophysica Acta* 990: 8–14.

Singh, O. V., S. Ghai, D. Paul, and R. K. Jain. 2006. Genetically modified crops: Success, safety assessment, and public concern. *Applied Microbiology and Biotechnology* 71: 598–607.

Soós, V., B. Jóri, and E. Páldi et al. 2005. Ferritin2 gene in paraquat-susceptible and resistant biotypes of horseweed *Conyza Canadensis* (L.) Cronq. *Journal of Plant Physiology* 163: 979–982.

Spence, M. J., M. T. Henzl, and P. J. Lammers. 1991. The structure of a *Phaseolus vulgaris* cDNA encoding the iron storage protein ferritin. *Plant Molecular Biology* 17: 499–504.

Strozycki, P. M., A. Skapska, K. Szczesniak, E. Sobieszczuk, J. F. Briat, and A. B. Legocki. 2003. Differential expression and evolutionary analysis of the three ferritin genes in the legume plant *Lupinus luteus*. *Physiologia Plantarum* 118: 380–389.

Surguladze, N., S. Patton, A. Cozzi, M. G. Fried, and J. R. Connor. 2005. Characterization of nuclear ferritin and mechanism of translocation. *Biochemical Journal* 388: 731–740.

Theil, E. C. 2000. Ferritin. In *Handbook of Metalloproteins*, A. Messerschmidt, R. Huber, T. Poulos, and K. Wieghardt (Eds.), Chichester: John Wiley and Sons, pp. 771–781.

Theil, E. C. and J. F. Briat. 2004. *Plant Ferritin and Non-heme Iron Nutrition in Humans*. Harvest Plus Technical Monograph 1. Washington, DC and Cali, International Food Policy Research Institute and International Center for Tropical Agriculture.

United Nations System Standing Committee on Nutrition (SCN). 2004. 5th Report on the World Nutrition Situation Nutrition for Improved Development Outcomes. SCN, Geneva.

United Nations Children's Fund/UNU/World Health Organization/MI Technical Workshop. 1999. Preventing Iron Deficiency in Women and Children. Technical Consensus on Key Issues. International Nutrition Foundation, New York, pp. 29–32.

Van Wuytswinkel, O. and J. F. Briat. 1995. Conformational changes and *in vitro* core formation modification induced by site directed mutagenesis of the specific N-terminus of pea seed ferritin. *Biochemical Journal* 305: 959–965.

Wardrop, A. J., R. E. Wicks, and B. Entsch. 1999. Occurrence and expression of members of the ferritin gene family in cowpeas. *Biochemical Journal* 337: 523–530.

Welch, R. M. 2005. Biotechnology, biofortification, and global health. *The Food Nutrition Bulletin* 26: 419–421.

Welch, R. M. and W. A. Norvell. 1999. Mechanisms of cadmium uptake, translocation and deposition in plants. In *Cadmium in Soils and Plants*, M. J. McLaughlin, and A. Singh (Eds.), Dordrecht: Kluwer Academic Publishers, pp. 125–150.

WHO/UNICEF/UNU. 2001. Iron deficiency anaemia: assessment prevention and control. Geneva, World Health Organization, (WHO/NHD/ 01.3). http://www.who.int/nutrition/publication/ida_assessment_prevention_control.pdf. (Accessed July 1, 2008.)

13 Progress in Understanding the Molecular Regulation of Iron Uptake in Strategy I Plants

Hong-Qing Ling, Juan Du, and Ning Wang

CONTENTS

13.1 INTRODUCTION

Iron is an essential nutritional element for all organisms. It functions as a component of many important enzymes and proteins involved in fundamentally biochemical processes. Iron (Fe) deficiency is one of the most serious malnutrition conditions threatening human health in the world because various physiological diseases, such as anemia, Wilson's, Parkinson's, and Menken's, are triggered by Fe deficiency

(Yuan et al. 1995). The biofortification of Fe content and availability in plant food products may be the most effective and economic method to fight Fe deficiency in human health.

In plants, Fe is one of the most common elements limiting plant growth because it exists predominantly in an oxidized ferric form [Fe(III)] in aerobic environments. The ferric Fe has an extremely low solubility at neutral or basic pH and is not readily available to plants. To meet Fe demand for growth and development, two effective Fe acquisition systems known as strategy I and strategy II have been evolved in higher plants (Römheld and Marschner 1986). Grasses use a chelating mechanism, called strategy II, to acquire Fe from the soil. Under Fe limitation, grasses (strategy II plants) synthesize an abundance of ferric chelators (phytosiderophores; i.e., mugineic acid family [MAs]), that are extruded into rhizosphere for chelating ferric Fe. Thereafter, the Fe(III)-MAs complex are taken up by Fe(III)–MAs transporter YS1 (Curie et al. 2001). MAs are biosynthesized from methionine through nicotianamine, and the genes encoding key enzymes for MAs biosynthesis have been isolated from graminaceous plants (Mori 1999). Except grasses, all other plants use the strategy I mechanism for effective Fe acquisition from soil. The core components of this strategy are (1) acidification of rhizosphere by enhanced extrusion of proton to increase solubility of ferric Fe, (2) activation of ferric-chelate reductase reducing Fe^{3+} to Fe^{2+} on root surface in the subapical region, and (3) induction of the high-affinity Fe^{2+}-transporter system to absorb ferrous Fe from soil into roots. Additionally, the strategy I plants also show morphological changes of roots, such as thickening of the subapical root zone and an increased formation of root hairs.

Iron deficiency in plants causes chlorosis and leads to decrease yield and food quality. Excess Fe is also toxic and affects the plant's normal metabolism by damaging the cell constituents with the formation of oxidative oxygen. Hence, Fe homeostasis in plants must be fine-regulated. In this chapter, we are focusing on reviewing the progress of molecular regulation mechanisms of Fe acquisition in the strategy I plants.

13.2 IRON UPTAKE GENES IN STRATEGY I PLANTS

Arabidopsis and tomato are two model plants for studying molecular mechanism of Fe uptake in strategy I plants. Most efforts have focused on the characterization of ferric chelate reductase and Fe transporter during the past decade. In this regard, great progress has been made, and the genes responsible for effective Fe acquisition from soil have been identified and well characterized.

13.2.1 Ferric-Chelate Reductase

Reduction of ferric to ferrous Fe is an obligatory step for Fe uptake in strategy I plants (Yi and Guerinot 1996). The reduction is carried out by a plasma membrane-bound ferric chelate reductase (Buckhout et al. 1989; Holden et al. 1991). Chelated ferric Fe, but not ionic Fe^{3+}, is the substrate for Fe reduction (Laulhere and Briat 1993). The Fe^{3+}-chelate reductase, one of the most extensively studied reductases on cell membrane, was experimentally addressed by analysis of Arabidopsis mutant *frd1* (ferric chelate reductase-defective mutant 1), which defects in Fe^{3+}-chelate reduction (Yi and Guerinot 1996). Loss of inducible reduction function of ferric Fe in roots of the *frd1* mutant led to growth impairment under Fe-limiting conditions. The corresponding gene was cloned and termed as *FRO2* by Robinson et al. (1999). A nonsense mutation in the first exon of *FRO2* in *frd1-1* and a missense mutation in *frd1-3* made these mutants fail to reduce ferric Fe and thus impair the effective Fe acquisition from soil. Further analysis revealed that there are eight *FRO* genes in *Arabidopsis thaliana* genome. They showed different expression patterns in different tissues and also displayed different Fe reduction activity when expressed in yeast (Wu et al. 2005). *FRO2* and *FRO3* are two genes mainly expressed in roots under Fe deficiency in Arabidopsis. The two genes may function in roots for Fe acquisition from soil. Considering that chlorosis and growth impairment occurred in the knockout mutant of *FRO2*, but not in the mutant of *FRO3*, *FRO2* is obviously more important than *FRO3* for Fe acquisition under Fe limitation. Overexpression of *FRO2* in Arabidopsis (Connolly et al. 2003) and heterologous expression in soybean (Vasconcelos et al. 2006) revealed that the transgenic plants were more tolerant of Fe deficiency. All data indicate that *FRO2* is a major ferric-chelate reductase for Fe acquisition in Arabidopsis.

The ferric chelate reductase *LeFRO1* from tomato (Li et al. 2004) and *PsFRO1* from pea (Waters et al. 2002) have been isolated by PCR-based homologous cloning. The two genes showed Fe-chelate reductase activity when expressed in yeast cells. Additionally, their expression intensity is dependent on the Fe status in roots. The data suggest that *LeFRO1* in tomato and *PsFRO1* in pea, same as *FRO2* in Arabidopsis, play a major role in ferric-chelate reduction for effective Fe acquisition in the two plants.

13.2.2 Iron (Fe) Transporters

In strategy I plants, Fe is mainly acquired from soil as ferrous ion into cells. For the transfer of ferrous Fe to occur across the barrier of cell membrane, Fe-specific transporter(s) are required. *IRT1*, an Fe-regulated transporter, was the first Fe transporter cloned in higher plants by functional expression in an Fe-inefficient yeast mutant strain *fet3fet4*, in which the high and low affinity Fe transportation systems were impaired (Eide et al. 1996). The *fet3fet4* cells expressing *IRT1* cDNA of *Arabidopsis thaliana* showed restoration of Fe^{2+} uptake activity, indicating *IRT1* was an Fe transporter. *IRT1* is a membrane protein containing eight transmembrane domains and a metal-binding domain. In Arabidopsis, it is expressed mainly in roots and the expression is induced by Fe deficiency (Eide et al. 1996; Vert et al. 2002). Expressing *IRT1* can complement the yeast strains defective in uptake of Fe, Mn and Zn to wild type. Further, the uptake of Mn, Zn, and Co is also enhanced when *IRT1* expression is induced by Fe deficiency. These observations implied that *IRT1* is a broad-range metal transporter (Korshunova et al. 1999). Additionally, the gene family of *IRT1* contains other two members *IRT2* and *IRT3* (Eide et al. 1996). The expression of *IRT2* is in roots (including root hairs, epidermis and cortex) of Arabidopsis and is upregulated by Fe deficiency. Heterologous expression of *IRT2* in yeast reveals that *IRT2* can transport Fe and zinc, but not manganese and cadmium (Vert et al. 2001). IRT3 shows homology with *IRT1*, but its function is still not characterized. Genetic analysis showed that *IRT1*, not *IRT2*, is an essential transporter for Fe uptake from soil and for plant growth in *Arabidopsis*. Knocking out *IRT1* by T-DNA insertion, the mutant plants (*irt1-1* and *irt1-2*) exhibited typical Fe deficiency symptoms causing severe leaf chlorosis and lethality in soil, and the defect functions of *IRT1* could be not complemented by overexpressing *IRT2* (Varotto et al. 2002).

Orthologs of *IRT1* and *IRT2* in tomato have been isolated from a cDNA library constructed from roots under Fe deficiency using *IRT1* as a probe, termed *LeIRT1* and *LeIRT2* (Eckhardt et al. 2001). Transcription of *LeIRT2* was unaffected by the Fe status of the plant, whereas *LeIRT1* was strongly upregulated under Fe deficiency, suggesting that *LeIRT1* might play a more important role in Fe acquisition from soil than *LeIRT2*.

13.2.3 NRAMP in Arabidopsis and Tomato

NRAMP, with a highly conserved 10-12 potential transmembrane hydrophobic core and a C-terminal long intracellular tail, is a protein family that has been implicated in the transportation of divalent metal ions (Cellier et al. 1995; Williams et al. 2000). In *Arabidopsis*, there are five NRAMP genes. Heterologous expression experiments in yeast cells showed that *AtNRAMP1*, not *AtNRAMP2*, is able to complement the defect functions of Fe transportation in *fet3fet4* yeast mutant. The expression of *AtNRAMP1* is upregulated under Fe deficiency. Overexpression of *AtNRAMP1* in Arabidopsis leads to an increased resistance to toxic Fe concentrations, suggesting that AtNRAMP1 might participate in Fe detoxification in plant cells (Curie et al. 2000). AtNRAMP3 and AtNRAMP4 are induced under Fe deficiency and target the vacuole. The two proteins function in the mobilization of vacuolar Fe for seed germination in low Fe conditions because the germination of *nramp3 nramp4* double mutant is arrested under low Fe conditions and is rescued by high Fe supply (Lanquar et al. 2005). In tomato, two *NRAMP* genes (*LeNRAMP1* and *LeNRAMP3*) were identified and characterized (Bereczky et al. 2003). The transcription abundance of *LeNRAMP1* was related to Fe status, whereas *LeNRAMP3* expressed constitutively, regardless of Fe supply. Both LeNRAMP1 and LeNRAMP3 are enabling to complement the defect functions of Fe transportation when expressed in yeast *fet3fet4* mutant, suggesting they are metal transporters involved in Fe homeostasis.

13.3 TRANSCRIPTION FACTORS INVOLVED IN THE POSITIVE CONTROL OF Fe UPTAKE

13.3.1 FER Is a Positive Regulator Controlling Fe Deficiency Responses in Tomato

FER, isolated by map-based cloning from tomato, is the first regulator characterized in Fe homeostasis in higher plants. It encodes a protein containing a highly conserved *bHLH* domain characteristic for the family of eukaryotic *bHLH* transcriptional regulatory proteins (Bienfait 1988). The elucidation of the biological functions of *FER* is based on the characterization of the mutant T3238*fer*, which is caused by a large DNA-fragment insertion at the end of the first exon of *FER* (Ling et al. 2002). The T3238*fer* mutant is not able to activate the whole physiological and morphological responses (i.e., the development of transfer cells, the enhancement of proton extrusion, the increase of ferric-chelate reductase activity, and the induction of the high affinity Fe^{2+}-transporter gene expression) to Fe deficiency when grown under Fe-limiting conditions (Ling et al. 1996). Obviously, *FER* is involved in the regulation of the Fe-deficiency responses of tomato. Reciprocal grafting experiments indicated that *FER* protein is active and is required only in roots and not in shoots (Ling et al. 2002). The expression of *FER* is regulated at the transcriptional level by Fe status. It was reported that the accumulation of *FER* protein was detected only under Fe deficiency, although the expression of a *FER* cDNA driven by cauliflower mosaic virus 35S promoter on the *fer* background is abundant under both Fe deficient and sufficient conditions (Brumbarova and Bauer 2005). Recently, Zhao and Ling (2007) observed that the transcript level of *FER* was dependent on environmental pH value (i.e., decreases at high pH and increases at low pH). Taken together, *FER* is fine-controlled at both the transcriptional and posttranscriptional levels.

As described above, the ferric chelate reductase *LeFRO1* and the ferrous transporter *LeIRT1* are two important components for the effective Fe uptake in tomato. Characterization of *FER* loss-function mutant T3238*fer* revealed that *FER* protein is involved in the activation of *LeFRO1* and *LeIRT1* transcription under Fe-limiting stress (Bereczky et al. 2003; Li et al. 2004; Zhao and Ling 2007). The induced transcription activity of *LeFRO1* and *LeIRT1* in roots of T3238*fer* was impaired under Fe deficiency (Li et al. 2004) and under high pH (Zhao and Ling 2007). Compared with wild type, the expression of *LeFRO1* in T3238*fer* is almost completely blocked, and the expression of *LeIRT1* is dramatically reduced, although its transcript is still detectable. Clearly, the expression activation of *LeFRO1* under Fe deficiency is dependent on FER, whereas the *LeIRT1* is only partially regulated by FER. As described above, *LeIRT2*, *LeNRAMP1* and *LeNRAMP3*, as metal transporters, are possibly involved in Fe homeostasis in tomato. In *fer* mutant, *LeNRAMP1* mRNA showed a significant decrease in roots compared to wild type, whereas the expression of *LeNRAMP3* and *LeIRT2* were not affected (Bereczky et al. 2003). Taking all together, it can be concluded that the expression of *LeFRO1*, *LeIRT1*, and *LeNRAMP1*, but not *LeIRT2*, and *LeNRAMP3*, is dependent on FER.

13.3.2 FIT, an Ortholog of Tomato FER, Is Involved in Controlling Iron Uptake in *Arabidopsis*

Arabidopsis FIT (other names are *AtbHLH29/FIT1/FRU*) encode a *bHLH* protein and share a high homology with tomato *FER* (42% identity and 72% similarity at the amino acid level) (Colangelo and Guerinot 2004; Jakoby et al. 2004; Yuan et al. 2005; Bauer et al. 2007). The heterologous expression of *FIT* in *fer* mutant T3238*fer* is able to completely complement the defect functions of *FER*, and the transgenic plants are enabled again to activate the Fe-deficiency responses as wild type under Fe-limitation stress, demonstrating that *FIT* is a functional ortholog of tomato *FER* in *Arabidopsis*, and can replace FER in controlling the effective Fe acquisition in tomato (Yuan et al. 2005). *FIT* mRNA is induced in outer cell layer of Fe deficient root (Colangelo and Guerinot 2004; Jakoby et al. 2004). Knocking out of *FIT* by

T-DNA insertion, the *FIT* mutants (*fit1-1* and *fit 1-2*) exhibited similar phenotype (chlorosis and lethality) as *FER* mutant T3238*fer* under normal culture conditions. In mutants, ferric chelate reductase activity is completely blocked (Colangelo and Guerinot 2004; Jakoby et al. 2004), and the expression of the major ferric-chelate reductase *FRO2* was not detected under Fe deficiency, whereas abundant *FRO2* mRNA appeared in wild type. These responses indicate that *FIT*, the same as *FER*, are involved in regulation of *FRO2* at the transcriptional level (Colangelo and Guerinot 2004; Jakoby et al. 2004). Based on this, *IRT1* transcript is only partially decreased and its protein is undetectable in the *fit1-1* mutant. Colangelo and Guerinot (2004) suggested that *FIT* regulates *IRT1* at both the transcriptional and post-transcriptional levels. In 35S:*FIT* transgenic lines, the expression pattern of *FRO2* and *IRT1* is unchanged compared to wild type, despite the constitutively expressed mRNA of *FIT* in transgenic lines (Colangelo and Guerinot 2004; Jakoby et al. 2004) that induces the enhancing expression of *FIT* alone can't trigger the expression of the major Fe uptake genes such as *FRO2* and *IRT1*.

13.3.3 Other Transcription Factors Putatively Involved in Fe Uptake

Recently, four *BHLH* proteins (*BHLH38*, *BHLH39*, *BHLH100*, *BHLH101*), comprising a subgroup of *BHLH Ib* (Heim et al. 2003), were identified in *Arabidopsis* (Wang et al. 2007). Their expressions are regulated by Fe status. The four genes share great similarity in sequence, and knocking out one of them doesn't show visible phenotype, indicating that the four genes have functional redundancy. Analysis of expression patterns of the four genes revealed that their response to Fe limitation is not dependent on *FIT*. Four proteins may play a role in Fe homeostasis in the upstream or in parallel of *FIT*.

13.4 CHLN IS REQUIRED IN DOWNREGULATION OF Fe DEFICIENCY RESPONSES IN TOMATO

CHLN, encoding a nicotianamine synthase responsible for synthesis of nicotianamine, is located on chromosome 1 of tomato and isolated by map-based cloning (Ling et al. 1999). Its mutant *chloronerva* is an auxotroph for nicotianamine caused by the single base change on nicotianamine synthase. Because the *CHLN* is not able to turn off the Fe deficiency responses under Fe-repletion conditions, the plant takes up more Fe from soil and accumulates it in shoots. Meanwhile, the mutant plants display morphological and physiological symptoms of Fe deficiency due to lack of nicotianamine. RT-PCR analysis showed that *LeFRO1* and *LeIRT1* lost their downregulation capability in roots of *chloronerva* under Fe-sufficient conditions, which indicates that *CHLN* is required for downregulation of the two genes. Biochemically, nicotianamine is a metal chelator, chelating ferrous Fe in plant cells to form Fe^{2+}-nicotianamine complex for Fe transportation and utilization. Reasonably, the Fe^{2+}-nicotianamine complex is thought to be a sensor for Fe status in plant cells (Scholz et al. 1992; Stephan and Scholz 1993). Under an Fe sufficiency condition, Fe^{2+}-nicotianamine may combine to an unknown repressor to downregulate the Fe-deficiency responses.

13.5 REGULATION MODEL OF Fe UPTAKE IN STRATEGY I PLANTS

As described above, the strategy I mechanism used in dicots and nongraminaceous plants for the effective Fe acquisition mainly include rhizosphere acidification, ferric-chelate reduction, and high affinity transportation of ferrous Fe. These responses are activated under an Fe deficiency and are turned off when sufficient Fe is present in cells. *FRO2* (ferric-chelate reductase) and *IRT1* (ferrous transporter) in *Arabidopsis* are two essential components for effective Fe uptake because if one of the two genes is knocked out, the mutants are not able to effectively acquire Fe from the soil envFement, and the plants exhibit chlorosis and lethality. Their corresponding genes in tomato are *LeFRO1* and *LeIRT1*. *FER* in tomato and its ortholog *FIT* in *Arabidopsis* are an essential regulator for controlling the transcription of Fe chelate reductase *FRO2/LeFRO1* and high affinity ferrous transporter *IRT1/LeIRT1,* as well other Fe-deficiency responses. *FER* and FIT are *bHLH* proteins functioning as dimer (heterodimer or homodimer) in activation of downstream genes. Overexpression of *FER* or *FIT* does not change expression patterns of Fe uptake genes such as *FRO2* and *IRT1* (Brumbarova and Bauer 2005; Colangelo and Guerinot 2004), and improves Fe uptake ability. Additionally, *FER* transcription intensity was not correlated with the transcriptional abundance of *LeFRO1* and *LeIRT1* in tomato under the culture condition with high pH (Zhao and Ling 2007). This suggests that *FER* and *FIT* might function as a heterodimer in activation of downstream genes. *AtbHLH38*, *AtbHLH39*, *AtbHLH100,* and *AtbHLH101* are four other *bHLH* proteins that are regulated by Fe (Wang et al. 2007). Presumably, some of the four *bHLH* proteins might work with *FIT* together to regulate the downstream genes for Fe uptake.

13.6 CONCLUSIONS

Malnutrition of Fe is a serious problem of human health in the world. The biofortification of Fe content and availability in plant products is an economical and effective method to improve Fe nutrition of human beings. The uptake, transport, and storage of Fe in plants are tightly controlled by many genes. Understanding their molecular regulation mechanisms is the first and basic step for improving content and availability in plant products by a biotechnology approach. As described above, many genes corresponding to Fe uptake and metabolism have been isolated and characterized. The identification and characterization of regulatory genes can construct the linkage of individual genes, and provide clues for understanding the complicated pathway of Fe acquisition, transportation, and storage. No doubt, the knowledge and genes will be useful for molecular breeding to enhance Fe content and availability in plant food products.

ACKNOWLEDGMENTS

This work was supported by the National Science Foundation of China (Grant No., 30530460 and 30521001).

REFERENCES

Bauer, P., H. Q. Ling, and M. L. Guerinot. 2007. FIT, the FER-like Fe deficiency induced transcription factor in *Arabidopsis. Plant Physiology Biochemistry* 45(5): 260–261.

Bereczky, Z., H. Y. Wang, V. Schubert, M. Ganal, and P. Bauer. 2003. Differential regulation of NRAMP and IRT metal transporter genes in wild type and Fe uptake mutants of tomato. *Journal of Biological Chemistry* 278(27): 24697–24704.

Bienfait, H. F. 1988. Proteins under the control of the gene for Fe efficiency in tomato. *Plant Physiology* 88: 785–787.

Brumbarova, T. and P. Bauer. 2005. Fe-mediated control of the basic helix-loop-helix protein FER, a regulator of Fe uptake in tomato. *Plant Physiology* 137(3): 1018–1026.

Buckhout, T. J., P. F. Bell, D. G. Luster, and R. L. Chaney. 1989. Fe-stress induced redox activity in tomato (*Lycopersicum esculentum* Mill.) is localized on the plasma membrane. *Plant Physiology* 90(1): 151–156.

Cellier, M., G. Prive, and A. Belouchi et al. 1995. Nramp defines a family of membrane proteins. *Proceedings of the National Academy of Sciences of the United States of America* 92(22): 10089–10093.

Colangelo, E. P. and M. L. Guerinot. 2004. The essential basic helix-loop-helix protein FIT1 is required for the Fe deficiency response. *Plant Cell* 16(12): 3400–3412.

Connolly, E. L., N. H. Campbell, N. Grotz, C. L. Prichard, and M. L. Guerinot. 2003. Overexpression of the FRO2 ferric chelate reductase confers tolerance to growth on low Fe and uncovers posttranscriptional control. *Plant Physiology* 133(3): 1102–1110.

Curie, C., J. M. Alonso, M. Le Jean, J. R. Ecker, and J. F. Briat. 2000. Involvement of NRAMP1 from *Arabidopsis thaliana* in Fe transport. *The Biochemical Journal* 347: 749–755.

Curie, C., Z. Panaviene, C. Loulergue, S. L. Dellaporta, J. F. Briat, and E. L. Walker. 2001. Maize yellow stripe1 encodes a membrane protein directly involved in Fe(III) uptake. *Nature* 409(6818): 346–349.

Eckhardt, U., A. Mas Marques, and T. J. Buckhout. 2001. Two Fe-regulated cation transporters from tomato complement metal uptake-deficient yeast mutants. *Plant Molecular Biology* 45(4): 437–448.

Eide, D., M. Broderius, J. Fett, and M. L. Guerinot. 1996. A novel Fe-regulated metal transporter from plants identified by functional expression in yeast. *Proceedings of the National Academy of Sciences of the United States of America* 93(11): 5624–5628.

Heim, M. A., M. Jakoby, M. Werber, C. Martin, B. Weisshaar, and P. C. Bailey. 2003. The basic helix-loop-helix transcription factor family in plants: A genome-wide study of protein structure and functional diversity. *Molecular Biology Evolution* 20(5): 735–747.

Holden, M. J., D. G. Luster, R. L. Chaney, T. J. Buckhout, and C. Robinson. 1991. Fe-chelate reductase activity of plasma membranes isolated from tomato (*Lycopersicon esculentum* Mill.) roots: Comparison of enzymes from Fe-deficient and Fe-sufficient roots. *Plant Physiology* 97(2): 537–544.

Jakoby, M., H. Y. Wang, W. Reidt, B. Weisshaar, and P. Bauer. 2004. FRU (BHLH029) is required for induction of Fe mobilization genes in *Arabidopsis thaliana. FEBS Letters* 577(3): 528–534.

Korshunova, Y. O., D. Eide, W. G. Clark, M. L. Guerinot, and H. B. Pakrasi. 1999. The IRT1 protein from *Arabidopsis thaliana* is a metal transporter with a broad substrate range. *Plant Molecular Biology* 40(1): 37–44.

Lanquar, V., F. Lelievre, and S. Bolte et al. 2005. Mobilization of vacuolar Fe by AtNRAMP3 and AtNRAMP4 is essential for seed germination on low Fe. *The EMBO Journal* 24(23): 4041–4051.

Laulhere, J. P. and J. F. Briat. 1993. Fe release and uptake by plant ferritin: Effects of pH, reduction and chelation. *The Biochemical Journal* 290 (Pt 3): 693–699.

Li, L., X. Cheng and H. Q. Ling. 2004. Isolation and characterization of Fe(III)-chelate reductase gene LeFRO1 in tomato. *Plant Molecular Biology* 54(1): 125–136.

Ling, H. Q., P. Bauer, Z. Bereczky, B. Keller, and M. Ganal. 2002. The tomato *FER* gene encoding a bHLH protein controls Fe-uptake responses in roots. *Proceedings of the National Academy of Sciences of the United States of America* 99(21): 13938–13943.

Ling, H. Q., G. Koch, H. Baumlein, and M. W. Ganal. 1999. Map-based cloning of chloronerva, a gene involved in Fe uptake of higher plants encoding nicotianamine synthase. *Proceedings of the National Academy of Sciences of the United States of America* 96(12): 7098–7103.

Ling, H. Q., A. Pich, G. Scholz, and M. W. Ganal. 1996. Genetic analysis of two tomato mutants affected in the regulation of Fe metabolism. *Molecular and General Genetics* 252(1–2): 87–92.

Mori, S. 1999. Fe acquisition by plants. *Current Opinion in Plant Biology* 2(3): 250–253.

Römheld, V. and H. Marschner. 1986. Evidence for a specific uptake system for Fe phytosiderophores in roots of grasses. *Plant Physiology* 80(1): 175–180.

Robinson, N. J., C. M. Procter, E. L. Connolly, and M. L. Guerinot. 1999. A ferric-chelate reductase for Fe uptake from soils. *Nature* 397(6721): 694–697.

Scholz, G., R. Becker, A. Pich, and U. W. Stephan. 1992. Nicotianamine—A common constituent of strategies I and II of Fe acquisition by plants: A review. *Journal of Plant Nutrition* 15: 1647–1665.

Stephan, U. W. and G. Scholz. 1993. Nicotianamine: Mediator of transport of Fe and heavy metals in the phloem. *Plant Physiology* 88: 522–529.

Varotto, C., D. Maiwald, P. Pesaresi, P. Jahns, F. Salamini, and D. Leister. 2002. The metal ion transporter IRT1 is necessary for Fe homeostasis and efficient photosynthesis in *Arabidopsis thaliana*. *Plant Journal* 31(5): 589–599.

Vasconcelos, M., H. Eckert, V. Arahana, G. Graef, M. A. Grusak, and T. Clemente. 2006. Molecular and phenotypic characterization of transgenic soybean expressing the *Arabidopsis* ferric chelate reductase gene, *FRO2*. *Planta* 224: 1116–1128.

Vert, G., J. F. Briat, and C. Curie. 2001. *Arabidopsis IRT2* gene encodes a root-periphery Fe transporter. *Plant Journal* 26(2): 181–189.

Vert, G., N. Grotz, and F. Dedaldechamp et al. 2002. IRT1, an *Arabidopsis* transporter essential for Fe uptake from the soil and for plant growth. *Plant Cell* 14(6): 1223–1233.

Wang, H. Y., M. Klatte, M. Jakoby, H. Baumlein, B. Weisshaar, and P. Bauer. 2007. Fe deficiency-mediated stress regulation of four subgroup Ib BHLH genes in *Arabidopsis thaliana*. *Planta* 226(4): 897–908.

Waters, B. M., D. G. Blevins, and D. J. Eide. 2002. Characterization of FRO1, a pea ferric-chelate reductase involved in root Fe acquisition. *Plant Physiology* 129(1): 85–94.

Williams, L. E., J. K. Pittman, and J. L. Hall. 2000. Emerging mechanisms for heavy metal transport in plants. *Biochimica et Biophysica Acta* 1465(1–2): 104–126.

Wu, H., L. Li, J. Du, Y. Yuan, X. Cheng, and H. Q. Ling. 2005. Molecular and biochemical characterization of the Fe(III) chelate reductase gene family in *Arabidopsis thaliana*. *Plant and Cell Physiology* 46(9): 1505–1514.

Yi, Y. and M. L. Guerinot. 1996. Genetic evidence that induction of root Fe(III) chelate reductase activity is necessary for Fe uptake under Fe deficiency. *Plant Journal* 10(5): 835–844.

Yuan, D. S., R. Stearman, A. Dancis, T. Dunn, T. Beeler, and R. D. Klausner. 1995. The Menkes/Wilson disease gene homologue in yeast provides copper to a ceruloplasmin-like oxidase required for Fe uptake. *Proceedings of the National Academy of Sciences of the United States of America* 92(7): 2632–2636.

Yuan, Y. X., J. Zhang, D. W. Wang, and H. Q. Ling. 2005. AtbHLH29 of *Arabidopsis thaliana* is a functional ortholog of tomato FER involved in controlling Fe acquisition in strategy I plants. *Cell Research* 15(8): 613–621.

Zhao, T. and H. Q. Ling. 2007. Effects of pH and nitrogen forms on expression profiles of genes involved in Fe homeostasis in tomato. *Plant, Cell and Environment* 30(4): 518–527.

14 Identification of Genes for Biofortification

Genetic and Molecular Analysis of Mineral Accumulation in Arabidopsis thaliana and Other Plant Species

Mark G.M. Aarts and Jian Wu

CONTENTS

14.1 IMPROVING THE MICRONUTRIENT STATUS OF PLANTS AND PLANT PRODUCTS

Many minerals, such as iron (Fe), zinc (Zn), manganese (Mn), nickel (Ni), and selenium (Se), are part of the essential micronutrient requirements of many organisms.

Deficiencies in minerals are therefore generally detrimental for proper development, growth, and ability to withstand biotic and abiotic stresses. Because plants provide a major contribution, or sometimes the only contribution, to the human diet, mineral deficiencies in plants will affect the mineral status of food and, thus, the mineral acquisition of humans. Mineral malnutrition among humans is prevalent all over the world but is especially dramatic in developing countries when people depend largely on staple crops for their daily supply of energy and micronutrients. Traditional approaches for reduction of mineral malnutrition have been the supplementation of minerals in tablets or through fortification of food. Examples of the latter are the supplementation of iodine to salt or the addition of Se to flour. However, in general, these approaches are only effective and economically feasible if there is a good infrastructure to educate people on the expected benefits and to provide access to fortified products and if the local governments are committed to support such actions over a sufficiently long period.

Recently, a new approach has been proposed, which is referred to as *biofortification* (Bouis 2002). This approach does not so much rely on the local infrastructure to repeatedly provide supplements or fortified products but aims at improving the nutritional value of food by improving the micronutrient status of the crops that are consumed daily (either in processed or fresh form). In order to sustainably reach a considerable part of the population, local crops need to be improved, either by breeding or through biotechnology, and seeds of improved varieties need to be distributed among local farmers. Provided that the micronutrient-improved crop varieties are at least comparable to existing varieties in terms of yield and quality, the superior nutritional value is expected to provide sufficient incentive for farmers to maintain such varieties and gradually replace the less nutritional ones.

A few considerations have to be made in order to develop proper varieties that can be used for biofortification. Plant products with enhanced mineral content will only be beneficial if the minerals are present in a bioavailable form. This is especially problematic for Fe because plants tend to produce sufficient antinutritional compounds such as phenolics and phytate to inhibit its bioavailability and to prevent maximum Fe absorption upon human consumption. In this respect, it is also of importance to consider the plant parts that should be targets for mineral improvement. In general, since grains are the most-consumed parts of plants, they seem the most appropriate plant organs to target for improved mineral content. However, not all parts of the grain are consumed. For example, rice grains are generally dehusked and polished to leave only the white endosperm for consumption. The endosperm is not particularly rich in Fe (or other minerals), although it still contains substantial quantities of the antinutrient phytate that will limit Fe bioavailability (Tuntawiroon et al. 1990). Thus, improving the bioavailable mineral content in rice should focus on enhancing mineral mobilization into the endosperm and on reducing the phytate content in endosperm.

Considerable refinement, and thus removal of mineral-rich parts, is not only typical for rice but for most of the cereals. In combination with the presence of phytate or other antinutrient compounds, this is one of the main reasons why cereals are generally poor contributors to the bioavailable mineral fraction in food (Welsh and Graham 1999). For legumes, most of them are consumed without too much refinement, but the high phytate content also inhibits mineral bioavailability (Sandberg 2002). Alternative organs are leaves, stems, and storage organs, such as tubers or roots. The main advantage of using leaves or stems as mineral supply organs is that most minerals are allocated to these sites by xylem transport only, whereas loading into storage organs requires phloem transport. The latter, which is also needed for grain loading, is more difficult to achieve by genetic modification and perhaps also by classical genetic breeding, because metal uptake, translocation, and remobilization all need to be considered.

In this chapter, we will focus on the possibilities of improving the bioavailable Fe and Zn status of crop products through biotechnology. To absorb and maintain a balance of potentially toxic metal ions, plants utilize poorly understood mechanisms involving a large number of membrane transporters and metal-binding proteins with overlapping substrate specificities and complex regulation. Therefore, the identification of genes involved in mineral accumulation and the molecular understanding of plant Fe and Zn homeostasis mechanisms are prerequisites for biofortification. Most of the research in this field has been performed using the model species *Arabidopsis thaliana* (Arabidopsis) and to a lesser extent Graminae, such as rice and maize. To date, several major strategies have been used to identify novel genes involved in metal homeostasis. These strategies and the approaches employing the identified genes for crop Fe and Zn biofortification will be discussed in this chapter.

14.2 MOLECULAR CONTROL OF IRON AND ZINC HOMEOSTASIS IN PLANTS

14.2.1 Iron Homeostasis

In general, plants acquire nearly all Fe and Zn through uptake via the roots. These elements are always found to be transported as ions, which will be Fe^{2+} or Fe^{3+} for iron and Zn^{2+} for Zn. (Ion charges have been omitted throughout the text. We will specifically indicate the charge only when a distinction needs to be made between ion charges.) For Fe uptake, plants use two different strategies, known as strategy I and II (Marschner 1995). In strategy I, which is the general strategy used by most plant families, Fe(III), which is generally the most common form of Fe in soil, is reduced to Fe(II) by ferric-chelate reductases embedded in the plasma membrane of root epidermal cells. In Arabidopsis, these enzymes are encoded by the *FRO* genes, of which there are at least eight described (Wu et al. 2005; Mukherjee et al. 2006). The reduced Fe is transported over the plasma membrane by Fe(II) transporters. In Arabidopsis, these are encoded by *IRT1* and *IRT2*, two genes belonging to the larger family of ZIP metal transporters (Guerinot 2000).

Graminae species use an additional uptake strategy (strategy II), by which Fe is not reduced but complexed to phytosiderophores as Fe(III). Phytosiderophores are mugineic acid (MA) compounds, a set of biochemically similar compounds derived from nicotianamine (NA), which is enzymatically synthesized from three molecules of S-adenosyl methionine by nicotianamine synthase (NAS). Genes known to be involved in the conversion of NA to MA compounds are nicotianamine amino transferase (NAAT) and deoxymugineic acid synthase (DMAS) (Higuchi et al. 1999).

Phytosiderophores are secreted by roots to chelate Fe(III). Fe(III)-MA chelates are taken up by specific transporters in the roots. For maize, one gene for such transporters has been cloned and found to be the *YS1* gene (Curie et al. 2001). Recently, a similar gene was cloned from barley (Murata et al. 2006). Studies in rice showed that Graminae are not exclusively using strategy II for Fe acquisition, because they have the full machinery available for operating strategy I Fe acquisition using *IRT*- and *FRO*-like genes (Ishimaru et al. 2006).

Transport of Fe to the stele of the root is assumed to occur symplastically. Once inside the stele, Fe will be loaded into the xylem. A transporter for this process has not been found yet, but genes encoding ferroportin- or IREG-like ("iron regulated exporter from gut") transporters have been suggested to encode such a candidate (Schmidt 2003). However, the only characterized ferroportin-like gene, *AtIREG2*, encodes a tonoplast-located metal transporter that was not able to transport Fe, but only Ni, when expressed in yeast (Schaaf et al. 2006). In addition to *AtIREG2*, there are two other ferroportin-like genes expressed in Arabidopsis, but they are not yet characterized. Furthermore, YSL2, which is located at the root endodermis and pericycle, has been suggested to be a candidate transporter for Fe(II)(-NA) loading of xylem (DiDonato et al. 2004). YSL proteins are discussed in more detail in the following text. Once Fe is loaded into the xylem, it will be complexed to citrate, which seems to be provided to the xylem via a MATE-efflux transporter. In Arabidopsis, this transporter is encoded by the *FRD3* gene (Durrett et al. 2007); a similar transporter has been identified in rice (Inoue et al. 2004). Iron is subsequently translocated via the xylem to the leaves, where it is unloaded and distributed to surrounding tissues.

Candidates for xylem unloading (i.e., loading into xylem parenchyma cells) are the YSL-transporters, orthologs of the maize Yellow Stripe1 protein involved in transport of MA–phytosiderophore-chelated Fe complexes (Curie et al. 2001). MA–phytosiderophores are typically produced in Graminae plants, but other species are able to make NA, the precursor of MA, which is also able to chelate metals (Stephan et al. 1996). In Arabidopsis, the *YSL1, 2,* and *3* genes have been characterized in some detail (DiDonato et al. 2004; Koike et al. 2004; Le Jean et al. 2005; Schaaf et al. 2005; Waters et al. 2006). These genes, and other members of the *YSL* gene family, all appear to be involved in the transport of metal–nicotianamine complexes across membranes. Expression of *YSL1*, *2* and *3* in vascular parenchyma cells suggests a function in sequestration of Fe and possibly other metals such as copper (Cu) from the veins to surrounding tissues in older leaves (Waters et al. 2006). Because *ysl1* mutants suffer from inadequate Fe in seeds (Le Jean et al. 2005; Waters et al. 2006), *YSL1* is also likely to play a determining role in seed Fe loading, a process that is driven by phloem transport of Fe (Briat 2007). Thus, *YSL* genes are thought to confer both loading and unloading of xylem, and loading of phloem by vascular parenchyma cells.

Transport by *YSL* transporters relies on the presence of NA. In the absence of NA, the *chloronerva* mutant is defective in NAS biosynthesis in the tomato (Ling et al. 1999) and in *NAAT*-overexpressing tobacco plants (Takahashi et al. 2003), the transport of Fe, and also Zn, is reduced or absent. The large number of *YSL* genes in Arabidopsis (nine members) suggests redundancy or development of different metal-binding specificities or tissue-specific expression patterns. *YSL* genes are also found in rice (at least 18 members), where they appear to play a similar role as in Arabidopsis (Koike et al. 2004). The OPT (oligopeptide transporter) family of transporters is structurally related to the *YSL* family, of which there are also nine members in Arabidopsis (Koh et al. 2002). Although it is not clear which compound is transported, the *AtOPT3* gene of Arabidopsis is involved in Fe homeostasis, because the knock-out mutant of this gene shows Fe deficiency symptoms even at high plant Fe levels (Stacey et al. 2008).

In plant cells, Fe is typically sequestered in vacuoles, where it is stored until further use. The mobilization of Fe into and out of vacuoles should also be mediated by specific transporters. Vacuolar export is conducted by NRAMP transporters. In Arabidopsis, the vacuolar Fe exporters AtNRAMP3 and 4 are particularly important for the remobilization of stored Fe from vacuoles (Thomine et al. 2003; Lanquar et al. 2005). Recently, the Arabidopsis *VIT1* gene was found to mediate Fe import into vacuoles. Although expression of this gene is found throughout the plant, expression is highest in developing embryos (Kim et al. 2006). In *vit1* mutants of Arabidopsis, proper localization of Fe in provascular strands of the embryo is disrupted, which disturbs mutant seedling growth under Fe-limited conditions similarly to that observed in *nramp3nramp4* double mutants (Lanquar et al. 2005).

Iron is not only stored in vacuoles but can also be stored in plastids. The major protein involved in such storage appears to be ferritin, which rapidly accumulates upon exposure to excess Fe. There are currently four *FER* genes known in Arabidopsis, which are expressed throughout the plant, but at different levels and with tissue-specific differences (Petit et al. 2001). One of the transporters involved in plastid Fe uptake has recently been identified in Arabidopsis to be encoded by the *PIC1* gene. This transporter belongs to an ancient family of permeases and is targeted to the inner envelope of chloroplasts (Duy et al. 2007).

14.2.2 Zinc Homeostasis

Zinc basically follows the same route as Fe in plant uptake and distribution; however, the genes involved are even less well known than for Fe. For Zn uptake, only one strategy is known, which relies on ZIP-like Zn uptake transporters. Zinc does not need to be reduced. Hence, *FRO* genes are not expected to play a role, although some *FRO* genes are induced in roots by Zn deficiency (van de Mortel et al. 2006). Initially, four *ZIP* genes were implicated in Zn uptake by the roots: *ZIP1, 2, 3*, and *4* (Grotz et al. 1998). These genes were induced by Zn deficiency, which was also confirmed in later microarray studies on Zn deficiency in Arabidopsis (van de Mortel et al. 2006). This study also revealed that other *ZIP* genes, such as *ZIP5, ZIP9, ZIP10, ZIP11, ZIP12*, and *IRT3*, are induced by Zn deficiency, and thus, they may also play a role in Zn uptake. Loading of Zn into the xylem may occur via the HMA family of metal transporters, and in Arabidopsis, specifically by HMA2 and HMA4 (Hussain et al. 2004). These are plasma-membrane-located Zn/Cd efflux transporters. Only *HMA2* expression is induced in Arabidopsis roots by Zn deficiency (van de Mortel et al. 2006). So far, one *HMA* gene, *OsHMA9*, has been characterized from rice, where it also seems to be involved in Zn as well as Cu efflux from the cytoplasm (Lee et al. 2007).

Similar to Fe, xylem transport of Zn is affected in the Arabidopsis *frd3* mutant (Rogers and Guerinot 2002), which suggests that citrate is of major importance for proper xylem transport and distribution of Zn into the leaves. Xylem unloading of Zn is not well understood. *HMA2* and *HMA4* are also expressed in leaf vascular parenchyma cells, and may play a role, but it is also conceivable that *YSL* transporters play a role. In addition, Zn deficiency in Arabidopsis also induces expression of the Zn transporter gene *ZIP4* in shoots (van de Mortel et al. 2006), which may be needed for proper Zn uptake. Similar to Fe, the phloem transport of Zn seems to be largely mediated by *YSL* transport of NA-chelated Zn, but Zn specific transporters are not known.

As Zn is also stored in vacuoles, tonoplast-specific Zn transporters must be active. Import of Zn into vacuoles can be accomplished by members of the Cation Diffusion Facilitator (CDF) family, which in Arabidopsis are encoded by the *MTP* genes. *MTP1*, and *MTP3*, in particular, appear to mediate Zn sequestration into vacuoles under excess-Zn conditions (Desbrosses-Fonrouge et al. 2005; Arrivault et al. 2006). *ZIF1* (Zn-induced facilitator) is another gene encoding a tonoplast-located Zn transporter in Arabidopsis, acting in addition to MTP1, in sequestering Zn in the vacuole. The ZIF protein is a member of the major facilitator superfamily of transporters but is not closely related to MTP1 (Haydon and Cobbett 2007). Zinc export from the vacuole may be mediated by NRAMP4, which in Arabidopsis transports Zn as well as Fe (Oomen et al. unpublished data). This observation is in contrast to NRAMP3, which does not transport Zn (Lanquar et al. 2005).

14.2.3 Regulation of Iron and Zinc Homeostasis

An important step for modifying Fe and Zn homeostasis would be to identify regulatory genes that control the Fe or Zn deficiency response by inducing expression of Fe and Zn uptake and remobilization transporters. Both Fe and Zn homeostasis are very tightly controlled, most likely involving many regulatory genes. For Zn, such regulators have not yet been described, but for Fe some transcription factors have been identified that control the expression of Fe uptake genes. Recently, Kobayashi et al. (2007) described cloning of the *IDEF1* gene of rice, which encodes an ABI3/VP-like transcription factor binding to Fe-deficiency-responsive element 1 (IDE1; CATGC) in the promoter of Fe-deficiency-responsive genes. One of these genes is the rice *OsIRO2* gene, which itself encodes a bHLH (basic helix–loop–helix) transcription factor, which directly regulates the expression of a large number of Fe-deficiency-responsive genes, involved in both strategy I and strategy II uptake. They all contain the bHLH binding site CACGTGG in their promoter (Ogo et al. 2007). *OsIRO2* is similar to *bHLH38* and *bHLH39*, which are two transcription factor genes that were recently found to be induced by Fe deficiency in leaves and roots of Arabidopsis together with two other bHLH genes (*bHLH100* and *bHLH101*) (Wang et al. 2007). Overexpression of these genes to determine which genes they may regulate has not yet been studied. Another bHLH transcription factor, encoded by the *FIT1* (Fe-deficiency-induced transcription factor) gene, previously also known as FRU and bHLH29, has been implicated in controlling the Fe deficiency response in Arabidopsis through the control of expression of *FRO2* and *IRT1* and some 70 additional genes (Colangelo and Guerinot 2004; Bauer et al. 2007). *FIT1* was identified on the basis of similarity to the bHLH transcription factor gene *FER* isolated from tomato (Ling et al. 2002).

14.3 ISOLATION OF IRON OR ZINC MINERAL HOMEOSTASIS MUTANTS

14.3.1 Phenotype Screening of Mutant Collections

The phenotypic screen for mutants affected in the process is generally a very rewarding exercise in trying to identify genes acting in specific processes. Surprisingly, this has not been very successful for the identification of mutants impaired in Fe or Zn homeostasis. Examples of Fe homeostasis mutants that have previously been found by phenotypic screening in crop species are the *fer* and *chloronerva* (*chn*) mutants of tomato, the *ys1* mutant of maize, and the *brz* mutant of pea (Guerinot and Yi 1994). The *fer* and *ysl* mutants have an Fe deficiency phenotype, and do not respond phenotypically to Fe supplementation to the soil. The *chn* and *brz* mutants also appear to be Fe-deficient, but they are accumulating Fe in the leaves. Supplying additional Fe to the soil has no effect on this phenotype. With the exception of the *BRZ* gene, all of these genes have been cloned, and their further characterization has significantly advanced our understanding of Fe homeostasis in plants. Already in the early 1990s, the Guerinot lab performed phenotypic screens to identify Fe uptake mutants in Arabidopsis. Arabidopsis is particularly rewarding for forward genetic screens owing to the availability of large T-DNA or transposon insertion mutant collections that allow rapid cloning of the mutated genes (www.arabidopsis.org). Although the obtained *frd* mutants proved very useful for further progress in understanding strategy I Fe uptake (Robinson et al. 1999; Rogers and Guerinot 2002), relatively few mutants were found (Yi and Guerinot 1996), indicating strong genetic redundancy of Fe homeostasis genes.

New strategies may be employed to identify additional mutants such as screening for sensitivity to excess Zn (Haydon and Cobbett 2007). The secondary mutagenesis of existing mutants may identify suppressor mutations or screens for mutants with an altered transcriptional response of selected Fe/Zn responsive promoters fused to an expression marker gene such as GUS or luciferase. Screens using the *IRT1* promoter of Arabidopsis have yielded several mutants, but cloning of new genes has not yet been reported. In our lab, we recently screened for mutants altered in their Zn deficiency response using the Zn-deficiency-responsive *ZIP4* promoter fused to GUS—this resulted in several mutants with either reduced response under inducing conditions or induced response under noninducing conditions (A.G.L. Assunção and M.G.M. Aarts, unpublished data).

Alternatively, high-throughput screens can be employed that are specifically dedicated for screening for mutants with an altered Fe or Zn leaf concentration. This screen has been initiated by Dr. D. Salt through the setup of the Purdue Ionomics Information Management System (PiiMS) (www.purdue.edu/dp/ionomics), a facility that is dedicated to high-throughput ICP-MS phenotyping of plants that largely focuses on analysis of P, Ca, K, Mg, Cu, Fe, Zn, Mn, Co, Ni, B, Se, Mo, Na, As, and Cd content in Arabidopsis shoots (Baxter et al. 2007). Previous screening of M2 mutant populations identified several mutants disturbed in leaf mineral element profile, one of which was a new *frd3* mutant (Lahner et al. 2003).

14.3.2 Reverse Genetics to Identify Mineral Homeostasis Mutants

It is becoming increasingly easier to obtain a fairly complete impression of all transcripts that are specifically expressed in different tissues upon exposure of the plant to Fe or Zn deficiency or excess with the availability of large-scale transcript analysis tools, such as whole-transcriptome microarray analysis or whole-transcriptome sequencing technologies (Busch and Lohmann 2007). Such an analysis has been performed for Fe in barley (Negishi et al. 2002), and was used to identify genes involved in Fe-deficiency response upon germination of rice (Nozoye et al. 2007). Microarray comparison of Fe-deficient Arabidopsis wild-type plants and *fit1* mutants also revealed a large set of genes induced by Fe deficiency (Colangelo and Guerinot 2004). For Zn, such analyses have been performed for Arabidopsis (Becher et al. 2004; Weber et al. 2004; van de Mortel et al. 2006), and for the Zn/Cd/Ni hyperaccumulator species *Thlaspi caerulescens* (van de Mortel et al. 2006) and the Zn/Cd hyperaccumulator *Arabidopsis halleri* (Becher et al. 2004; Weber et al. 2004; Talke et al. 2006). Although most attention in these papers has been on further characterization of genes already known to be involved in Fe or Zn homeostasis, they potentially report on a very efficient way to identify new Fe or Zn homeostasis genes. This approach has not yet been utilized to its full extent, particularly for crop species, and much more knowledge on mineral homeostasis is to be gained from such analysis, particularly when studying different organs or tissues and different exposure times to low/high Fe/Zn supply.

Once potentially new Fe or Zn homeostasis genes are identified, their exact functions need to be established. Such functional analysis is based on experimental data and basically involves overexpression or ectopic expression of the gene, analysis of tissue-specific gene expression, and determining the effect of knocking-out the expression of the gene (Somerville and Somerville 1999). Many molecular genetic tools have been developed specifically for functional gene analysis, especially in Arabidopsis, maize, and rice. A recent development for Arabidopsis gene function research is the public availability of microarray data sets in one database such as Genevestigator (https://www.genevestigator.ethz.ch/) (Grennan 2006), which allows the visualization of gene expression of any particular gene under different conditions or at different developmental stages or in different tissues. An additional useful tool is the Arabidopsis coexpression tool (ACT; www.arabidopsis.leeds.ac.uk/act/), which makes it possible to identify coexpressed genes from a large collection of microarrays. Although most of these tools were initially developed for the model species Arabidopsis, with the availability of more and more genome sequence data and other genomic data from additional plant species, more tools will become available to make use of all these data (e.g., PlantGDB [www.plantgdb.org] for comparative genomics of plant species [Duvick et al. 2008] or for use in Gramene [www.gramene.org], a curated database for genetic and genomic data of major Graminae crop species [Liang et al. 2008]).

14.4 YEAST FUNCTIONAL COMPLEMENTATION TO IDENTIFY IRON OR ZINC TRANSPORTERS

The initial focus of plant mineral homeostasis research was largely dedicated, for practical reasons, to the analysis of cation transporters in the absence of a large collection of mineral homeostasis plant mutants (Mäser et al. 2001). In yeast, several mutants impaired in proper mineral uptake had been found (Eide 1998). These proved to be excellent tools for complementation screens by expressing plant proteins in yeast mutants growing on medium favoring selection of genes complementing the mutant phenotype (Hall and Williams 2003). Commonly used yeast (*Saccharomyces cerevisiae*) mutants are as follows: (1) the *fet3fet4* mutant impaired in low- and high-affinity Fe uptake (Eide 1998), (2) the *smf1* mutant, which is sensitive to high concentrations of the divalent cation chelator EGTA owing to a mutation in an *NRAMP* gene (Supek et al. 1996), (3) the *zrc1cot1* double mutants of Cation Diffusion Facilitator genes impaired in storage of Zn in the vacuole (MacDiarmid et al. 2000), and (4) the *zrt1zrt2* mutant in which the low- and high-affinity Zn uptake transporters are mutated (Zhao and Eide 1996a,b).

To screen for plant genes involved in mineral uptake or vacuolar transport, a mineral-uptake-deficient or mineral-excess-sensitive yeast mutant strain was transformed with a cDNA library constructed from RNA of relevant plant tissues (e.g., mineral-deficient roots). The transformants were then screened for complementation of the yeast mutant phenotype. Clones expressing cDNAs that restored the mutant growth were isolated, confirmed, and sequenced. However, although the genes identified by yeast functional complementation screening confer mineral transport or tolerance at the single-cell yeast level, it is possible that these genes do not participate in the same process in plants, particularly if more complex mechanisms based on the operation and interplay of multiple cell types, tissues, and organs are functioning (Papoyan and Kochian 2004). Therefore, a series of characterizations is still needed to confirm the function of the protein encoded by the cloned cDNAs, such as yeast mineral uptake studies, heterologous or ectopic expression in other plant species, and transcriptional response to differences in mineral supply conditions.

14.5 COMPARATIVE GENOMICS OF METAL HYPERACCUMULATOR SPECIES

With the development of tools to analyze genome or transcriptome differences between samples, it is also possible to compare plant genotypes even if they do not belong to the same species. Comparative genomics can be very useful for identifying genes involved in plant Fe or Zn homeostasis, particularly if there are closely related species that show substantial phenotypic differences in mineral homeostasis. For actual genome comparison, the full genome sequence should be available for the compared species. This is not yet the case for many species, so this may not yet be very useful for the identification of genes that can be applied in biofortification. For transcript comparison, the availability of full genome sequences is not a requirement. In many cases, cDNA or oligonucleotide microarrays, containing probes for one species, can be used for heterologous transcript analysis of another species, on the condition that there is sufficient cross-hybridization of heterologous genes to their orthologous probes (van de Mortel and Aarts 2006).

Heavy metal hyperaccumulator species are characterized by their ability to accumulate extremely high amounts of certain metals, often corresponding with a strong preference to store these metals in the leaves (Baker and Brooks 1989). For Fe, such species are not known, but there are several Zn hyperaccumulators, some of which, like *Arabidopsis halleri* and *Thlaspi caerulescens*, have been extensively studied at the physiological and transcriptional level. These are remarkable species, which most likely initially evolved metal tolerance to grow on metal-rich soils and subsequently developed metal hyperaccumulation to deter herbivores (Coleman et al. 2005). Zinc accumulation in *A. halleri* and *T. caerulescens* reaches extremely high levels (up to 3% of dry weight), especially when grown on soils rich in Zn. Even under normal soil conditions, Zn concentrations of 0.5% of dry weight are not uncommon (Assunção et al. 2003; Macnair and Callow 2003). The transcriptome analyses of Zn hyperaccumulators have so far only been performed on species closely related to Arabidopsis, because relevant heterologous custom-made or commercial microarray platforms are only available for Arabidopsis (Becher et al. 2004; Weber et al. 2004; Filatov et al. 2006; Hammond et al. 2006; Talke et al. 2006; van de Mortel et al. 2006). Transcript profiling has provided a broad and unbiased picture of the molecular basis for metal homeostatic networks in Zn hyperaccumulating plant species. All of these studies showed that a large number of genes are differentially expressed in response to changes in Zn and Fe exposure. This observation helped to elucidate the molecular mechanism of Zn homeostasis and provides an additional tool to establish how the relevant genes are regulated at the transcription level. This also suggested additional genes that may potentially be involved in the Zn homeostasis network, because they show a conserved strong transcriptional difference when comparing Zn hyperaccumulating species with nonaccumulating relatives (Broadley et al. 2007). Many of these genes, such as *MTP1*, *MTP8*, *MTP11*, *IRT3*, and *HMA3* encode Zn transporters, and their seemingly coordinated "over"-expression in Zn hyperaccumulators suggests the presence of a regulatory mechanism in which few transcription factors control the expression of many Zn homeostasis genes in response to signals provided by the root and shoot Zn-status-sensing receptors.

At the genomic level, the altered transcription regulation of Zn homeostasis genes in a Zn hyperaccumulator can be explained in several ways: (1) the presence or specificity of some cis-elements or trans-factors controlling transcription is altered; (2) the affinity of the Zn-sensing compounds that provide the signals to eventually downregulate transcript levels of Zn acquisition genes is reduced; and (3) the Zn requirement is high due to the sequestration of Zn inside or outside particular leaf cells in such a way that Zn is no longer available to the surrounding tissue; and thus conferring a constant "Zn-deficiency" state on the hyperaccumulator plant. The identification of genes involved in the transcriptional regulation and signal transduction in Zn homeostasis will be very important for genetic engineering approaches for biofortification. So far, this approach seems to be most appropriate for biofortification through increased xylem transport and sequestration in leaves. Although seed Zn concentrations are also somewhat elevated in Zn hyperaccumulators compared to regular species, this accumulation is often only about 10-fold higher compared to the generally > 100-fold higher accumulation in leaves.

14.6 QTL ANALYSIS ON NATURAL VARIATION

A final important way to identify genes needed for biofortification is the genetic analysis of natural variation in plant populations. Phenotypic variation is central to evolutionary adaptation due to natural selection to local conditions among organisms of one species. Quantitative trait locus (QTL) analysis is a powerful tool for the dissection of natural variation (Alonso-Blanco and Koornneef 2000). QTL analysis unravels the genetic variation that exists between different accessions or genotypes of one species and is generally performed on genetic segregating populations (Koornneef et al. 2004). QTL is an unbiased investigation of the genes affecting a certain trait, meaning that genes encoding structural as well as regulatory proteins involved in the process of investigation can be identified. It is not an easy procedure, as several steps are needed for the identification of genes for a certain trait when using QTL analysis.

The first step is to establish statistically significant genomewide evidence for linkage or association between genetic markers and the desired trait. Although QTL analysis only identifies chromosomal regions containing one or more genes relevant for the process under investigation, it can be used as a starting point for the identification and cloning of such genes (Paran and Zamir 2003). After fine mapping of the identified QTL intervals using a large recombinant population (e.g., a F2), near-isogenic lines (NILs) or other introgression lines, which differ only by a small introgression of one genotype in the background of another, then candidate genes can be assigned to the QTL after determining the genomic sequence of the fine mapped locus. This is a straightforward process for species with a fully sequenced genome. For other species, the genomic sequence has to be obtained after screening genomic DNA libraries, such as Bacterial Artificial Chromosome (BAC) libraries, using the closest flanking molecular markers as probes (Tanksley et al. 1995). Sometimes it is possible to use the information of related species with a sequenced genome to obtain information on possible candidate genes based on genomic micro colinearity, but the frequent observation of small rearrangements in related genomes makes this approach unreliable (Bennetzen and Ma 2003). With the availability of genome sequence, candidate genes may be confirmed by phenotyping knock-out or overexpression mutants. This is not always conclusive, in which case the best evidence of cloning the right gene is provided by reciprocal transfer of the allele of one parent into the genetic background of the other (El-Din El-Assal et al. 2001).

Mineral accumulation is a typical quantitative trait and thus suitable for QTL analysis. In Arabidopsis, we previously showed substantial genetic variation for 8 different minerals among 25 different accessions (Vreugdenhil et al. 2004). A genetically well-characterized recombinant inbred line (RIL) population based on a cross between accessions Landsberg *erecta* (Ler) and Cape Verde Islands (Cvi) was used to identify QTLs for the accumulation of these minerals in seeds. Many QTLs were found at different loci in the genome. Some QTLs for different minerals colocalize, indicating common loci controlling accumulation of several minerals. For Fe and

Zn, two and four QTLs were, respectively, identified. Two of these could be confirmed using NILs, and cloning of the genes involved is in progress (A. Ghandilyan, J. Du, and M.G.M. Aarts, unpublished data). The same population was used for QTL analysis of phytate concentrations in seeds, which identified one major locus on the top of chromosome 3 (Bentsink et al. 2003). Cloning of this gene is also in progress. Because the seed phytate QTL does not colocate with a Zn or Fe QTL, it looks as though it is possible to breed for varieties with high seed Fe and Zn concentrations and low phytate concentrations.

In addition to the Ler x Cvi population, other populations have been used for mineral QTL analysis (Baxter et al. 2007). The PiiMS ionomics facility is also very well suited for association mapping or linkage disequilibrium (LD) mapping, and is an additional genetic approach for gene identification based on the phenomenon that closely linked genomic DNA stretches remain connected over many generations, and may spread through a population as haplotypes. Analysis of many densely genotyped and properly phenotyped accessions of one species allows detection of such associations between markers and closely linked alleles (and subsequent identification of the gene involved), if these alleles contribute in a distinctive way to the phenotype of the plant (Yu and Buckler 2006). In Arabidopsis, LD-mapping only detects genes within 10 kb of a genetic marker, which means a very high marker density is needed for saturated mapping. Recently, Kim et al. (2007) described LD-mapping in Arabidopsis and designed an Affymetrix genotyping array with which 250,000 SNPs can be detected.

Although Arabidopsis is not a crop, the species is very valuable in studying the genetics of Fe and Zn accumulation. Of course, this can lead to the cloning of new genes controlling mineral homeostasis, but perhaps even more important is that the species facilitates the study of environmental and developmental effects on Fe and Zn content of various plant organs. Unpublished work in our laboratory on three additional Arabidopsis RIL populations grown under different environmental conditions showed that there are many Zn and Fe QTLs in Arabidopsis affecting leaf or seed mineral concentrations or both (A. Ghandilyan et al., unpublished data). This unpublished study also showed that there is a strong effect of the environment on the expression of QTLs and there are considerable QTL × QTL interactions. It will be very important to apply this information in breeding for high-yielding crop varieties with stable high Fe and Zn content when grown at different locations.

In addition to Arabidopsis, QTL analysis has also been employed for other species, such as the Zn hyperaccumulators *T. caerulescens* and *A. halleri* (Assunção et al. 2006; Deniau et al. 2006; Filatov et al. 2007), but only a few crop species have been studied. Guzmán-Maldonado et al. (2003) described QTLs for Fe and Zn content in common bean (*Phaseolus vulgaris*). Analysis of a segregating rice population identified several QTLs for plant growth under Zn deficiency, which leads to low Zn accumulation (Wissuwa et al. 2006). When examining a segregating doubled haploid (DH) population of Chinese cabbage (*Brassica rapa*), we identified leaf mineral QTLs, including two for Zn and one for Fe (Wu et al. 2008).

Recently, a remarkable observation was made by Uauy et al. (2006). They cloned the *Grain Protein Content B1 (Gpc-B1)* locus from wild emmer wheat and found a gene encoding for a NAC-like transcription factor underlying the trait. Surprisingly, this gene was found to also control senescence of wheat and, indirectly, remobilization of minerals to seeds. Wild-type alleles of the gene cause early senescence of wild wheat plants, which increases the remobilization of proteins and nutrients, such as Fe and Zn from the leaves to the grain. In domesticated durum wheat and modern hexaploid bread wheat varieties, a mutant allele of this gene is present, rendering the transcription factor nonfunctional. This effect delays senescence by a few days and thus reduces the grain protein, Zn, and Fe content by 10–15%. Studies in Arabidopsis showed that remobilization of minerals from leaves may not be a very significant contribution to seed loading. The major part of seed minerals might originate from other than storage sources, indicating that maintaining adequate mineral uptake throughout the development of the plant will be needed to ensure sufficient seed mineral loading (Waters and Grusak 2008). This observation also indicates that focus on mineral homeostasis genes alone may not cover the full spectrum of genetic variation contributing to seed mineral content, and that general growth and development of plants are likely to have important pleiotropic effects on leaf and seed mineral contents. Thus, breeding for improved yield may have unexpected and unintended negative effects on leaf and seed mineral contents, and vice versa, breeding for improved mineral contents can have unintended negative pleiotropic effects on other favored traits. These possibilities should be taken well into account when designing a breeding strategy for improving leaf or seed mineral contents.

14.7 STRATEGIES FOR IMPROVED IRON OR ZINC BIOFORTIFICATION THROUGH GENETIC MODIFICATION

In the past, several attempts have been made to improve the Fe or Zn biofortification potential of plants by introducing genes into crops by genetic modification. One of the first successful experiments was the expression of the soybean ferritin gene in rice endosperm (Goto et al. 1999). This action increased the Fe content of rice seeds by about threefold, which is a respectable achievement. However, at this high Fe-loading capacity, the Fe uptake capacity appears to be limiting, as the rest of the plant gradually becomes depleted of Fe (Qu et al. 2005). Overexpression of genes controlling Fe deficiency response, such as the *OsIRO2* and/or *OsIDEF1* genes from rice (Kobayashi et al. 2007; Ogo et al. 2007), will induce overexpression of Fe uptake genes. Even individual overexpression of such genes, for example, the iron-deficiency-responsive promoter-driven yeast Fe(III)-chelate reductase gene, may be effective in enhancing the ability to withstand Fe deficiency (Ishimaru et al. 2007), but this does not necessarily lead to increased Fe uptake under Fe-sufficient conditions.

Iron homeostasis in a plant is very tightly controlled. Expression of proteins such as IRT1 and FRO2 in Arabidopsis is controlled at the transcript level and also at the post-translational level in such a way that these proteins are only present under Fe-deficient conditions (Connolly et al. 2003). Post-translational control of IRT-like proteins has also been reported for yeast (Gitan and Eide 2000), and this may reflect a general and conserved mechanism in eukaryotes to tightly control Fe homeostasis through uptake control. Whether such regulation is present in all plants has not been shown yet, but it seems likely. Unless this method of regulation is understood better and ways are found to modify it, it will be difficult to increase the total plant content of Fe or Zn.

A possible way of circumventing this tight control may be by creating strong Fe sinks in seeds, leaves, or storage organs in such a way that the plant actually experiences Fe deficiency in the root, automatically prompting the endogenous Fe uptake system to become active. This system appears to be one of the possible mechanisms that are considered to explain the exceptionally high Zn accumulation found in Zn hyperaccumulators (Assunção et al. 2003). A remarkable similarity between *A. halleri* and *T. caerulescens* is the strong expression of Zn uptake and translocation genes, even at high Zn supply, as if the plant either does not sense the high Zn accumulation in the leaves, or indeed suffers from Zn deficiency in the roots because of the strong accumulation elsewhere (van de Mortel et al. 2006; Talke et al. 2006). It is not known yet if Zn uptake transporters also experience post-translational control of expression, but if so, this seems less tightly controlled than for Fe uptake. Ectopic constitutive expression of the *T. caerulescens ZNT1* Zn uptake transporter gene in Arabidopsis increased the leaf Zn concentration two- to threefold, irrespective of the Zn supply (Talukdar, Aarts et al., unpublished data).

In metal hyperaccumulators, most of the metals accumulate in the leaves, thus creating a Zn sink. Creating such a sink through biotechnology could potentially be achieved by leaf-specific overexpression of vacuolar Zn uptake transporters. This approach was tried by van der Zaal et al. (1999) and Haydon and Cobbett (2007), who used CaMV 35S-promoter-mediated overexpression of the Arabidopsis *MTP1* (*ZAT*) gene and the *ZIF1* gene, respectively, both known vacuolar Zn uptake genes. Although this effort increased the tolerance of transgenic Arabidopsis to high Zn, it did not noticeably increase Zn accumulation in leaves under normal Zn supply. One of the reasons could be the use of the CaMV 35S promoter, which is not specific for leaves but also increases expression in roots, thus creating a sink in the roots rather than the leaves. Specific promoters have not been tried much, except for the iron-deficiency-induced promoter driving the expression of the *IDEF1* gene in rice (Kobayashi et al. 2007), and the endosperm-specific glutelin promoter used to drive the ferritin expression in rice (Goto et al. 1999).

The preferred sites for Fe or Zn accumulation are not the leaves, unless for vegetable species, but rather endosperm or embryo tissue in seeds or storage tissues in roots or tubers (e.g., cassava, sweet potato, or potato). It seems even more difficult to increase Fe and Zn accumulation in these sites through genetic modification as a preferred Fe or Zn transport route because the plant needs to be designed requiring highly specific tissue and developmental-stage-specific expression of Fe or Zn export and import transporters. This most likely will also involve increasing expression of nicotianamine synthase genes and the specific YSL transporters that transport Fe- or Zn-NA chelates needed for phloem-mediated loading of storage organs such as seeds and tuberous roots. Although most of the components needed to create such a route are identified, it will still be a difficult challenge to successfully accomplish this.

In addition to increasing the plant organ Fe or Zn concentration, improving bioavailability of the plant Fe or Zn for humans is still a major concern. The dominant mineral antinutrient in seeds is phytate. Although natural or induced phytate biosynthesis mutants have been identified for barley, maize, and rice (Raboy 2002), the mutations often have pleiotropic effects on the general performance of the plant. Recently, Shi et al. (2007) described a way to overcome this general drawback, by specifically silencing the *LPA1* gene of maize, encoding for a phytate transporter, in maize embryos. The resulting transgenic plants showed none of the pleiotropic effects known for *lpa1* mutants. Silencing of the homologous gene in soybean had the same effect, which means this approach may be very effective also for other cereals and legumes (Shi et al. 2007).

14.8 CONCLUSIONS

In this chapter, approaches were described that can be used for identification and application of genes for Fe and Zn biofortification in crops. However, it is apparent that many aspects of metal accumulation in plants are yet to be understood. An important process to be studied in more detail is the phloem-mediated loading of Fe and Zn into seeds. Biofortified seeds are the most efficient plant source to enhance the nutrient content in the human diet, simply because seeds are a substantial component of the human diet. In addition to understanding the seed Fe and Zn loading, it is necessary to determine how to improve the bioavailability of seed-stored Fe and Zn by decreasing the concentration of the antinutrient phytate without seriously impairing the vigor of next generation seedlings. Using leaves instead of seeds as a biofortified food source is perhaps less effective, because vegetables do not always form a substantial part of the human diet. However, it will be worthwhile to obtain more information on the presence of antinutrients in leaves because it seems easier to improve leaf Zn content compared to seed Zn content, as it does not require phloem transport. Some plants, but not all, are rich in polyphenols that have been shown to limit bioavailability of Fe (Hurrell et al. 1999). Perhaps leafy vegetables with lower polyphenol content can be used for Zn biofortification or the polyphenol biosynthesis can be altered.

Another important process that is still puzzling is the signal transduction of Fe or Zn deficiency in shoots, leading to increased uptake by roots. There are only few reports describing the identification of transcription factors that regulate the expression of metal transporters and metal-binding proteins, as outlined in this chapter. Although it is not completely unthinkable, increasing the Fe or Zn content in the whole plant or even in specific organs in a commercially viable crop is unlikely to be achieved by introduction of single genes, but only by specific alteration of a set of genes. Thus, it will be extremely important to identify all of the transcription factors through which downstream sets of Fe or Zn homeostasis genes can be controlled.

REFERENCES

Alonso-Blanco, C. and M. Koornneef. 2000. Naturally occurring variation in Arabidopsis: An underexploited resource for plant genetics. *Trends in Plant Science* 5: 22–29.

Arrivault, S., T. Senger, and U. Krämer. 2006. The Arabidopsis metal tolerance protein AtMTP3 maintains metal homeostasis by mediating Zn exclusion from the shoot under Fe deficiency and Zn oversupply. *The Plant Journal* 46: 861–79.

Assunção, A. G. L., B. Pieper, J. Vromans, P. Lindhout, M. G. M. Aarts, and H. Schat. 2006. Construction of a genetic linkage map of *Thlaspi caerulescens* and quantitative trait loci analysis of zinc accumulation. *New Phytologist* 170: 21–32.

Assunção, A. G. L., H. Schat, and M. G. M. Aarts. 2003. *Thlaspi caerulescens*, an attractive model species to study heavy metal hyperaccumulation in plants. *New Phytologist* 159: 351–60.

Baker, A. J. M. and R. R. Brooks. 1989. Terestrial higher plants which hyperaccumulate metalic elements—a review of their distribution, ecology and phytochemistry. *Biorecovery* 1: 81–126.

Bauer, P., H.-Q. Ling, and M. L. Guerinot. 2007. FIT, the FER-like iron deficiency induced transcription factor in Arabidopsis. *Plant Physiology and Biochemistry* 45: 260–61.

Baxter, I., M. Ouzzani, S. Orcun, B. Kennedy, S.S. Jandhyala, and D.E. Salt. 2007. Purdue ionomics information management system: An integrated functional genomics platform. *Plant Physiology* 143: 600–11.

Becher, M., I. N. Talke, L. Krall, and U. Krämer. 2004. Cross-species microarray transcript profiling reveals high constitutive expression of metal homeostasis genes in shoots of the zinc hyperaccumulator *Arabidopsis halleri*. *The Plant Journal* 37: 251–68.

Bennetzen, J. L. and J. Ma. 2003. The genetic colinearity of rice and other cereals on the basis of genomic sequence analysis. *Current Opinion in Plant Biology* 6: 128–33.

Bentsink, L., K. Yuan, M. Koornneef, and D. Vreugdenhil. 2003. The genetics of phytate and phosphate accumulation in seeds and leaves of *Arabidopsis thaliana*, using natural variation. *Theoretical and Applied Genetics* 106: 1234–43.

Bouis, H. E. 2002. Plant breeding: A new tool for fighting micronutrient malnutrition. *Journal of Nutrition* 132: 491S–94.

Briat, J. F. 2007. Iron dynamics in plants. *Advances in Botanical Research* 46: 137–180.

Briat, J. F., C. Curie, and F. Gaymard. 2007. Iron utilization and metabolism in plants. *Current Opinion in Plant Biology* 10: 276–82.

Broadley, M. R., P. J. White, J. P. Hammond, I. Zelko, and A. Lux. 2007. Zinc in plants. *New Phytologist* 173: 677–702.

Busch, W. and J. U. Lohmann. 2007. Profiling a plant: Expression analysis in Arabidopsis. *Current Opinion in Plant Biology* 10: 136–41.

Colangelo, E. P. and M. L. Guerinot. 2004. The essential basic helix-loop-helix protein FIT1 is required for the iron deficiency response. *Plant Cell* 16: 3400–12.

Coleman, C. M., R. S. Boyd, and M. D. Eubanks. 2005. Extending the elemental defense hypothesis: Dietary metal concentrations below hyperaccumulator levels could harm herbivores. *Journal of Chemical Ecology* 31: 1669–81.

Connolly, E. L., N. H. Campbell, N. Grotz, C. L. Prichard, and M. L. Guerinot. 2003. Overexpression of the FRO2 ferric chelate reductase confers tolerance to growth on low iron and uncovers posttranscriptional control. *Plant Physiology* 133: 1102–10.

Curie, C., Z. Panaviene, C. Loulergue, S. L. Dellaporta, J. F. Briat, and E. L. Walker. 2001. Maize *yellow stripe1* encodes a membrane protein directly involved in Fe(III) uptake. *Nature* 409: 346–49.

Deniau, A., B. Pieper, W. Ten Bookum, P. Lindhout, M. Aarts, and H. Schat. 2006. QTL analysis of cadmium and zinc accumulation in the heavy metal hyperaccumulator *Thlaspi caerulescens*. *Theoretical and Applied Genetics* 113: 907–20.

Desbrosses-Fonrouge, A.-G., K. Voigt, A. Schroder, S. Arrivault, S. Thomine, and U. Krämer. 2005. *Arabidopsis thaliana* MTP1 is a Zn transporter in the vacuolar membrane which mediates Zn detoxification and drives leaf Zn accumulation. *FEBS Letters* 579: 4165–74.

DiDonato, R. J., L. A. Roberts, T. Sanderson, R. B. Eisley, and E. L. Walker. 2004. Arabidopsis *Yellow Stripe-Like2 (YSL2):* A metal-regulated gene encoding a plasma membrane transporter of nicotianamine-metal complexes. *The Plant Journal* 39: 403–14.

Durrett, T. P., W. Gassmann, and E. E. Rogers. 2007. The FRD3-mediated efflux of citrate into the root vasculature is necessary for efficient iron translocation. *Plant Physiology* 144: 197–205.

Duvick, J., A. Fu, U. Muppirala et al. 2008. PlantGDB: A resource for comparative plant genomics. *Nucleic Acids Research* 36: D959–65.

Duy, D., G. Wanner, A. R. Meda, N. von Wiren, J. Soll, and K. Philippar. 2007. PIC1, an ancient permease in Arabidopsis chloroplasts, mediates iron transport. *Plant Cell* 19: 986–1006.

Eide, D. J. 1998. The molecular biology of metal ion transport in *Saccharomyces cerevisiae. Annual Review of Nutrition* 18: 441–69.

El-Din El-Assal, S., C. Alonso-Blanco, A. J. M. Peeters, V. Raz, and M. Koornneef. 2001. A QTL for flowering time in Arabidopsis reveals a novel allele of *Cry2. Nature Genetics* 29: 435–40.

Filatov, V., J. Dowdle, N. Smirnoff, B. Ford-Lloyd, H. J. Newbury, and M. R. Macnair. 2007. A quantitative trait loci analysis of zinc hyperaccumulation in *Arabidopsis halleri. New Phytologist* 174: 580–90.

Filatov, V., J. Dowdle, N. Smirnoff, B. Ford-Lloyd, H. J. Newbury, and M. R. Macnair. 2006. Comparison of gene expression in segregating families identifies genes and genomic regions involved in a novel adaptation, zinc hyperaccumulation. *Molecular Ecology* 15: 3045–59.

Gitan, R. S. and D. J. Eide. 2000. Zinc-regulated ubiquitin conjugation signals endocytosis of the yeast ZRT1 zinc transporter. *Biochemical Journal* 346: 329–36.

Goto, F., T. Yoshihara, N. Shigemoto, S. Toki, and F. Takaiwa. 1999. Iron fortification of rice seed by the soybean *Ferritin* gene. *Nature Biotechnology* 17: 282–86.

Grennan, A. K. 2006. Genevestigator. Facilitating web-based gene-expression analysis. *Plant Physiology* 141: 1164–66.

Grotz, N., T. Fox, E. Connolly, W. Park, M. L. Guerinot, and D. Eide. 1998. Identification of a family of zinc transporter genes from arabidopsis that respond to zinc deficiency. *Proceedings of the National Academy of Sciences of the United States of America* 95: 7220–24.

Guerinot, M. L. and Y. Yi. 1994. Iron: Nutritious, noxious, and not readily available. *Plant Physiology* 104: 815–20.

Guerinot, M. L. 2000. The ZIP family of metal transporters. *Biochimica et Biophysica Acta (BBA)—Biomembranes* 1465: 190–98.

Guzmán-Maldonado, S. H., O. Martinez, J. A. Acosta-Gallegos, F. Guevara-Lara, and O. Paredes-Lopez. 2003. Putative quantitative trait loci for physical and chemical components of common bean. *Crop Science* 43: 1029–35.

Hall, J. L. and L. E. Williams. 2003. Transition metal transporters in plants. *Journal of Experimental Botany* 54: 2601–13.

Hammond, J. P., H. C. Bowen, P. J. White et al. 2006. A comparison of the *Thlaspi caerulescens* and *Thlaspi arvense* shoot transcriptomes. *New Phytologist* 170: 239–60.

Haydon, M. J. and C. S. Cobbett. 2007. Transporters of ligands for essential metal ions in plants. *New Phytologist* 174: 499–506.

Higuchi, K., K. Suzuki, H. Nakanishi, H. Yamaguchi, N.-K. Nishizawa, and S. Mori. 1999. Cloning of nicotianamine synthase genes, novel genes involved in the biosynthesis of phytosiderophores. *Plant Physiology* 119: 471–80.

Hurrell, R. F., M. Reddy, and J. D. Cook. 1999. Inhibition of non-haem iron absorption in man by polyphenolic-containing beverages. *British Journal of Nutrition* 81: 289–95.

Hussain, D., M. J. Haydon, Y. Wang et al. 2004. P-type ATPase heavy metal transporters with roles in essential zinc homeostasis in Arabidopsis. *Plant Cell* 16: 1327–39.

Inoue, H., D. Mizuno, M. Takahashi, H. Nakanishi, S. Mori, and N. K. Nishizawa. 2004. A rice FRD3-like (*OsFRDL1*) gene is expressed in the cells involved in long-distance transport. *Soil Science and Plant Nutrition* 50: 1133–40.

Ishimaru, Y., S. Kim, T. Tsukamoto et al. 2007. Mutational reconstructed ferric chelate reductase confers enhanced tolerance in rice to iron deficiency in calcareous soil. *Proceedings of the National Academy of Sciences of the United States of America* 104: 7373–78.

Ishimaru, Y., M. Suzuki, T. Tsukamoto et al. 2006. Rice plants take up iron as an Fe^{3+}-phytosiderophore and as Fe^{2+}. *The Plant Journal* 45: 335–46.

Kim, S. A., T. Punshon, A. Lanzirotti et al. 2006. Localization of iron in Arabidopsis seed requires the vacuolar membrane transporter VIT1. *Science* 314: 1295–98.

Kim, S., V. Plagnol, T. T. Hu et al. 2007. Recombination and linkage disequilibrium in *Arabidopsis thaliana*. *Nature Genetics* 39: 1151–55.

Kobayashi, T., Y. Ogo, R. Nakanishi Itai et al. 2007. The transcription factor IDEF1 regulates the response to and tolerance of iron deficiency in plants. *Proceedings of the National Academy of Sciences of the United States of America* 104: 19150–55.

Koh, S., A. M. Wiles, J. S. Sharp, F. R. Naider, J. M. Becker, and G. Stacey. 2002. An oligopeptide transporter gene family in Arabidopsis. *Plant Physiology* 128: 21–29.

Koike, S., H. Inoue, D. Mizuno et al. 2004. OsYSL2 is a rice metal-nicotianamine transporter that is regulated by iron and expressed in the phloem. *The Plant Journal* 39: 415–24.

Koornneef, M., C. Alonso-Blanco, and D. Vreugdenhil. 2004. Naturally occurring genetic variation in *Arabidopsis thaliana*. *Annual Review of Plant Biology* 55: 141–72.

Lahner, B., J. Gong, M. Mahmoudian et al. 2003. Genomic scale profiling of nutrient and trace elements in *Arabidopsis thaliana*. *Nature Biotechnology* 21: 1215–21.

Lanquar, V., F. Lelièvre, S. Bolte et al. 2005. Mobilization of vacuolar iron by AtNRAMP3 and AtNRAMP4 is essential for seed germination on low iron. *The EMBO Journal* 24: 4041–51.

Lee, S., Y.-Y. Kim, Y. Lee, and G. An. 2007. Rice P1b-type heavy-metal ATPase, OsHMA9, is a metal efflux protein. *Plant Physiology* 145: 831–42.

Le Jean, M., A. Schikora, S. Mari, J.-F. Briat, and C. Curie. 2005. A loss-of-function mutation in *AtYSL1* reveals its role in iron and nicotianamine seed loading. *The Plant Journal* 44: 769–82.

Liang, C., P. Jaiswal, C. Hebbard et al. 2008. Gramene: A growing plant comparative genomics resource. *Nucleic Acids Research* 36: D947–53.

Ling, H.-Q., P. Bauer, Z. Bereczky, B. Keller, and M. Ganal. 2002. The tomato *FER* gene encoding a bHLH protein controls iron-uptake responses in roots. *Proceedings of the National Academy of Sciences of the United States of America* 99: 13938–43.

Ling, H.-Q., G. Koch, H. Baumlein, and M. W. Ganal. 1999. Map-based cloning of *chloronerva*, a gene involved in iron uptake of higher plants encoding nicotianamine synthase. *Proceedings of the National Academy of Sciences of the United States of America* 96: 7098–103.

MacDiarmid, C. W., L. A. Gaither, and D. Eide. 2000. Zinc transporters that regulate vacuolar zinc storage in *Saccharomyces cerevisiae*. *The EMBO Journal* 19: 2845–55.

Macnair, M. R. and J. A. Callow. 2003. The hyperaccumulation of metals by plants. In *Advances in Botanical Research*. London: Academic Press, pp. 63–105.

Marschner, H. 1995. Mineral nutrition of higher plants. 2nd ed. London: Academic Press.

Mäser, P., S. Thomine, J. I. Schroeder et al. 2001. Phylogenetic relationships within cation transporter families of Arabidopsis. *Plant Physiology* 126: 1646–67.

Mukherjee, I., N. Campbell, J. Ash, and E. Connolly. 2006. Expression profiling of the Arabidopsis Ferric Chelate Reductase (FRO) gene family reveals differential regulation by iron and copper. *Planta* 223: 1178–90.

Murata, Y., J. F. Ma, N. Yamaji, D. Ueno, K. Nomoto, and T. Iwashita. 2006. A specific transporter for iron(III)-phytosiderophore in barley roots. *The Plant Journal* 46: 563–72.

Negishi, T., H. Nakanishi, J. Yazaki et al. 2002. cDNA microarray analysis of gene expression during Fe-deficiency stress in barley suggests that polar transport of vesicles is implicated in phytosiderophore secretion in Fe-deficient barley roots. *The Plant Journal* 30: 83–94.

Nozoye, T., H. Inoue, M. Takahashi et al. 2007. The expression of iron homeostasis-related genes during rice germination. *Plant Molecular Biology* 64: 35–47.

Ogo, Y., R. Nakanishi Itai, H. Nakanishi et al. 2007. The rice bHLH protein OsIRO2 is an essential regulator of the genes involved in Fe uptake under Fe-deficient conditions. *The Plant Journal* 51(3): 366–77.

Papoyan, A. and L. V. Kochian. 2004. Identification of *Thlaspi caerulescens* genes that may be involved in heavy metal hyperaccumulation and tolerance. Characterization of a novel heavy metal transporting ATPase. *Plant Physiology* 136: 3814–23.

Paran, I. and D. Zamir. 2003. Quantitative traits in plants: Beyond the QTL. *Trends in Genetics* 19: 303–06.

Petit, J. M., J. F. Briat, and S. Lobréaux. 2001. Structure and differential expression of the four members of the *Arabidopsis thaliana* ferritin gene family. *Biochemical Journal* 359: 575–82.

Qu, L., T. Yoshihara, A. Ooyama, F. Goto, and F. Takaiwa. 2005. Iron accumulation does not parallel the high expression level of ferritin in transgenic rice seeds. *Planta* 222: 225–33.

Raboy, V. 2002. Progress in breeding low phytate crops. *Journal of Nutrition* 132: 503S–05.

Robinson, N. J., C. M. Procter, E. L. Connolly, and M. L. Guerinot. 1999. A ferric-chelate reductase for iron uptake from soils. *Nature* 397: 694–97.

Rogers, E. E. and M. L. Guerinot. 2002. FRD3, a member of the multidrug and toxin efflux family, controls iron deficiency responses in Arabidopsis. *Plant Cell* 14: 1787–99.

Sandberg, A.S. 2002. Bioavailability of minerals in legumes. *British Journal of Nutrition* 88: S281–S285.

Schaaf, G., A. Honsbein, A. R. Meda, S. Kirchner, D. Wipf, and N. von Wiren. 2006. *AtIREG2* encodes a tonoplast transport protein involved in iron-dependent nickel detoxification in *Arabidopsis thaliana* roots. *Journal of Biological Chemistry* 281: 25532–40.

Schaaf, G., A. Schikora, J. Haberle et al. 2005. A putative function for the Arabidopsis Fe-phytosiderophore transporter homolog AtYSL2 in Fe and Zn homeostasis. *Plant and Cell Physiology* 46: 762–74.

Schmidt, W. 2003. Iron solutions: acquisition strategies and signaling pathways in plants. *Trends in Plant Science* 8: 188–93.

Shi, J., H. Wang, K. Schellin et al. 2007. Embryo-specific silencing of a transporter reduces phytic acid content of maize and soybean seeds. *Nature Biotechnology* 25: 930–37.

Somerville, C. and S. Somerville. 1999. Plant functional genomics. *Science* 285: 380–83.

Stacey, M. G., A. Patel, W. E. McClain et al. 2008. The Arabidopsis AtOPT3 protein functions in metal homeostasis and movement of iron to developing seeds. *Plant Physiology* 146: 589–601.

Stephan, U. W., I. Schmidke, V. W. Stephan, and G. Scholz. 1996. The nicotianamine molecule is made-to-measure for complexation of metal micronutrients in plants. *BioMetals* 9: 84–90.

Supek, F., L. Supekova, H. Nelson, and N. Nelson. 1996. A yeast manganese transporter related to the macrophage protein involved in conferring resistance to mycobacteria. *Proceedings of the National Academy of Sciences of the United States of America* 93: 5105–10.

Takahashi, M., Y. Terada, I. Nakai et al. 2003. Role of nicotianamine in the intracellular delivery of metals and plant reproductive development. *Plant Cell* 15: 1263–80.

Talke, I. N., M. Hanikenne, and U. Krämer. 2006. Zinc-dependent global transcriptional control, transcriptional deregulation, and higher gene copy number for genes in metal homeostasis of the hyperaccumulator *Arabidopsis halleri*. *Plant Physiology* 142: 148–67.

Tanksley, S. D., M. W. Ganal, and G. B. Martin. 1995. Chromosome landing: A paradigm for map-based gene cloning in plants with large genomes. *Trends in Genetics* 11: 63–68.

Thomine, S., F. Lelievre, E. Debarbieux, J. I. Schroeder, and H. Barbier-Brygoo. 2003. AtNRAMP3, a multispecific vacuolar metal transporter involved in plant responses to iron deficiency. *The Plant Journal* 34: 685–95.

Tuntawiroon, M., N. Sritongkul, L. Rossander-Hultén et al. 1990. Rice and iron absorption in man. *European Journal of Clinical Nutrition* 44: 489–97.

Uauy, C., A. Distelfeld, T. Fahima, A. Blechl, and J. Dubcovsky. 2006. A *NAC* gene regulating senescence improves grain protein, zinc, and iron content in wheat. *Science* 314: 1298–301.

van de Mortel, J. E. and M.G. M. Aarts. 2006. Comparative transcriptomics—Model species lead the way. *New Phytologist* 170: 199–201.

van de Mortel, J. E., L. Almar Villanueva, H. Schat et al. 2006. Large expression differences in genes for iron and zinc homeostasis, stress response, and lignin biosynthesis distinguish roots of *Arabidopsis thaliana* and the related metal hyperaccumulator *Thlaspi caerulescens*. *Plant Physiology* 142: 1127–47.

van der Zaal, B. J., L. W. Neuteboom, J. E. Pinas et al. 1999. Overexpression of a novel Arabidopsis gene related to putative zinc-transporter genes from animals can lead to enhanced zinc resistance and accumulation. *Plant Physiology* 119: 1047–56.

Vreugdenhil, D., M. G. M. Aarts, M. Koornneef, H. Nelissen, and W. H. O. Ernst. 2004. Natural variation and QTL analysis for cationic mineral content in seeds of *Arabidopsis thaliana*. *Plant, Cell and Environment* 27: 828–39.

Wang, H.-Y., M. Klatte, M. Jakoby, H. Bäumlein, B. Weisshaar, and P. Bauer. 2007. Iron deficiency-mediated stress regulation of four subgroup Ib bHLH genes in *Arabidopsis thaliana*. *Planta* 226: 897–908.

Waters, B. M., H.-H. Chu, R. J. DiDonato et al. 2006. Mutations in Arabidopsis *yellow stripe-like1* and *yellow stripe-like3* reveal their roles in metal ion homeostasis and loading of metal ions in seeds. *Plant Physiology* 141: 1446–58.

Waters, B. M. and M. A. Grusak. 2008. Whole-plant mineral partitioning throughout the life cycle in *Arabidopsis thaliana* ecotypes Columbia, Landsberg *erecta*, Cape Verde Islands, and the mutant line *ysl1ysl3*. *New Phytologist* 177: 389–405.

Weber, M., E. Harada, C. Vess, E. von Roepenack-Lahaye, and S. Clemens. 2004. Comparative microarray analysis of *Arabidopsis thaliana* and *Arabidopsis halleri* roots identifies nicotianamine synthase, a ZIP transporter and other genes as potential metal hyperaccumulation factors. *The Plant Journal* 37: 269–81.

Welsh, R. M. and R. D. Graham. 1999. A new paradigm for world agriculture: meeting human needs: productive, sustainable, nutritious. *Field Crops Research* 60: 1–10.

Wissuwa, M., A. M. Ismail, and S. Yanagihara. 2006. Effects of zinc deficiency on rice growth and genetic factors contributing to tolerance. *Plant Physiology* 142: 731–41.

Wu, H., L. Li, J. Du, Y. Yuan, X. Cheng, and H.-Q. Ling. 2005. Molecular and biochemical characterization of the Fe(III) chelate reductase gene family in *Arabidopsis thaliana*. *Plant and Cell Physiology* 46: 1505–14.

Wu, J., Y. Yuan, X. Zhang, et al. 2008. Mapping QTLS for mineral accumulation and shoot dry biomass under different Zn nutritional conditions in Chinese cabbage (*Brassica rapa* L. ssp. *pekinensis*). *Plant and Soil* (in press).

Yi, Y. and M. L. Guerinot. 1996. Genetic evidence that induction of root Fe(III) chelate reductase activity is necessary for iron uptake under iron deficiency. *The Plant Journal* 10: 835–44.

Yu, J. and E. S. Buckler. 2006. Genetic association mapping and genome organization of maize. *Current Opinion in Biotechnology* 17: 155–60.

Zhao, H. and D. Eide. 1996a. The *Zrt2* Gene encodes the low affinity zinc transporter in *Saccharomyces cerevisiae*. *Journal of Biological Chemistry* 271: 23203–10.

Zhao, H. and D. Eide. 1996b. The Yeast *Zrt1* Gene encodes the zinc transporter protein of a high-affinity uptake system induced by zinc limitation. *Proceedings of the National Academy of Sciences of the United States of America* 93: 2454–58.

15 Uptake of Trace Elements by Soybean in Typic Quartzipsamment Soil Treated with Different Phosphorous Fertilizers in Brazil

Francisco Nildo da Silva,
Antonio Eduardo Furtini Neto,
and Andrew C. Chang

CONTENTS

15.1 INTRODUCTION

The development of high-yielding soybean (*Glycine max* L.) has led to a substantial increase in the area under cultivation in many parts of Brazil, mainly in the Cerrado (savanna); nevertheless, one of the major constraints for crop production in those areas is that Cerrado soils are quite deficient in phosphorus (P). Studies have confirmed that 65.1% of all tropical soils are strongly deficient in P, whereas 26.6% are moderately deficient (Roche et al. 1980). Consequently, the application of phosphate fertilizers is an essential and necessary agricultural fertilization practice to achieve higher productivities, especially with the growth of those intensively cultivated crops in Cerrado soils (Novais and Smyth 1999).

The existence of several large phosphate mines in the region makes the use of crude phosphate an attractive option. Other phosphate fertilizers produced from Morocco, Israel, and the United States are also widely used in Brazil owing to their high P solubility. The phosphate rocks used for fertilizer production usually contain heavy metals and have become one of the most important sources of metal contamination in agricultural soils (Mortvedt 1987). Huton and Symon (1986) indicated that phosphate rocks in the United Kingdom have an average of 7.7 mg arsenic (As) per kg. In the European Common Market about 300 tons of cadmium (Cd) are added to agricultural soils each year (Huton and Symon 1986), whereas phosphate rocks from the western United States contributed nearly 100 g Cd ha^{-1} y^{-1} to the soil (Mulla et al. 1980). In Brazil, the Cd and chromium (Cr) contents in phosphate rocks are 2–7 mg kg^{-1} and 4–26 mg kg^{-1}, respectively, which are relatively lower than the concentrations in other countries, such as in the United States (i.e., 5–200 mg Cd kg^{-1} and 7–500 mg Cr kg^{-1}). However, the lead (Pb) concentrations in Brazilian phosphate rocks are relatively high (28–36 mg kg^{-1}).

Heavy metal contamination in soil influences the uptake and accumulation of nutritionally important elements in food or crop tissues, and thus affects the nutritional quality of food products, with potential implications for human health (Senesi et al. 1999). The addition of P to soils as commercial P fertilizers or ground phosphate rocks has been shown to result in increased concentrations of metals in surface soils (Charter et al. 1993; Kpomblekov and Tabatabai 1994; Camelo et al. 1997). There are no federal laws in Brazil that regulate contaminants in fertilizers, except for some state regulations such as those in the state of São Paulo. Those regulations deal with As, Cd, and Pb concentrations in Brazilian fertilizers because of public health concerns.

Phosphorus fertilizers contain a broad range of Cd, ranging from <10 to > 100 mg Cd kg^{-1}, depending on the source/type of phosphate rock. Continuous input of P fertilizers at high application rates resulted in more Cd accumulated in agricultural soils than in Cd removed by harvested plants (i.e., phytoextraction). According to FAO, the amount of phosphatic rock imported by Brazil has increased tenfold over the past 10 years. A similar situation prevails in some South American countries, such as Argentina. According to Camelo et al. (1997), annual fertilizer usage in Argentina has been increased from 500,000 tons in 1992 to approximately 1.6 million tons in 1996, which was primarily used for wheat and corn cultivation compared with the soybean and corn production in Brazil. A large proportion of the fertilizers used in Argentina is from imported countries, and at present, heavy metal content in the fertilizers has not been monitored. Furthermore, according to McLaughlin et al. (1996), the annual Cd input from fertilizers and phosphate rock is approximately 40 tons in Australian soils; the total output of Cd accumulated in agricultural products (such as wool, cotton, milk, or dairy products) is approximately 0.5 ton per year. Clearly, Cd levels are slowly accumulating in agricultural soils.

Indeed, the long-term use of phosphate fertilizers is an important anthropogenic source of heavy metal contamination in soil (Mortvedet 1987; Nriagu 1990; Singh 1990). The amounts of heavy metals added to the soil from P fertilizers depend on the origin of crude phosphates (Soler and Soler 1996). The accumulation of heavy metals, such as Cd, in the soil and associated environmental risks are important matters of concern in Brazil (Compos et al. 2005). Importantly, the toxic effects of Cd to humans, as observed in Japan, suggest that soil Cd contamination from P fertilizer uses could be an important matter of concern to environmental management and food production (Friberg and Vather 1983). Thus, the aims of this study were (1) to measure the contents of Cd, As, and Pb in different phosphate fertilizers used in Brazil and (2) to determine the uptake of Cd, As, and Pb by soybean cultivated in Brazilian soils supplied with different phosphate fertilizers.

15.2 MATERIALS AND METHODS

15.2.1 Greenhouse Study

The experimental soil was classified as Typic Quartzipsamment under the USDA soil taxonomy. Typic Quartzipsamment soils are very permeable and have low carbon content, poor water retention, and low levels of fertility and available P. Therefore, we amended the soil with phosphorous fertilizers to show the uptake of the trace elements by soybean under different application rates and sources of phosphorous fertilizers. The experimental soil had the following chemical characteristics: pH, 5.0; organic matter, 8 g dm^{-3}; CEC, 34 mmolc dm^{-3}; and trace element concentrations of Cd, 0.02; As, 0.36; and Pb, 3.05 mg kg^{-1}. The soils were collected at a depth of 0–20 cm and amended with two types of phosphorous fertilizers and three types of phosphate rocks at four doses (0, 100, 300, and 600 mg kg^{-1} total P).

A completely randomized experimental design with four replicates was conducted in a 5×4 factorial arrangement, consisting of five sources of P [i.e., triple superphosphate (ST), thermophosphate (TM), two imported reactive phosphates (FR1 and FR2), and one domestic natural phosphate (FN1)] and four application rates of total P (i.e., 0, 100, 300, and 600 mg kg^{-1}). The quantity of P applied from each source was calculated on the basis of total P_2O_5 content in each fertilizer (Table 15.1). The P fertilizers were applied to the soil before the soybean seeds were sown.

Dolomitic lime was used to neutralize the soil acidity at an amount sufficient to increase base saturation to 60%. All experimental soils received the same fertilization as follows: 50 mg N kg^{-1}, 60 K mg kg^{-1} (potassium nitrate), 0.5 B mg kg^{-1} (boric acid), 1.5 mg Cu kg^{-1} (copper sulfate), 5 mg Zn kg^{-1} (zinc sulfate), 5 mg Fe kg^{-1} (iron sulfate), and 3.6 mg Mn kg^{-1} (manganese sulfate). Each pot was filled with 4 kg

TABLE 15.1
Concentrations of As, Cd, Pb, and P (as P_2O_5) in Different Phosphorous Fertilizers

	Metal Concentration (mg kg^{-1})			P Concentration (%)	
Fertilizer Type	As	Cd	Pb	Total P	Citric Acid P
Super triple	— [a]	22.33	5.51	46.1	40.0
Thermo	77.42	0.29	64.29	18.1	17.6
Phosphates reactive imported 1	17.68	18.89	36.50	29.0	9.8
Phosphates reactive imported 2	—	66.80	28.66	32.0	9.0
National natural phosphates	148.64	1.97	233.83	22.7	4.3

[a] below the detection limit.

amended soils, wetted with distilled water to 40% of the water-holding capacity of each treated soil, and incubated for 1 week.

Six soybean seeds were sown at 0.5 cm depth. The seedlings were thinned to three plants per pot. One plant was harvested at 60 days after germination to measure leaf concentrations of As, Cd, and Pb. The two other plants were harvested at the end of the soybean-growing season. The beans were analyzed for concentrations of As, Cd, and Pb.

15.2.2 Chemical Analysis of Heavy Metals

15.2.2.1 Phosphate Fertilizers and Soil

Concentrations of As, Cd, and Pb were determined by following the USEPA 3051 method. According to the method, a 0.5 g sample of fertilizers or soil received 9.0 mL HNO_3, 3.0 mL HF, 2.0 mL HCl, 2.0 mL H_2O_2, and 4.0 mL deionized water. The mixture was microwaved at 1200 W for a total of 15 min. The samples were heated at 180 ± 5°C for 10 min. The chemical analysis was carried out at University of California at Riverside. Concentrations of Cd, As, and Pb in the digestion solution were determined using a PerkinElmer Atomic Absorption Spectrometer with a graphite furnace atomizer (GFAAS).

Standards were prepared to calibrate the instrument by diluting stock solutions of each element in 1% Optima nitric acid. A 50 μg L^{-1} standard solution was used for the As and Pb analyses, and a 5 μg l^{-1} standard solution was used for Cd analysis. The autosampler was capable of making the serial dilutions required to establish multiple points for the calibration curve. The 1% Optima nitric acid solution was analyzed to provide a calibration blank. The analysis of each element also required the preparation of matrix modifiers to be added to each sample before analysis. The As analysis required the addition of 5 μL 0.1% Pd solution and 5 μL of 0.06% $Mg(NO_3)_2$ solution to a 20 μL aliquot of sample or standard. The lead and cadmium analyses required the addition of 5 μL 1% $NH_4H_2PO_4$ solution and 5 μL of 0.06% $Mg(NO_3)_2$ solution to a 20 μL aliquot of sample or standard.

15.2.2.2 Plant Tissues

Plant samples were acid-digested at University of California at Riverside using a CEM Mars 5 microwave system that is capable of monitoring both temperature and pressure of a representative sample during digestion. The acid digestion used HP-500 Plus vessels and PFA (perfluoro alkoxy ethylene) liners. According to the method, a 0.5 g plant sample received 4.0 mL HNO_3, 4.0 mL H_2O_2, and 2.0 mL deionized water. The mixture was microwaved for a total of 15 min (see the same procedure described earlier). The digestion solutions were transferred to volumetric flasks with a final volume of 25 mL, and then filtered. Concentrations of the three metals in the digested solutions were determined using AAS, following the same procedure for the chemical analysis of soil and fertilizer samples described earlier.

15.2.3 Quality Control and Data Analysis

Two standard reference materials [the San Joaquin Valley Soil (NIST 2709) and the peach leave (NIST 1547)] were used in this study to determine the accuracy of the chemical analyses. The metal concentrations determined in this study were comparable to the NIST-certified values, or within ± 20% of the certified value. To ensure the precision of the chemical analysis, each sample was digested and analyzed in duplicate. The data were analyzed by the ANOVA procedure at the 5% probability level by the Scott Knott multiple range test.

15.3 RESULTS AND DISCUSSION

15.3.1 Effects of Phosphorous Fertilizers on Biomass Production and Phosphorus accumulation in Soil and Plant

The soybean biomass production values are presented in Table 15.2 for the experiment. Total soybean dry matter varied significantly among the fertilizer treatments. For example, the thermophosphate treatment had the highest biomass production (40.25 g per pot), whereas the national natural phosphate treatment had the lowest (18.94 g per pot).

The P concentrations in soybean leaves increased as application rates of the fertilizer in the soil increased, and varied among the fertilizers. Phosphorus absorption was quite different depending on the P source that was applied. The amount of P accumulated in the beans increased almost sixfold with the triple superphosphate, compared to the reactive phosphate from Morocco.

This study showed a good response for some phosphate fertilizers under the experimental conditions described. Roche et al. (1980) reported that 65.1% of all tropical soils are strongly deficient in P, whereas 26.6% are moderately deficient. In the present study, the sharp growth response to the use of phosphate fertilizer indicates severe P deficiency in the Brazilian soil used in this experiment. In most cases, tropical soils are very poor or deficient in P for plant growth or agricultural productions.

TABLE 15.2

Biomass Production of Soybean Cultivated in the Typic Quartzipsamment Soil Treated with Different Phosphorous Fertilizers

Fertilizer Treatment	Shoot Height (cm)	Shoot Diameter (cm)	Biomass Production (g DW) Root	Bean	Total Biomass
Phosphate imported	36.45	0.59	9.17	7.65	25.98
Phosphate imported	37.09	0.59	9.07	8.37	25.87
National Phosphate	33.31	0.53	6.16	7.30	18.94
Super triple	43.47	0.63	8.52	10.50	33.11
Thermophosphate	45.26	0.65	10.51	13.53	40.25
CV%	10.36	7.90	23.82	24.45	19.45

15.3.2 Buildup of Heavy Metals in Soil/Plant

15.3.2.1 Arsenic, Cadmium, and Lead in Phosphate Fertilizers

The chemical analysis of the heavy metals in each P fertilizer is given in Table 15.1. The data show that different types of P fertilizers have various chemical compositions of As, Cd, and Pb. Concentrations of Cd are higher in reactive phosphate fertilizers than in natural phosphate. The Pb concentration was significantly higher in natural phosphate from Brazil (FN) than the other types of fertilizers. Also, the concentration of As was the highest in national natural phosphate from Brazil.

Raven and Loeppert (1997) determined the trace element composition in 24 fertilizers and soil amendments, and found that phosphate rocks have the highest concentrations of trace elements among the tested fertilizers. In the present study, high Pb concentrations were found on the natural phosphate rock from Brazil (FN), followed by the reactive phosphate fertilizer 2 (FR2). Therefore, applications of these P fertilizers or soil amendments would contribute a significant amount of heavy metals to the soil, and result in an increase of the heavy metal load in the treated soils with long-term application (Williams and David 1973; Mulla et al. 1980; McLaughlin et al. 1997; Raven and Loeppert 1997; Lavado et al. 2001; Campos et al. 2005). Because of the high contents of trace elements in the phosphate rock, the addition of these P fertilizers to soils may result in the buildup of trace elements at levels exceeding their natural abundances. Kpomblekou-A and Tabatabai (1994) measured 12 phosphate rocks for trace and macroelement composition, and showed that the trace element contents of the phosphate rocks analyzed varied greatly, with similarly high Cd and Pb concentrations found in reactive phosphate fertilizers (FR1 and FR2) in this study.

The findings from this study suggest that concentrations of As in the various P fertilizers were relatively low, and As as a contaminant would not likely cause any significant environmental and human health problems in Brazilian soils and plants. In this experiment we observed a high Fe contents in the soil, which allows As to form As–Fe complexes in the soil. On the other hand, several studies have noted that

P soil amendments can increase the availability of As to plants (Creger and Peryea 1994; Alam et al. 2001). The chemical similarities between arsenate and phosphate result in the competition between the two elements for binding sites in soils. Therefore, the use of phosphate fertilizers might potentially increase the solubility and mobility of As in the treated soil. For example, Peryea (1991) found that As released by P amendments tended to remain in a readily soluble form, thereby making it more available for plant uptake.

15.3.2.2 Cadmium, Arsenic, and Lead Accumulated in Soybean

Concentrations of As, Cd, and Pb in soybean tissues are presented in Table 15.3. There were statistically significant treatment effects on metal accumulation in leaves and beans ($P < 0.05$). The highest Cd concentration in soybean leaves was found in the soil treated with phosphates reactive imported 2 (FR2), whereas the lower Cd concentration was observed with the thermophosphate treatment (Table 15.3). With regard to the concentrations of Cd, As, and Pb in soybean beans, As in beans was below the ICP-AES detection limit due to low plant uptake and/or strong adsorption of As in the soil. The highest concentration of Cd accumulated in the beans was 0.23 mg kg^{-1}, when the soil was supplied with phosphates reactive imported 2 (FR2). The highest bean Pb concentration of 52 mg kg^{-1} was observed in the soil amended with super triple (ST) fertilizer (Table 15.3).

Loganathan et al. (1995) studied Cd distribution in hill pastures as influenced by 20 years of phosphate fertilizer applications and sheep grazing. They concluded that a strong relationship was obtained between soil Cd and P in the 0–30 mm layer of a high fertilization farm, and thus demonstrated the link between phosphate fertilizer applied and Cd accumulation. In a recent long-term field study at seven locations across the Canadian prairies, Grant et al. (2007) evaluated the impact of application of monoammonium phosphate fertilizer on Cd accumulation in durum wheat and flax. They found that Cd concentrations in the seeds of both crops increased with the

TABLE 15.3

Concentrations of As, Ca, and Pb in Soybean Cultivated in Typic Quartzipsamment Soils Treated with Different Phosphorous Fertilizers

Fertilizer Treatment	As (mg kg^{-1})		Cd (mg kg^{-1})		Pb (mg kg^{-1})	
	Leaf	Bean	Leaf	Bean	Leaf	Bean
Super triple	0.11	—[a]	0.47	0.12	0.22	0.52
Thermo	0.07	—	0.11	0.02	0.43	0.21
Phosphates reactive imported 1	0.02	—	0.25	0.12	0.54	0.10
Phosphates reactive imported 2	0.12	—	0.60	0.23	0.50	0.47
National natural phosphates	0.02	—	0.11	0.02	0.19	0.13

[a] below the detection limit.

application of P fertilizers, even when the fertilizer contained only trace concentrations of Cd. According to their study, the impact of the fertilizers on Cd concentration in plant tissues became more apparent with repeated applications over the time period. The authors suggested the possibility of reducing Cd concentrations in crops by reducing P application rates and Cd concentrations in P fertilizers.

In a different study, Osztoics et al. (2003) examined the uptake of Cd, Cr, and Sr by spring barley as affected by P fertilizers in a pot experiment, and found that Cd concentration in plant tissues was also relatively higher in a sandy soil when compared to Podzol and Chernozem soils. Shinonaga et al. (1999) experimentally determined the mobility of other trace elements in soybean plants, and they concluded that rubidium showed the highest mobility in soybean among those radionuclides, followed by selenium and cobalt, which were classified as "mobile," whereas Mn, As, Sr, and Re were classified in a group with "medium mobility." Thus, the soybean plant may take up unwanted trace elements.

In the present study, the highest bean Cd concentration was 0.23 mg kg^{-1}. Lavado et al. (2001) found a Cd concentration of 0.85 mg kg^{-1} in soybean in an experiment in which they determined nutrient and heavy metal concentrations and metal distributions in soybean, corn, and wheat as affected by different tillage systems in the Argentine pampas. Within this context, Mortvedt (1987) established that the Cd input from the use of phosphate fertilizers to the soil in the range from 0.3 to 1.2 g ha^{-1} year did not increase Cd concentrations in the plants. According to the same author, when the Cd content was lower than 10 mg kg^{-1} in the phosphate fertilizer, the Cd uptake by the plant was not significant. Singh (1990) found that the application of 40 mg Cd kg^{-1} in the N-P-K fertilizer applied at the rate of 30 mg P kg^{-1} did not increase the Cd concentration in oats. However, the same fertilization at the application rate of 90 mg P kg^{-1} increased the Cd concentration in the plant. These findings are important and demonstrate that the Brazilian government must think about regulating the amount of trace elements in phosphorus fertilizers, and determine the types of P fertilizers that can be used in Brazilian farmlands.

The accumulation of Pb in the soybean reached a high of 0.52 mg kg^{-1} when super triple was used as fertilizer, while only a concentration of 0.10 mg kg ^{1} was measured with the fertilizer phosphates reactive import 1 (Table 15.4). The accumulation of Pb in beans might be affected both by the Pb concentration in the fertilizers and by the chemical behavior of Pb in the treated soil. Lead is more easily adsorbed in the soil, has a stronger complexing capacity with organic matter, and precipitates with carbonates, phosphates, and sulfates in several states of oxidation (McBride and Spiers 2001).

15.3.2.3 Regression Models Describing Effects of the Fertilizers on Heavy Metals in Soybean

According to the analysis of variance, there was a significant interaction effect between the fertilizer treatments and the Cd and Pb contents in plant tissues (Table 15.4). This interaction probably occurred because each source had different contents of the metals. Using the bean Cd and Pb contents (mg kg^{-1}) as the dependent variable for the application rate and the type of fertilizer applied, regression equations were obtained

TABLE 15.4

Regression Equations Adjusted for the Concentrations of Cd and Pb in Beans of Soybean as the Dependent Variable of the Application Rate and the Type of Phosphorous Fertilizer Applied in the Typic Quartzipsamment Soil

Fertilizer Treatment	Regression Equation	R2	Significance
	Cd in Beans		
Phosphates reactive imported 1	$y = 0.032 + 0.00033x$	0.95	$P < 0.01$
Phosphates reactive imported 2	$y = 0.041 + 0.00077x$	0.99	$P < 0.01$
Super triple	$y = 0.032 + 0.00036x$	0.99	$P < 0.01$
	Pb in Beans		
National natural phosphates	$y = 0.204 + 0.00028x$	0.80	$P < 0.01$
Phosphates reactive imported 1	$y = 0.067 + 0.00013x$	0.97	$P < 0.01$
Super triple	$y = 0.405 + 0.047x^{0.5} - 0.00198x$	0.83	$P < 0.01$
Thermo	$y = 0.257 + 0.014x^{0.5} - 0.00091x$	0.90	$P < 0.01$

and fitted to the linear and quadratic root models (Table 15.4). For example, the estimated concentration of Pb in soybean beans was higher when triple superphosphate was applied to the soil; the Pb concentration of 0.65 mg kg^{-1} was obtained when the P application rate of 262 mg kg^{-1} was selected. Little information is available on the mechanisms of Pb uptake by plants. This may result from both the difficulty in measuring Pb at the low concentrations commonly found in the soil solution and the absence of detailed studies on transport processes within the soil and uptake processes from soil to roots. Some glasshouse experiments have shown the increased uptake of Pb when soil pH is decreased (McLaughlin et al. 1996).

Concentrations of Cd in the Typic Quartzipsamment soil and the plant tissues were affected by the different phosphate fertilizers supplied. Almost all regression equations had significant R^2 values, the only exception being the thermosphosphate (not shown in Table 15.4), for which a significant regression equation was not found. The Cd content in the bean tissues increased with increasing soil Cd input. The level of such increase was higher with triple superphosphate than with other types of fertilizers, in spite of the fact that the FR2 source had a higher Cd concentration.

Roberts et al. (1994) determined the Cd status of soils, plants, and grazing animals in New Zealand, and found that increased total soil P is a reflection of fertilizer use history. Therefore, the usage of phosphatic fertilizers is implicated in Cd accumulation in pasture soil. The overall average total Cd content in the top 7.5 cm layer of the soil was 0.20 μg Cd g^{-1} with a range of 0.02–0.77 μg g^{-1}, whereas the average total Cd content for pasture soil was 0.4 μg g^{-1} with a range of 0.04–1.53 μg g^{-1}.

The highly significant correlation ($r = 0.62$, $P < 0.001$, $n = 397$) between the total P and the total Cd across all study sites was obtained in Australia (Merry 1988), suggesting that Cd enrichment in pastoral soils would be partially related to P fertilizer applications. Other authors (Williams and David 1973; Rothbaum et al. 1986) have previously attributed the Cd enrichment to the use of superphosphate. According to Roberts et al. (1994), this observation is not unreasonable considering the long history of superphosphate fertilizer usage. Natural superphosphates typically contain 34–69 $\mu g\ g^{-1}$ (Rothbaum et al. 1986). The same situation can occur in Brazilian soils owing to the high concentrations of Cd and Pb in the FR2 phosphate and in the FN phosphate, respectively. Similar results were also obtained by Campos et al. (2005) for the reactive phosphate fertilizers that contained high levels of Cd (e.g., 145 ± 13 mg kg^{-1}) and for natural phosphate fertilizer that contained a high level of Pb (234 ± 9 mg kg^{-1}).

15.4 CONCLUSIONS

The application of P fertilizers may play an important role in influencing the quality of food crops and result in various environmental and health problems. Such negative impacts are highly dependent on the origin and chemical composition of the fertilizer. The use of P fertilizers will not only increase crop yields, but they may also contribute to the slow buildup of toxic heavy metals in edible plant tissues. This outcome is possible if long-term application of some fertilizers contain heavy metals, and if specific food crops accumulate heavy metals.

This study shows that the unintended contribution of heavy metals (such as Cd, As, and Pb) to agricultural soils from phosphate fertilizers, and the possibility of toxic metal accumulation in soybean tissues. The metal contamination in soils from application of P fertilizers might be significant, but this will always be element- and fertilizer-specific. The highest Pb concentration in beans (0.52 mg kg^{-1}) was found with the triple superphosphate treatment, whereas the highest bean Cd concentration (0.23 mg kg^{-1}) was observed with the application of phosphates reactive imported 2. Because only trace amounts of As were found in both P fertilizers, As would not likely cause significant environmental problems in Brazilian soils and/or in soybean beans. However, it would be prudent to monitor the quality of any agricultural soil and associated crop where P fertilizers have been used for an extended period of time.

ACKNOWLEDGMENTS

The senior author thanks Drs. A. E. Furtine Neto, A. C. Chang, and L.R.G. Guilherme for their assistance to this study conducted at the University of California at Riverside. The financial support and the research facilities provided by the Federal University of Lavras and the scholarship award from CAPES both in the United States and Brazil are gratefully acknowledged.

REFERENCES

Alam, M. G. M., S. Tokunaga, and T. Maekawa. 2001. Extraction of arsenic in a synthetic arsenic-contaminated soil using phosphate. *Chemosphere* 43: 1035–1041.

Camelo, L. G. L., S. R. Miguez, and L. Marbán. 1997. Heavy metals input with phosphate fertilizers used in Argentina. *Science of the Total Environment* 204: 245–250.

Campos, M. L., F. N. Da Silva, A. E. Furtine Neto, et al. 2005. Determinação de cádmio, cobre, cromo, níquel, chumbo e zinco em fosfatos de rocha. *Pesquisa Agropecuaria Brasileira, Brasília* 40(4): 361–367.

Charter, R. A., M. A. Tabatabai, and J. W. Schafer. 1993. Metal contents of fertilizers marketed in Iowa. *Communications in Soil Science and Plant Analysis* 24: 961–972.

Creger, T. L. and F. J. Peryea. 1994. Phosphate fertilizer enhances arsenic uptake by apricot liners grown in lead-arsenate-enriched soil. *HortScience* 29(2): 88–92.

Friberg, L. F. and M. Vahter. 1983. Assessment of exposure to lead and cadmium through biological monitoring of a UNEP/WHO global study. *International Journal of Environmental Research* 30: 95–123.

Grant, C. A., L. Eastley, D. Flaten, E. Gawalko, and M. Tenuta. 2007. Toxicology and risk assessment: Cadmium in crop residues and phosphate fertilizer influences crop cadmium concentration. In *Biogeochemistry of Trace Elements: Environmental Protection, Remediation and Human Health*, Y. Zhu, N. Lepp, and R. Naidu (Eds.), Beijing: Tsinghua University Press, pp. 607–608.

Huton, M. and C. Symon. 1986. The quantities of cadmium, lead, mercury and arsenic entering the UK environment from human activities. *Science of the Total Environment* 57: 129–150.

Kpomblekou-A, K. and M. A. Tabatabai. 1994. Metal contents of phosphate rocks. *Communications in Soil Science and Plant Analysis* 25(17/18): 2871–2882.

Lavado, R. S., C. A. Porcelli, and R. Alvarez. 2001. Nutrient and heavy metal concentration and distribution in corn, soybean and wheat as affected by different tillage systems in the Argentine Pampas. *Soils and Tillage Research* 62: 55–60.

Loganathan, P., A. D. Mackay, J. Lee, and M. J. Hedley. 1995. Cadmium distribution in hill pastures as influenced by 20 years of phosphate fertilizer application and sheep grazing. *Australian Journal of Soil Research* 33: 859–871.

McLaughlin, M. J., K. G. Tiller, R. Naidu, and D. P. Stevens. 1996. Review: The behavior and environmental impact of contaminants in fertilizers. *Australian Journal of Soil Research* 34: 1–54.

McLaughlin, M. J., P. Simpson, and N. Fleming. 1997. Effect of fertilizer type on cadmium and fluorine concentrations in clover herbage. *Australian Journal of Experimental Agriculture* 37: 1019–1026.

Merry, R. H. 1988. Investigations on cadmium in South Australia: Rainfall, soils, cereals, pastures and soil-plant interactions. *National Symposium on Cadmium Accumulations in Australian Agriculture*, J. Simpson and B. Curnow (Eds.), Canberra; Australian Government Publication Service, pp. 62–79.

Mortvedt, J. J. 1987. Cadmium levels in soils and plants from some long-term soil fertility experiments in United States of America. *Journal of Environmental Quality* 16: 137–142.

Mulla, D. J., A. L. Page, and T. J. Gange. 1980. Cadmium accumulations and bioavailability in soils from long-term phosphorus fertilization. *Journal of Environmental Quality* 9: 408–412.

McBride, M. B. and G. Spiers. 2001. Trace elements content of selected fertilizers and dairy manures as determined by ICP–MS. *Communications in Soil Science and Plant Analysis* 32(1, 2): 139–156.

Nriagu, J. O. 1990. Global metal pollution. Poisoning the biosphere? *Environment* 32: 7: 7–33.

Novais, F. R. and T. J. Smyth. 1999. *Fosforo em solo e planta em condicoes tropicais*, R. F. Novais (Ed.), Viçosa: Universidade Federal de Viçosa Press, pp. 399.

Osztoics, E., P. Csatho, M. Magyar, T. Németh, and A. Osztoics. 2003. Cd, Cr and Sr uptake of spring barley as affected by P-fertilizers in a pot experiment. *Proceedings of the 7th International Conference on the Biogeochemistry of Trace Elements*, Uppsala, Sweden, G. R. Gobran and N. Lepp (Eds.), 1: 74–75.

Peryea, F. J. 1991. Phosphate-induced release of arsenic from soils contaminated with lead arsenate. *Soil Science Society of America Journal* 55: 1301–1306.

Raven, K. P. and R. H. Loeppert. 1997. Heavy metals in the environment: trace element composition of fertilizers and soil amendments. *Journal of Environmental Quality* 26(55): 1–557.

Roberts, A. H. C., R. D. Longhurst, and M. W. Brown. 1994. Cadmium status of soils, plants, and grazing animals in New Zealand. *New Zealand Journal of Agricultural Research* 37: 119–129.

Roche, P., L. Griere, D. Babre, H. Calba, and P. Fallavier. 1980. *Le Phosphore dans les Milieux Intertropicaux: Appréciation des Niveaux de Carence et des Besoins en Phosphore.* Paris, France: IMPHOS

Rothbaum, H. P., R. L. Gogirel, A. E. Johnston, and G. E. G. Mattingley. 1986. Cadmium accumulation in soils from long-continued applications of superphosphate. *Journal of Soil Science* 37: 99–107.

Senesi, G. S., G. Baldassare, and N. Senesi. 1999. Trace elements inputs into soils by anthropogenic activities and implications for humans health. *Chemosphere* 39(2): 343–377.

Shinonaga T., S. Ambe, and I. Yamaguchi. 1999. Uptake and distribution of trace elements in maturing soybean. *Biological Trace Element Research* 68: 235–248.

Singh, B. R. 1990. Cadmium and fluoride uptake by oats and rape from phosphate fertilizers in two different soils. *Norwegian Journal of Agriculture Science* 4: 239–249.

Soler, S. J. and R. J. Soler. 1996. Cadmium in organic fertilizers. *Fertilizer Environment* 541–545.

Williams, C. H. and D. J. David. 1973. The effect of superphosphate on the cadmium content of soils and plants. *Australian Journal of Soil Research* 11: 43–56.

16 Determination and Importance of Trace Elements in Chinese Medicinal Herbs Used for Healthcare and Medical Treatments

Chien-Yi Chen, Liyang Zhang, and Guangming Huo

CONTENTS

16.1 INTRODUCTION

One of the first forms of biofortification was the use of herbal-type crops. For over 4000 years, the Chinese people have effectively used medicinal herbs for medical treatments, including use as diuretics, pain relievers, health additives, and to relieve stasis (Ou 1993; Yen 1980). The therapeutic effect of Chinese medicinal herbs depends on their chemical constituents, most of which are organic compounds and trace elements that are key components of an essential enzyme system or other proteins (Mutaftchiev and Tzachev 2003). Many studies on the contents of trace elements in these herbs have been published (Chen and ChangLai 2001; Wang et al. 1996; Chen et al. 1993; Liu et al. 1991; Ou 1993; Chiao 1992; Chen and Lin 1988; Yen 1980). The roles of trace elements, iodine (I), iron (Fe), and zinc (Zn), are of particular interest to scientists in Chinese medicine. For example, I is known as an essential nutrient for humans, and it participates in the synthesis of thyroid hormones (Underwood 1973), whereas deficiencies of Fe and Zn have been associated with depressed growth and poor health in Taiwanese children. Apparently, some diseases are clearly associated with deficiencies of trace elements in the human body. In addition, the levels of toxic elements such as arsenic (As), antimony (Sb), and samarium (Sm) in Chinese medicinal herbs are also of prime interest in toxicological studies, in addition to their possible roles in human health (Ou 1993; Mertz 1981; Yen 1980).

Reliable multielement analytical techniques are essential for evaluating the contents of trace elements in Chinese herbal medicines to provide information to understand herb treatments used for particular conditions (Chen and ChangLai 2001; Chou et al. 1993). Trace elements are sometimes problematic to analyze because of possible spectral interference from polyatomic compounds when using graphite furnace, atomic absorption and emission spectrometry, and inductively coupled plasma spectrometry. In recent years, use of neutron activation analysis (NAA) has steadily increased as a sensitive nondestructive technique owing to its high sensitivity for providing a qualitative and quantitative determination of trace elemental concentrations in herbs (Chen et al. 2007; Chen 2002, 2004; Chen and Pan 2001; Wei et al. 1998; Wang et al. 1996; Chen et al. 1993; Chou et al. 1993; Liu et al. 1991). Because no digestion of herb samples is required for NAA, its usage for trace element determination is free of blank values. For most applications, delayed γ-radiation from the radioactive products was detected by NAA.

16.2 EXPERIMENTAL APPROACH

16.2.1 Selection and Description of Medicinal Herbs for Study

Table 16.1 shows the 12 medicinal herbs and the constituents that we evaluated for use as diuretic treatments. The parts used include rhizomes of two species, foliage of two species, stems of two species, seeds of five species, and one fungal cryptogam. To ensure that we had representative material, the 12 medicinal herbs [endorsed by the Taiwan Agricultural Research Institute (TARI)] were purchased from various local pharmaceutical stores throughout Taiwan; these herbs had all been imported from various regions of Mainland China (Wang et al. 1996; Chen et al. 1993; Ou 1993; Yen 1980). The medicinal herbs, lichen from the International

TABLE 16.1
Chinese Medicinal Herbs Commonly Used for Diuretic Treatments in Taiwan Selected for This Study

	Chinese Name	Scientific Name	Class
Rhizomes			
1.	Fangchi	*Aristolochia fangchi*	Aristolochiaceae
2.	Zexie	*Alismatis plantago-aquatica*	Alismataceae
Foliage (or Herbage)			
3.	Yinchenhao	*Artemisiae capillaries*	Compositae
4.	Yuxingcao	*Heartleaf houttuynia*	Saururaceae
Stem			
5.	Dengxincao	*Junci medulla*	Juncaceae
6.	Mutong	*Hocquartiae manshuriensis*	Lardizabalaceae
Fungal Cryptogam			
7.	Zhuling	*Polyporus umbellatus*	Polyporaceae
Seeds			
8.	Cheqianzi	*Plantaginis semen*	Plantaginaceae
9.	Dongguazi	*Benincasae semen*	Cucurbitaceae
10.	Lianzi	*Nelumbinis semen*	Nelumbo
11.	Shanchihzi	*Gardeniae fructus*	Rubiaceae
12.	Yiyiren	*Coicis lachrymal*	Gramineae

Atomic Environmental Agency (IAEA-336), and tomato leaves from the National Institute of Standard and the Technology (NIST) Standard Reference Material (NIST-SRM 1570a), were analyzed. In addition, NIST samples, standards, and background interference were utilized as quality controls for testing under identical conditions by neutron flux irradiation.

16.2.2 Neutron Activation Analysis

The detailed analytical approaches by neutron activation analysis technique have been described by Chen et al. (2003) and Chen and ChangLai (2001). In this study, approximately 20 samples (including lichen and a clean PE as background bag) were bundled and placed into the same container, and fixed tightly inside the conventional rabbit (a larger vial) for exposure to long and short irradiations in an open-pool reactor at the National Tsing Hua University. The γ-spectra were obtained using a calibrated HPGe (FWHM = 2.5 keV, 15% relative efficiency). The γ-rays were analyzed by Micro SAMPO90 software on a personal computer connected to a System-100 multichannel analyzer board for spectral acquisition.

The contents of at least 20 of the following trace elements were measured in the herbs; aluminum (Al), bromide (Br), chloride (Cl), iodine (I), magnesium (Mg), manganese (Mn) and vanadium (V) [(identified by short irradiation periods)]; and arsenic (As), cobalt (Co), chromium (Cr), iron (Fe), potassium (K), lanthanum (La), sodium (Na), rubidium (Rb), antimony (Sb), scandium (Sc), selenium (Se), samarium (Sm), and zinc (Zn) (identified by long irradiation periods).

16.3 TRACE ELEMENT COMPOSITIONS IN THE HERBS USED FOR DIURETIC TREATMENT

16.3.1 CONCENTRATIONS OF 20 TRACE ELEMENTS

Table 16.2 lists the trace element content found in the 12 Chinese medicinal herbs used as diuretic treatments (Chen et al. 1993; Ou 1993; Yen 1980). The trace element content varied over significant ranges in different herbs mainly because of their different growing environments in China. Among all trace elements, K was present in greater quantities than all other elements tested. Concentrations of I were determined using epithermal NAA. Iodine content in the 12 medicinal herbs is of great interest to the medicine community because of its major influence on human health (Chen 2002; Mertz 1981). Concentrations of I ranged from 0.19 ± 0.07 mg kg^{-1} in mutong to 0.54 ± 0.12 mg kg^{-1} in dengxinco. The contents of Mn ranged from 34.6 ± 2.0 mg kg^{-1} in mutong to 138 ± 23 mg kg^{-1} in dongguazi (commonly used to activate immune systems (Ou 1993). The maximum concentration of Mg was 174 mg kg^{-1} in shanchizi.

Concentrations of Sb ranged from a high of 380 ± 80 μg kg^{-1} in cheqianzi to a low of 65 ± 5 μg kg^{-1} in fangchi. In another example, the maximum concentration of Br was 3.50 ± 1.6 mg kg^{-1} in cheqianzi, whereas its concentration was negligible in both dengxincao and mutong. The Sc and V contents were verified in all species; the Sc and V concentrations varied from 5 ± 1 to 210 ± 7 μg kg^{-1}, and 0.10 ± 0.03 to 7.3 ± 1.2 mg kg^{-1}, respectively. The Al concentration was 3600 ± 1700 mg kg^{-1} in zhuling, which was approximately 36 times greater than that measured in shanchizi (100 ± 20 mg kg^{-1}). Chloride concentrations ranged from 6590 ± 500 mg kg^{-1} in yiyiren to 120 ± 20 mg kg^{-1} in dongguazi, and Na concentrations ranged from 1680 ± 540 mg kg^{-1} in dongguazi to 90 ± 10 mg kg^{-1} in mutong. The As concentration of 1.55 ± 0.09 mg kg^{-1} measured in zhuling was almost 11.9 times greater than 0.13 ± 0.03 mg kg^{-1} detected in yiyiren. The Zn content was 83 ± 4 mg kg^{-1} in zexie (commonly used to ease labor, and to promote conception). This Zn concentration in zexie was almost 14 times greater than that measured in zhuling, and 2.0 times greater than that measured in fangchi.

Selenium concentrations did not exceed 0.26 mg kg^{-1} (in cheqianzi) among the herbs. An accurate determination of Se is important because Se deficiency directly causes Keshan and Kashi–Beck's diseases in low-Se areas (Alaejos and Romero 1995). Cobalt, a constituent of vitamin B_{12}, was found to be as high as 560 ± 30 μg kg^{-1} in cheqianzi (commonly used to relieve diarrhea, remove body heat, and to tranquilize). The Co concentration was almost eight times greater than the Co concentrations of 63 ± 12 μg kg^{-1} measured in lianzi (used as a diuretic and/or to remove damp heat) (Yen 1980; Ou 1993; Chen et al. 1993). The contents of trace elements of measured Chinese medicinal herbs described previously are consistent with other published data (Wang et al. 1996; Chen et al. 1993; Chiao 1992; Chen and Lin 1988).

TABLE 16.2
Concentrations of Trace Elements in Chinese Medicinal Herbs Consumed in Taiwan

	Herbs:											
Element	Cheqianzi	Dongguazi	Dengxincao	Fangchi	Lianzi	Mutong	Shanchizi	Yinchenhao	Yuxingcao	Yiyiren	Zhuling	Zexie
K (%)	2.68 ± 9.0	1.06 ± 6.0	0.88 ± 0.08	1.01±1.3	0.22±0.09	0.29±0.09	1.77±1.4	2.73±0.10	2.46±0.06	0.17±0.042	0.55±0.032	1.51±0.11
Cl (mg/kg)	1600±130	120±20	910±60	940±140	730±200	970±110	780±70	1000±220	130±10	6590±500	580±120	660±100
Na (mg/kg)	440±110	1680±540	150±20	250±110	110±10	90±10	160±20	210±20	510±30	160±10	340±20	230±20
Al (mg/kg)	820±60	130±10	2610±200	130±10	1200±400	240±180	100±20	320±19	260±20	550±13	3600±1700	360±30
Fe (mg/kg)	268±20	580±60	380±120	180±14	58±6	121±7	64±15	154±30	1520±90	65±12	220±40	174±13
Mn (mg/kg)	51±9	138±23	52±10	70.2±3.2	129±55	34.6±2.0	38±15	123±33	60±13	41±16	93±4	100±30
Zn (mg/kg)	42.6±0.6	38.6±2.8	29.3±7.0	41.1±2.3	23.5±1.2	27.6±1.6	11.9±1.0	21.9±4.0	38±4	32.5±2.0	5.88±0.60	83±4
Rb (mg/kg)	28.0±2.6	10.8±0.8	11.2±0.9	9.55±0.30	5.24±0.30	2.44±0.40	7.30±2.4	3.0±0.2	11.7±0.7	2.23±0.30	4.93±0.60	4.67±0.30
Mg (mg/kg)	20.3±2.5	103±1.3	<MDC	51.4±3.7	210±36	62.3±7.8	174±28	5.10±0.54	<MDC	105±15	33.5±3.7	1.78±0.25
Br (mg/kg)	3.5±1.6	0.72±0.16	<MDC	2.73±0.35	2.05±0.13	<MDC	1.72±0.19	2.09±0.31	2.79±0.79	3.05±0.25	0.73±0.16	0.67±0.15
V (mg/kg)	1.08±0.14	0.11±0.03	3.63±0.60	0.12±0.03	2.11±0.50	0.69±0.11	7.3±1.2	2.90±0.40	0.28±0.08	1.52±0.30	1.90±0.10	0.10±0.03
As (mg/kg)	0.72±0.12	0.61±0.19	0.29±0.05	0.17±0.04	0.89±0.19	0.95±0.18	0.245±0.09	0.31±0.04	1.33±0.35	0.13±0.03	1.55±0.09	0.28±0.02
I (mg/kg)	0.43±0.08	<MDC	0.54±0.12	0.32±0.07	0.39±0.06	0.19±0.07	<MDC	0.41±0.09	<MDC	0.26±0.09	<MDC	0.43±0.10
Cr (mg/kg)	0.29±0.03	1.10±0.10	1.36±0.23	0.15±0.02	0.29±0.03	0.60±0.04	0.65±0.05	0.61±0.11	0.88±0.10	0.17±0.08	0.20±0.02	0.89±0.20
Se (mg/kg)	0.26±0.05	0.25±0.03	<MDC	0.24±0.03	<MDC	0.22±0.02	<MDC	<MDC	<MDC	0.12±0.02	<MDC	<MDC
Co (μg/kg)	560±30	490±30	530±30	510±30	63±12	127±17	85±8	85±8	370±150	70±7	101±20	379±20
Sb (μg/kg)	380±80	170±30	130±30	65±5	850±4	140±5	85±10	135±18	98±10	90±10	182±20	100±10
La (μg/kg)	290±40	630±10	<MDC	1190±120	<MDC	210±3	510±110	280±30	3960±200	<MDC	2900±550	610±70
Sc (μg/kg)	140±35	90±4	110±3	19±2	5±1	12±1	90±10	69±3	210±7	6±1	5±1	16±1
Sm (μg/kg)	55±4	80±4	45±5	54±8	15±3	19±3	15±4	<MDC	360±40	25±3	380±45	25±3

Note: < MDC: less than minimum detection concentration.
Data are mean values on a dry weight basis ± standard deviation.

Some differences in concentrations of trace elements are to be expected in the various medical herbs, because these herbs were purchased from various local pharmaceutical stores throughout Taiwan.

16.3.2 Children's Maximum Daily Intake

According to Ou and other local medical doctors in Taiwan (Ou 1993), the mean body weight of 3-year-old children is less than 15 kg. Taiwanese children are recommended to consume herbs at a rate of about 6 g per day (Wang et al. 1996). For this reason, attention must be given to the content of some trace elements, for example, As and Sb, which can be toxic when consumed in sufficiently large amounts over a long period of time. Only one sample exceeded MDI standards. Dengxincao had an I content that exceeded 0.50 mg kg^{-1}. The maximum daily intake (MDI) of I for 1-year-old and 10-year-old children is 65 and 110 mg kg^{-1}, respectively, as recommended by the Department of Health, Republic of China.

The National Academy of Sciences has listed essential and toxic elements for human nutrition (Underwood 1973). Table 16.3 lists the MDI of these elements by children with the suggested dose. These values indicate that the average Taiwanese child probably exceeds the daily dietary intake of Al, Fe, and Sc (Liu et al. 1991). The concentrations of Fe, Mn, and Zn are significantly lower than the U.S.-recommended dietary allowances (RDA) for essential elements, and also lower than the WHO/FAO provisional tolerable daily intakes (PTDI) (WHO 1987). The average daily ingestion of Fe and As would account for MDIs of Fe ranging from 350 to 9120 mg kg^{-1}, and 0.78 to 9.3 mg kg^{-1}, respectively, which is also below the 30 mg kg^{-1} of PTDI value for As recommended by WHO/FAO.

The MDI of Cr is estimated to range from 0.90 to 8.16 mg kg^{-1} in Taiwanese children, which is below the U.S. RDA (Food and Nutrition Board 1989). The recommended amount of Cr still falls within the PTDI recommend by FAO/WHO for children with a body weight of 15 kg. Notably, Taiwanese children also take in large amounts of essential elements, such as Al, Cl, Cr, Fe, Mn, Se, and Zn. Finally, the MDIs of K, Cl, Na, Al, Mn, Rb, Mg, Br, V, I, Se, Co, Sb, La, and Sm correspond closely with the levels recommended by WHO/RDA. The MDIs of Fe, Na, and Se were well below the maximum tolerable intakes suggested by expert groups (Food and Nutrition Board 1989; WHO 1987).

In summary, the levels of metals consumed from these tested medical herbs are safe, and they are still recommended for providing children and adults readily available amounts of many nutritious trace elements. The long-standing application of these herbs for diuretic treatments supports modern research (Chen et al. 2007). Using other medicinal herbs, for example, ginseng, may inexpensively provide humans with essential trace elements and other constituents for maintaining their health nutrition (see the following case study).

TABLE 16.3
Maximum Daily Intakes of Elements from Chinese Medicinal Herbs Used for Children's (15 kg Body Weight) Healthcare and Medical Treatments in Taiwan

Element	Dongguazi	Dengxincao	Yuxingcao	Zhuling	Zexie	RDA[a]	PTI[b]	Liu et al.[c]
K(%)	63.6	52.6	148	33.0	90.6			170
Cl (mg/kg)	720	5,460	780	3,480	3,960			3,500,000
Na (mg/kg)	10,080	900	3,060	2,040	1,380			2,800,000
Al (mg/kg)	780	15,700	1,560	21,600	2,160			9,200
Fe (mg/kg)	3,480	2,280	9,120	1320	1,044	15,000		8,000
Mn (mg/kg)	828	312	360	558	600	2,000		2,800
Zn (mg/kg)	232	176	228	35.3	498	15,000		7,300
Rb (mg/kg)	64.8	67.2	70.2	29.6	28.0			1,700
Mg (mg/kg)	61.8	ND[d]	ND[d]	201	10.7			
Br (mg/kg)	4.32	ND[d]	16.7	4.38	4.02			4,200
V (mg/kg)	0.66	21.8	1.68	11.4	0.60			
As (mg/kg)	3.66	1.74	7.78	9.30	1.52		30	
I (mg/kg)	ND[d]	3.24	ND[d]	ND[d]	2.58			
Cr (mg/kg)	6.60	8.16	5.28	1.20	5.36	50		92
Se (mg/kg)	1.50	ND[d]	ND[d]	ND[d]	ND[d]			110
Co (μg/kg)	2,940	3,180	2,220	610	2,270			
Sb (μg/kg)	1,020	780	590	1,090	600			
La (μg/kg)	3,780	ND[d]	23,800	17,400	3,660			
Sc (μg/kg)	540	660	1260	30	96			200
Sm (μg/kg)	480	270	2,160	2,280	150			

[a] U.S. recommended dietary allowances (10th ed.) for adults.

[b] Provisional tolerable daily intakes (WHO/FAO) for children with body weight 15 kg.

[c] Analyzing daily dietary intake for citizen in Taiwan; see Liu et al. 1991.

[d] ND = not detected.

16.4 CASE STUDY: *PANAX GINSENG* AND *GANODERMA LUCIDUM*

16.4.1 Germanium Contents

Elemental germanium (Ge) is a hard, lustrous, gray-white, and brittle metalloid with atomic number 32 and atomic weight 72.6. Germanium has two chemical forms in the environment: inorganic Ge, such as GeO_2 and organic Ge. Germanium occurs in many different plants, and concentrations of Ge vary significantly from species to species (Xie 1998). The plants containing exceptional quantities of Ge are those often used as Chinese medicinal herbs (Asai 1981). The two most commonly consumed Chinese medicinal herbs worldwide are *Panax ginseng* and *Ganoderma lucidum* owing to their various biological functions. In many Asian communities *Panax ginseng* and *Ganoderma lucidum* have particularly been used as important ingredients for antitumor, immunomodulatory, cardiovascular, respiratory, antihepatotoxic, or antinoeiceptive (pain-reducing) treatments (Chang 2006). As a case study in this chapter, we have focused on the biological functions of Ge in these two medicinal herbs.

Panax ginseng is the one of most important tonic medicinal herbs. It has been widely used in traditional Chinese medicine for the past 20 centuries (Chiou 2000), and oftentimes used to enhance human stamina and help people recuperate people from physical stress (Hong et al. 2002), and was even used as an aphrodisiac agent in ancient oriental medicine. It is believed that ginseng contains a biologically functional ingredient that can mitigate metabolic disorders related to such illnesses, as diabetes, diarrhea, nausea, rheumatism, insomnia, and leprosy. In ancient times herbalists interpreted the shape of ginseng as resembling a human figure. They thought that this shape was a sign that ginseng could be valuable for healing the entire human. In Chinese ginseng is referred to as the "man root" or "human-body shaped root." Indeed, ginseng as an important medicinal herb has been the subject of many Asian legends and folk stores in China, Japan, or Korea. In recent years, beverages blended with ginseng have been produced and are popular and available in North American markets.

Unlike ginseng, *Ganoderma lucidum* is an edible fungal species and has a long medicinal use history in Far Eastern Asia. The *Ganoderma lucidum* fruiting body has gained worldwide popularity in recent years as a dietary supplement. Chang and Buswell (1999) estimated that total world market value for *Ganoderma*-based natural healthcare products was 1628 million U.S. dollars in 1995, including $350 million in China, $215 million in Taiwan, $600 million in Korea, $300 million in Japan, and $60 million in Hong Kong (Chang 2006). According to the estimations by Rai (2003) and Lai et al. (2004), the current *G. lucidum* production worldwide is approximately 5000–6000 tonnes, with more than half of that produced in China. However, the most recent survey by China Edible Fungi Association showed that the production of *Ganoderma* mushroom in China was 36700 and 49200 tonnes in 2002 and 2003, respectively. Clearly, such large market demands cannot be met from wild culture and collections. Therefore, many countries have developed a great variety of cultivation practices for this medicinal herb, including solid fermentation technology (described later) (Zhang 2002; Zhang et al. 2005, 2006).

Earlier studies on a *Ginseng* and *Ganoderma* found that these two species contained various chemical compounds that have specific biological functions. The identified functional chemical compounds include polysaccharides, triterpenes, adenosines, and immunomodulatory proteins. Ginseng contains complex carbohydrates called saponins or ginsengsines, and may possess anti-inflammatory, antioxidant, and anticancer elements. High levels of Ge have been observed in these plant tissues (Yu 1997). In this regard, Asai (1981) measured the Ge content of those plants reputed to have beneficial effects in the treatment of malignant tumors, and he found that plants generally regarded as conducive to good human health contained fairly large quantities of Ge. For this reason, we believe that Ge might play an important role as an active compound in these medicinal herbs.

The distribution of Ge in plant is not homogenous (Xie 1998). For instance, in ginseng grown in Chinshan, Korea, Ge mainly accumulates in the coarse root, and then in the stem and leaf. The highest Ge concentration (up to 4,000 mg kg^{-1}) was observed in Ginseng coarse root, whereas the peripheral root hairs contained much less Ge.

16.4.2 Plant Germanium and Biological Functions

The levels of Ge in plant tissues vary depending on soil. In general, Ge plays an important role in the biological metabolisms of plants. For example, ginseng would be just as susceptible to the thousands of viruses and bacteria existing in the soil and would soon rot if it were not for its extremely high Ge content (Asai 1981). In addition to ginseng, germanium is also present in the structure of various mushrooms such as *Cortinellus shiitake*, *C. champignon*, and *C. kawaradake* (Asai 1981).

Many of the inorganic Ge salts are, however, toxic. Asai (1981) identified the organic Bis-β-carboxyethyl Ge sesquioxide that is commonly named as Ge-132, as a form of organically bound Ge that is virtually nontoxic to humans. When Ge is intravenously administered, it is excreted very rapidly, with about 60% lost in the urine within the first hour, and 90% lost within three hours. This organically bound Ge is apparently not metabolized at all by tissues, and its levels peak in the tissues within 6 hours following oral (sublingual) administration, and it is not retained significantly by any organ except possibly the vitreous humor of the eye. Ge-132 effects are "dose dependent"; meaning that the previously described effect become more significant as the dosage increases. Ultimately, the response plateaus, but increased dosage causes no harm. The healthier the person (as with healthy children), the less benefit Ge confers, and vice versa. There are many different treatment modalities for Ge, many of which, though differing, may be useful.

16.4.2.1 Immunomodulatory Effect of Germanium

In both humans and animals Ge-132 has reportedly been shown to (1) increase gamma interferon in the blood, (2) activate macrophages and natural killer cells, (3) raise blood hemoglobin levels and reduce white cell counts, and (4) stimulate immunomodulation activity in the B cell system (Asai 1981; Chen 1996; Tang and Hu 1997). This element, therefore, may be an excellent adjuvant for (aids the operation of) immunochemotherapeutic agents. The effects of Ge-132 on various immune parameters are almost identical to that of known γ interferon immunomodulating activity (Asai 1981). In addition, studies on immune-suppressed animals and on patients with malignancies or rheumatoid arthritis suggest that Ge-132 would normalize the function of T cells, B lymphocytes, antibody-dependent cellular cytotoxicity, natural killer cell activity, and numbers of antibody-forming cells. Apparently, organic germanium has some "normalizing" influence on the immune system (Kamen 1987).

16.4.2.2 Immunostimulant Effects

The effects of 2-carboxyethyl Ge sesquioxide (Ge-132) and spirogermanium in oncology and mechanisms of antineoplastic action were evaluated by Tang and Hu (1997). Today, Ge-132 and spirogermanium compounds are useful as antineoplastic agents in the prevention and cure of different antineoplastic-agents (Tang and Hu 1997). A key characteristic of Ge is its ability to "normalize" many physiological functions. For example, it has the ability to lower blood pressure in hypertensive people. This normalizing effect may extend to a broad variety of physiological/biochemical functions, including pH, potassium, calcium, chloride, triglycerides, cholesterol, bilirubin, uric acid, and blood hemoglobin. Our bodies have many physical/chemical systems that are in a state of "homeostasis," which is the ability to restore systems to their "normal" state. A substance that assists body systems in restoring normality is called an *adaptogen*. Germanium may be a type of adaptogen. In effect, analogous to thyroid hormones, Ge-132 may improve the overall metabolism of living cell and body organs. A most interesting characteristic of Ge is its ability to relieve a great deal of pain by inhibiting the natural body enzyme that in turn inhibits production of the body's endorphins.

16.4.3 Foods High in Organic Germanium

Germanium occurs naturally in peat and coal deposits, and has accumulated to levels as high as 10 mg kg^{-1} from the remains of once-living organisms (Chen 1996). In addition to ginseng and the previously mentioned mushrooms, many other medicinal plants contain high amounts of Ge. These include shelf fungus (*Trametes cinnabarina*, 800–2000 mg kg^{-1}), garlic (*Allium sativum*, 750 mg kg^{-1}), sushi (*Angelica pubescens*, 260 mg kg^{-1}), *Bandai moss Aralia cordata* (260 mg kg^{-1}), Japanese waternut (*Trapa japonica*, 240 mg kg^{-1}), Comfrey (*Symphytum officinale*, 150 mg kg^{-1}), boxthorn seed (*Lycium chinense*, 125 mg kg^{-1}), wisteria (*Wisteria floribunda*, 110 mg kg^{-1}), and pearl barley (*Fructus coicis* lacryma-jobi, 75 mg kg^{-1}) (Asai 1981).

16.4.4 Biofortified Products by Solid Fermentation

In addition to trace elements, such as Ge, there are also many other functional chemical compounds in *Panax ginseng* and *Ganoderma lucidum* (Tan and Vanitha 2004). In an attempt to industrially produce some of these chemical compounds found in medicinal herbs or fungi, the solid fermentation technique has recently been developed (Zhuang et al. 2004, 2007) by combining nutrient substrates with pharmaceutical fungi to produce functional fungus-like substances. The use of this technology could enhance the production of the targeted active ingredients from medicinal herbs or fungi, including secondary metabolites resulting from biotransformation processes. Zhang et al. (2006) have investigated the improvement on the immune function and growth of chicken that were fed with products produced from the solid fermentation with medicinal fungi, including *Ganoderma lucidum*. In a 30-day experiment, the growth of the chicken was enhanced by about 10–15%, compared to the control, and the antibody titer for avian influenza (A1, H5 subtype) increased by a factor of 2–4 (Zhang et al. 2005). Similarly, using a solid fermentation technology would be necessary for mass-producing those active compounds naturally found in medicinal herbs such as *Panax ginseng* and *Ganoderma lucidum*, which benefit human health.

16.5 CONCLUSIONS

Medicinal herbs have been used for centuries to treat diseases and to promote health quality. The therapeutic effects of medicinal herbs depend on their chemical constituents, including trace elements and pharmaceutical compounds. These studies showed the wide variance in the composition of 20 different trace elements in commonly available herbs in Taiwan and discussed some of the analytical techniques used to accurately determine their concentrations. Information was also presented on germanium content in the most traditional Chinese herb (*Panax ginseng*) and in *Ganoderma lucidum* (a traditional pharmaceutical fungus) and its role in immunomodulatory and immunostimulant effects. Clearly, medicinal herbs benefit human and even animal health. Future research endeavors in the area of biofortification should also include medicinal herbs or fungi as potential sources of nutrients or other beneficial compounds.

ACKNOWLEDGMENT

The authors wish to thank Dr. Liu, S.Y. of TARI, Medical Doctors, Tsai H.S., and Bai S.L. for identifying Chinese medicinal herbs and many fruitful discussions.

REFERENCES

Asai, K. 1981. Miracle Cure: Organic Germanium *(Gebundene ausgabe)*. Japan Publications Trading Company.

Alaejos M. S. and C. D. Romero. 1995. Analysis of selenium in body fluids: A review. *Chemical Review* 95: 227–257.

Chang, S. T., and J. A. Buswell. 1999. Ganoderma lucidum (Curt.:Fr.) P. Karst. (Aphyllophoromycetideae): A mushrooming medicinal mushroom. *International Journal of Medicinal Mushrooms* 1: 139–146.

Chang, S. T. 2006. *Ganoderma lucidum* (a leader of medicinal mushrooms): A prominent source for the healthcare market in the 21st century. In *Proceedings of the First Symposium on Development of China's Medicinal Fungi Industry*, 15–28. Mycological Society of China.

Chen, G. S. 1996. Study of organic germanium-132. *Environment and Exploitation* 2: 1–5.

Chen, C. Y. 2004. Trace elements in Taiwanese health food, Angelica keiskei, and other products. *Food Chemistry* 84(4): 545–549.

Chen, C. Y. 2002. Elemental analysis of Taiwanese health food Angelica keiskei with INAA. *Journal of Radioanalytical and Nuclear Chemistry* 252(3): 551–558.

Chen, C. Y., D. B. Lin, and W. K. Chen. 2007. Elements in the sera of preschool children living in central Taiwan. *Chinese Journal of Chemistry* 25: 515–520.

Chen, C. Y. and L. K. Pan. 2001. Trace elements of Taiwanese *dioscorea* spp. using neutron activation analysis. *Food Chemistry* 72: 255–260.

Chen, C. Y. and S. P. ChangLai. 2001. Determination of several trace elements in Chinese herbs with epithermal neutron activation analysis. *Analytical Science* (supplement) 17: i649–i652.

Chen, C. Y., Y. Y. Wei, S. P. ChangLai, and L. K. Pan. 2003. Concentration of bromine and iodine in the Chinese medical herbs via epithermal neutron activation analysis. *Journal of Radioanalytical and Nuclear Chemistry* 257(2): 405–410.

Chen, H. C. and S. M. Lin. 1988. Determination of trace element elements in certain crude drugs (part I). *Kaohsing Journal of Medical Science* 4: 259–272.

Chen, K. S., C. L. Tseng, and T. H. Lin. 1993. Trace elements in natural drugs determined by INAA. *Journal of Radioanalytical and Nuclear Chemistry* 170: 265–280.

Chiao, G. P. 1992. *Trace Element and Chinese Medicine Drugs.* Shanghai (in Chinese).

Chiou, W. F., W. C. Jan, C. J. Chou, and C. F. Chen. 2000. Non major ginsenosides contribute to the relaxation of *Panax ginseng* in rabbit *corpus cavernosum. China Medical Journal* 11(4): 197–204.

Chou F. I., H. D. Lin, J. C. Wei, A. Y. Wang, and J. G. Lo. 1993. Simplified measurement of protein-bound iodine with epithermal neutron activation analysis. *Nuclear Medicine and Biology* 20(5): 631–636.

Food and Nutrition Board, U.S. 1989. *Recommended Dietary Allowance*, 10th ed., World Health Organization, World Health Organization Technical Report, No. 776, 1989.

Hong, B., Y. JI, J. Hong, K. Nam, and T. Ahn. 2002. A double-blind crossover study evaluating the efficacy of Korean red Ginseng in patients with erectile dysfunction: A preliminary report. *The Journal of Urology* 168(5): 2070–2073.

Kamen, B. 1987. *Germanium: A New Approach to Immunity.* Nutritional Encounter, Inc., California. pp. 38.

Liu, S. M., C. Chung, J. T. Chuang, C. F. Wang, and N. K. Aras. 1991. Daily dietary intake of minor and trace elements by upper social groups in Taiwan. *Journal of Radioanalytical and Nuclear Chemistry* 150(2): 397–415.

Lai, T., Y Gao, and S. Zhou. 2004. Global marketing of medicinal Ling Zhi mushroom *Ganoderma lucidum* (W.Curt.:Fr.) Lloyd (Aphyllophoromycetideae) products and safety concerns. *International Journal of Medicinal Mushrooms* 6: 104–109.

Mertz, W. 1981. The essential trace elements. *Science* 218(18): 1332–1338.

Mutaftchiev, K. and K. Tzachev. 2003. Determination of manganese (II) in some medicinal plants and their decoctions by a kinetic spectrophotometric method. *Phytochemical Analysis* 14: 160–163.

Ou, M. 1993. *Chinese-English Manual of Common-Used Prescriptions in Traditional Chinese Medicine.* 2nd ed. Ngai Kwong Printing (H.K.) Co. Ltd., Hong Kong.

Rai R. D. 2003. Successful cultivation of the medicinal mushroom Reishi, *Ganoderma lucidum. India Mushroom Research* 12(2): 87–91.

Tang, R. H. and S. W. Hu. 1997. Germanium and antineoplastic agents. *Studies of Trace Elements and Health* 3: 57–58.

Tan, B. K. and J. Vanitha. 2004. Immunomodulatory and antimicrobial effects of some traditional Chinese medicinal herbs: A review. *Current Medicinal Chemistry* 11(11): 1423–1430.

Underwood, E. J. 1973. *Toxicants Occurring Naturally in Foods.* National Academy of Sciences, pp. 43.

Wang, C. F., M. J. Duo, and E.E. Chang. 1996. Essential and toxic elements in the Chinese medicine. *Journal of Radioanalytical and Nuclear Chemistry* 211: 333–347.

Wei, Y.Y., Chung, C., and C. W. Wu. 1998. Optimal analytical conditions for determining elements in gastric tissue and blood of cancer patients using neutron activation. *Journal of Analytical Chemistry* 53(7): 776–782.

World Health Organization. 1987. *Evaluation of Certain Food Additives and Contaminants*, 30th Report of the Joint FAO/WHO Expert Committee on Food Additives, Geneva, Switzerland.

Xie, Y. Q. 1998. A study on environmental ecology of germanium. *Trace Elements Science* (2): 23–25.

Yen, K. Y. 1980. *The Illustrated Chinese Material Medical Crude Drugs*. 1st ed. Southern Materials Center, Inc., Taiwan (in Chinese).

Yu, Z. X. 1997. The pharmacological effects and clinical application methods of *Ganoderma lucidum*. *Journal of Yunnan Normal University* 17(2): 65–68.

Zhang, L. Y. 2002. Research on the approaches of measurement of biomass of *Monascus purpureus* in solid fermentation. *Journal of Nanjing Xiaozhuang College* 18(4): 6–8.

Zhang, L. Y., Y. F. Zhou, W. Y. Yu, D. L. Zhang, and Y. K. Wang. 2005. The study on medicinal fungi fermentation substance to regulate endogenous hormone secretion and immunity in broiler chicken. *Journal of Yangzhou University: Agricultural and Life Science Edition* 26(4): 38.

Zhang, L. Y., Y. F. Zhou, D. L. Zhang, Y. P. Jiang, and Y. Zhuang. 2006. Application of mycology in the field of veterinary medicine and of mycoplasm to control avian influenza. *Edible Fungi of China* 25(2): 3–6.

Zhuang, Y., Y. M. Chi, S. B. Chen, R. N. Ding, and Z. D. Min. 2004. Preparation of medicinal fungal new type bi-directional solid fermentation engineering and Huai Qi fungal substance. *Chinese Pharmaceutical Journal* 39(3): 175–178.

Zhuang, Y., X. M. Xie, and L. Y. Zhang. 2007. The origin, development and its advantage and potential of the bi-directional solid fermentation for medicinal fungi. *Edible Fungi of China* 26(2): 3–6.

17 Field-to-Fork Perspective *Soil Nutrition/Biofortification as the Root of Human Health and Well-Being*

Arden B. Andersen

CONTENTS

17.1 INTRODUCTION

17.1.1 Nutritional Quality of Food

Fundamental nutritional imbalances in humans reside squarely in diet and food. However, food, which constitutes diet, is not the same as food 70 years ago, because of its depletion in nutrient density. Conventional agriculture can be represented by the N-P-K model, which has resulted in significant environmental and human health problems. Food nutrient quality is directly and intimately correlated to the nutrient quality of the soil environment upon or in which that food is grown. Hippocrates, thousands of years ago, profoundly stated, "Let food be your medicine." Like all common sense declarations, it remains as profound and appropriate today. True human health begins with soil fertility and food nutrition (Kirschenmann 2006).

The fundamental premise of this brief chapter is to present a human perspective on recognizing the link between human health and soil fertility. International Bibliographic Information on Dietary Supplements (IBIDS) database includes over 760,000 bibliographic records as of 2007, concerning the link between dietary supplements and human and animal health (IBIDS 2007). This pharmaceutical paradigm further neglects the obvious and profound need to address the nutrient quality of food itself, which should be the consumer's best source of sufficient therapeutic levels of vitamins, minerals, and antioxidants.

As foods have become more refined, resulting in lower nutrient density at their origin, consumers have created a growing demand for nutritional supplements to make up the deficiencies. Here lies the crux of the current problem with food quality; it has been consistently declining in nutritional value. The USDA and U.K. Ministry of Agriculture data showed that foods produced today have less nutritional value than they had in the late 1940s (Thomas 2003).

Increasingly, people are seeking answers in the marketplace to accommodate their need and desire for more nutritious and healthful food. This search is demonstrated by the demand for organic food, which continues to grow at 20% per year (Dimitri and Greene 2002). The outcome of importance is nutrient density of the food. The consumers' challenge is not in their willingness to change or in the lack of opportunities for them to do so, but rather with finding quality food to fulfill the dietary and nutritional prerequisites of a healthy lifestyle. This pharmaceuticalization of food starkly demonstrates that nutrition is at the root of the problem. It further raises the problem of challenging human and animal immune systems. What we have, worldwide, is a serious deficiency of essential nutrients in our soils, and subsequently, in our foods (Thomas 2003).

17.2 BIOLOGICAL LINKS WITH NUTRITION

Agronomists do acknowledge a direct association between plant disease and nutrition under a few conditions, for example, blossom end rot in tomatoes and better pit in apples, which are related to calcium deficiency. However, they find it difficult to take this concept to its logical conclusion regarding insect pests, disease pathogens (viruses, fungi, and bacteria), and weeds. Chaboussou (2004) additionally demonstrated that disease organisms that attack plants act similar to insect pests in that they proliferate where reducing sugars, free amino acids, and free nitrogen compounds abound, and not where nonreducing sugars, complex carbohydrates, and complete proteins predominate. In fact, these animals will seek out plants that satisfy inherent nutritional needs as these needs arise. This is seen plainly in the administration of free-choice supplements for grazing livestock (Dettloff 2004). Animals that fail to acquire such nutritional needs become weakened and ill, succumbing to diseases, parasites, and predators (Konarev 1996). William Albrecht, while Department Head at the University of Missouri, recognized that animal health was directly correlated to feed quality, which was in turn directly correlated to soil nutritional health. He determined a desired base saturation model required to improve crop quality, which is correlated with animal health (Albrecht 1975). Comprehensive nutritional management (biofortification) is required to achieve such a state of health (Chaboussou 2004).

17.2.1 Calcium: Foundation Nutrient

The beginning of an appropriate nutritional balancing program starts with laying the foundation. Too often, calcium (Ca) is addressed in soil fertility, not as a nutrient, but rather as a liming material. Soil pH, in this model, is used as the major determining factor for or against liming. This approach perpetuates the neglect of calcium, the nutrient, and the consequent soil compaction, grass weeds, limited crop yields, and poor-quality foods (Andersen 2006; Skow and Walters 1995). Calcium is the key ubiquitous signaling entity in living plants. Every metabolic process from nutrient transport to ATP production, from protein production to cell division requires sophisticated communication within the plant so that the various living processes proceed to their logical conclusion. All living cells, in this case plant cells, use a network of signal transduction pathways to conduct developmental programs, obtain nutrients, control metabolism, and deal with their environment. In studying the "language" of these signaling systems in plants, scientists have identified messengers such as Ca^{2+}, lipids, pH, and cyclic GMP (cGMP). For lipids and cGMP to be formed, calcium ion messengers must be present. No single messenger has been demonstrated to respond to more stimuli than cytosolic free calcium ions. In other words, calcium ions may be part of the primary messenger system throughout the plant that is so important for living functions to occur (Sanders et al. 1999).

17.3 CONCLUSIONS

The N-P-K model of fertilization used today and in the past has created environmental, economic, and public health calamities that will be felt for generations. The recent global call for carbon sequestration serves as a further impetus to the need for change in agricultural management.

A new standard is evolving for food quality, driven by consumer demand. This standard is a standard derived from soil and crop management. It is the standard of nutrient density, food value, safety, and integrity. This means real accountability from producer to retailer for the actual value of the food to consumer health and nutrition. This requires a significant shift in fertilization management in the field, a reversal no less, from mining the soil to regenerating it. We must integrate biofortification from field to fork, and not simply as a genetic tag-on. In doing so, we will be practicing true preventive medicine, helping both those starving for lack of food and those who are starving to death on full stomachs.

REFERENCES

Albrecht, W. A. 1975. *The Albrecht Papers*, Volumes I and II. Kansas City: Acres USA.

Andersen, A. B. 2006. *Science in Agriculture*, Austin, TX: Acres USA.

Chaboussou, F. 2004. *Healthy Crops: A New Agricultural Revolution*. Charlbury, U.K.: Jon Carpenter Publishing.

Dettloff, P. 2004. *Alternative Treatments for Ruminant Animals*. Austin. TX: Acres, USA.

Dimitri, C. and C. Greene. 2002. Recent Growth Patterns in the U.S. Organic Foods Market. U.S. Department of Agriculture, Economic Research Service, Market and Trade Economics Division and Resource Economics Division. Agriculture Information Bulletin Number 777. Washington, DC, 20036 September 2002, http://www.ers.usda.gov/Publications/AIB777/ (accessed September 20, 2007).

Kirschenmann, F. 2006. Farming, Food and Health. Gleanings. Leopold Center for Sustainable Agriculture, Iowa State University, Summer 2006 http://www.glynwood.org/resource/foodandhealth.pdf (accessed September 22, 2007).

Konarev, A. I. V. 1996. Interaction of insect digestive enzymes with plant protein inhibitors and host-parasite coevolution. *Euphytica*, 92(1/2): 89–94.

Sanders, D., C. Brownlee, and J. Harper. 1999. Communicating with calcium. *Plant Cell* 11: 691–706.
Skow, D. and C. Walters Jr. 1995. *Mainline Farming for Century 21.* Austin, TX: Acres USA.
Thomas, D. 2003. A study on the mineral depletion of the foods available to us as a nation over the period 1940 to 1991. *Nutrition and Health* 17: 85–115.

Index

C

D

E

F

K

L

M

N

O

P

S

T

U

W

X

Y

Z